75 Structure and Bonding

Long-Range Electron Transfer in Biology

With Contributions by
P. Bertrand B. E. Bowler J. Chang H. B. Gray
B. H. Hoffman A. Kuki A. G. Mauk G. McLendon
M. J. Natan J. M. Nocek A. L. Raphael A. G. Sykes
M. J. Therien S. A. Wallin

With 73 Figures and 23 Tables

Springer-Verlag
Berlin Heidelberg New York London
Paris Tokyo Hong Kong Barcelona

ISBN 3-540-53260-9 Springer-Verlag Berlin Heidelberg New York
ISBN 0-387-53260-9 Springer-Verlag New York Berlin Heidelberg

Library of Congress Cataloging-in-Publication Data
Long-range electron transfer in biology/editor, G. A. Palmer: with contributions by P. Bertrand . . . [et al.].
p. cm. – (Structure and bonding; 75)
ISBN 0-387-53260-9 (U.S.: alk. paper)
1. Metalloproteins. 2. Electron transport. I. Palmer, Graham A.
II. Bertrand, P. (Patrick), 1947- . III. Series.
QD461.S92 vol. 75 [QP552.M46] 541.2'2 s – dc20 [574.19'245] 90-26896 CIP

Typesetting: Macmillan India Ltd, Bangalore-25; Printing: Colordruck, Berlin; Bookbinding: Lüderitz & Bauer, Berlin
51/3020-5 4 3 2 1 0 – Printed on acid-free paper

Preface

The possibility that there might be long-range electron transfer between redox-active centers in enzymes was first suspected by biochemists working on the mechanism of action of metalloenzymes such as xanthine oxidase which contain more than one metal-based redox center. In these enzymes electron transfer frequently proceeds rapidly but early spectroscopic measurements, notably those by electron paramagnetic resonance, failed to provide any indication that these centers were close to one another.

However, it took the seminal experiments on the temperature-independent light-induced oxidation of cytochromes in photosynthetic bacteria by Devault and Chance in 1966 to persuade physical scientists that long-range electron transfer in biological systems might be a real phenomenon. The subsequent theoretical contributions of Hopfield and Jortner placed a more rigorous focus on the problem and triggered a substantial effort towards defining the physico-chemical basis for this phenomenon. This effort proceeds unabated today, as this issue of *Structure and Bonding* testifies.

This field has progressed rapidly in the last decade and consequently it appeared worthwhile to ask a number of the individuals who have participated in these advances to provide their perspective, both on the current state of our knowledge and on a review of the most recent results from their respective laboratories.

In an interdisciplinary area such as this it is more than natural that the character of the individual articles differs widely. We start at one extreme with the contributions of Bertrand and Kuki which contain a great deal of theoretical and state-of-the-art physics. We then proceed to those of Gray and Hoffman who cleverly exploit protein chemistry, continue on to Mauk and McLendon who take advantage of the tools of modern genetic engineering and end with Sykes who describes the utility of drawing on Nature's engineering in unravelling the details of metalloprotein redox reactivity.

By bringing together such disparate approaches is hoped that this volume will serve as a convenient starting point for someone, regardless of background, who is interested in acquiring some appreciation of the origins of this field, of our current state of knowledge and of the breadth of approaches that have been successful in bringing this field to its current level of insight.

Houston, February 1991 Graham Palmer

Table of Contents

Application of Electron Transfer Theories to Biological Systems

Patrick Bertrand

Laboratoire d'Electronique des Milieux Condensés URA CNRS 784,
Université de Provence, Centre de St Jérôme, boîte 241, 13397 MARSEILLE Cedex 13, France

In biological systems, the mechanisms of conversion and storage of energy involve sequences of oxido-reduction reactions in which electrons are transferred along a chain of redox centers embedded in a protein medium. The theoretical interpretation of the kinetics of these transfers pertains to Quantum Mechanics, and was developed by chemists and physicists. However, owing to the fundamental importance of these processes, many biochemists are also concerned with these theories and their practical application to biological systems. This introductory chapter is an attempt to clarify the physical basis of current theoretical interpretations of biological electron transfers. It comprises an account of the standard formalism appropriate for non-adiabatic processes, and a detailed review of different approaches which have been developed to apply this formalism to the analysis of kinetic data. Important advances in this field have resulted, on the one hand, from precise theoretical calculations based on molecular structures, and on the other hand, from implementation of elaborate experimental methods based on efficient chemical and biochemical techniques. This topic is illustrated by many examples taken from the recent literature which concern redox proteins as well as photosynthetic systems.

1 Introduction

The basis of electron transfer theories were established by Marcus at the beginning of the 1960s. Originally, Marcus' work was concerned with bimolecular reactions between small inorganic molecules in solution, for which transfer occurs within a transient complex without disruption of the first coordination sphere (i.e. an outer sphere process). However, the residual interactions between the centers which result from the overlap of their orbitals were thought to be strong enough for the rate of intracomplex electron transfer to be independent of them (an adiabatic process). When the temperature is sufficiently high so that the nuclear motions of the system can be described classically, the rate expression takes an activated form [1, 2, 3]:

$$k = \nu \exp(-(\Delta G^0 + \lambda)^2 / 4\lambda k_B T), \tag{1}$$

where ν is a characteristic frequency for the nuclear motion, ΔG^0 is the redox free energy change, and λ the reorganization energy. By using some simplifying assumptions, Marcus derived the well known cross-relations which were successfully tested in a number of redox reactions between inorganic complexes in solution [4].

The field of application of the theory was considerably enlarged after the contributions first by Levich [5] and Dogonadze et al. [6], and later by Kestner et al. [7] and Hopfield [8]. These authors have shown that when the residual interactions between the centers are weak enough, owing to their large separation, the expression of the electron transfer rate depends explicitly on these interactions through an electronic factor (a non-adiabatic process). This expression also involves a nuclear factor, whose value is determined by the nuclear motions of the system. These motions are treated semi-classically by Hopfield, and quantum-mechanically by Jortner [7] but it is remarkable that the expressions obtained for the nuclear factor give in both cases an activated-type behavior similar to Eq. (1) in the high-temperature limit. The theoretical interpretation of the temperature dependence of the photo-induced cytochrome oxidation in the bacterium *Chromatium* [8, 9] was the starting point of an impressive number of theoretical and experimental studies devoted to applications of electron transfer theories to biological systems.

In these systems, the redox centers are constituted by relatively large prosthetic groups which are held to the proteins by a complex folding of the peptide chain. This organization prevents the close approach of the centers, and results generally in long-range electron transfers. The weak residual interactions between the orbitals of the centers then place these processes in the non-adiabatic regime, so that the reaction rate depends on both an electronic and a nuclear factor. These two factors contribute to many basic properties of biological electron transfer systems. For example, in electron transport chains coupled to energy conversion in mitochondria and in photosynthetic systems, redox proteins are often organized in membrane-bound complexes where specific electron transfer processes are coupled to proton transfers. Since the electronic factor decreases very rapidly when the intercenter distance increases, a quasi-linear ordering of the centers within these complexes helps to achieve a directional electron transfer without short circuits. In photosynthetic systems, the attainment of high quantum yields requires very high forward rates and much slower recombination rates, particularly in the first steps of charge separation. In this case, the role of the nuclear factors is essential. Soluble redox proteins are also involved in some steps of these electron transport chains, and in many other electron transfer processes which are coupled to enzymatic reactions. These proteins participate in bimolecular reactions, whose specific character is ensured by electrostatic interactions between complementary side chains at the interface of the two partners. The conformation of the transient complex in which electron transfer takes place is determined by these interactions, and does not necessarily optimize either the nuclear factor, or the electronic factor; the latter is expected to be very sensitive to structural details. In this case, the selectivity of the reaction may result in a relatively slow transfer rate.

Although electron transfers in biological systems are generally expected to be non-adiabatic, it is possible for some intramolecular transfers to be close to the adiabatic limit, particularly in proteins where several redox centers are held in a very compact arrangement. This situation is found for example in cytochromes c_3 of sulfate-reducing bacteria which contain four hemes in a 13 kDa molecule [10, 11], or in *Escherichia coli* sulfite reductase where the distance between the siroheme iron and the closest iron of a 4Fe-4S cluster is only 4.4 Å [12]. It is interesting to note that a very fast intramolecular transfer rate of about $10^9\ s^{-1}$ was inferred from resonance Raman experiments performed in *Desulfovibrio vulgaris* Miyazaki cytochrome c_3 [13].

In early applications of the theory, some key parameters of the models, such as the reorganization energy λ or the parameter α that characterizes the decrease of the electronic factor with distance, were considered as adjustable or were fixed rather arbitrarily. Such treatments were not fully convincing, since they sometimes allowed the interpretation of a specific set of experimental data with different sets of parameters, and the validity of the theory could not even be considered as really tested. Since the beginning of the 1980s, important advances have been made in the theory, concerning, for instance, the influence of the nature of the intercenter medium on the electronic factor, and the effective calculation of the reorganization energy. The recent synthesis of model systems, in which a donor and an acceptor are separated by a bridge of known geometry, has facilitated a detailed test of theoretical predictions. Moreover, the fruitful discussions between experimentalists and theoreticians that began in 1979 at the Philadelphia Conference [14], were the starting point for an impressive number of experimental studies intended to characterize electron transfer reactions in biological systems. These studies, which have very remarkably exploited the new possibilities given by chemical modification and site-directed-mutagenesis techniques, are largely reported in several of the chapters of the present volume. In addition, the X-ray crystallographic structure of two bacterial photosynthetic reaction centers has been recently determined at a resolution of about 3 Å [15–21]. This allows precise theoretical calculations to be directly compared to the numerous kinetic data available on these systems, and many papers on this topic have already been published.

In biological systems, electron transfer kinetics are determined by many factors of different physical origin. This is especially true in the case of a bimolecular reaction, since the rate expression then involves the formation constant K_f of the transient bimolecular complex as well as the rate of the intracomplex transfer [4]. The elucidation of the factors that influence the value of K_f in redox reactions between two proteins, or between a protein and organic or inorganic complexes, has been the subject of many experimental studies, and some of them are presented in this volume. The complexation step is essential in ensuring specific recognition between physiological partners. However, it is not considered in the present chapter, which deals with the intramolecular or intracomplex steps which are the direct concern of electron transfer theories.

We will first review in Sect. 2 the physical basis of the theory for a non-adiabatic process, and show how standard simplifying assumptions result in the definition of an electronic and a nuclear factor for the reaction rate. The effective calculation of the electronic fator requires a realistic choice of the wavefunctions describing the initial and final electronic states of the system, and different models have been proposed for this. However, all these models lead to very similar concepts concerning the contribution of the medium orbitals to the value of the electronic factor. The validity of these concepts is confirmed by experimental data obtained in model systems. Next, we will examine in Sect. 3 how this general formalism can be used to obtain information from kinetic experiments performed in biological systems. The theoretical expression of the rate involves many independent parameters, and their determination requires very careful investigations. A convenient and often used procedure consists in the study of rate variations as a function of one parameter, the others being maintained constant. In practice, this last condition is very difficult to achieve in biological systems, and a definite interpretation of the results must await further experiments. The different approaches which have been used are illustrated by many examples concerning redox proteins and photosynthetic systems taken from the recent literature. Some recent theoretical studies devoted to the primary step of charge separation in bacterial photosynthetic reaction centers are also briefly presented.

2 The Physical Basis

Exhaustive reviews dealing with the applications of electron transfer theories to biological systems have been published recently [4, 22], and should be consulted for a general presentation of electron transfer processes as well as detailed mathematical developments. Shorter reviews are also available [23, 24]. In this section, we review the physical basis of the formalism generally used in the case of non-adiabatic processes, following a presentation similar to that given in Refs [7, 25, 26], with some modifications. In this formalism, the electron transfer rate is defined as the Boltzmann average of transition probabilities between two states represented by Born-Oppenheimer wave functions. This leads to the definition of a nuclear factor which determines the temperature dependence, and an electronic factor which plays a central role in the case of long-range electron transfers. The principal models that have been proposed to calculate these factors are then reviewed. We distinguish those in which the nuclear motions coupled to the process are represented by a set of harmonic oscillators, and others where some motions are treated classically. Turning to the electronic factor, we first examine the principle of semi-empirical determinations, and then present one-electron and many-electron theoretical calculations. We show that the validity of the bridge-assisted electron transfer concept is supported by experimental data obtained in model systems. At the end of this section, we indicate the various improvements to the theory which have been proposed over the last years.

All through this chapter, we have avoided the terminology "electron transfer by tunneling" which is rather confusing though it often appears in the literature [14]. The nature of the different tunneling effects involved in electron transfer processes is discussed in the previously cited reviews [4, 22, 23].

2.1 *Expression of the Electron Transfer Rate for a Non-adiabatic Process*

Electron transfer processes induce variations in the occupancy and/or the nature of orbitals which are essentially localized at the redox centers. However, these centers are embedded in a complex dielectric medium whose geometry and polarization depend on the redox state of the system. In addition, a finite delocalization of the centers' orbitals through the medium is essential to promote long-range electron transfers. The electron transfer process must therefore be viewed as a transition between two states of the whole system. The expression of the probability per unit time of this transition may be calculated by the general formalism of Quantum Mechanics.

2.1.1 *Calculation of the Transition Probability*

Let us consider an electron transfer system, whose Hamiltonian may be written:

$$\mathscr{H}(\mathrm{r}, \mathrm{Q}) = \mathrm{H}(\mathrm{r}, \mathrm{Q}) + \mathrm{T_N},$$

where r and Q represent the whole sets of electronic and nuclear coordinates respectively, H is the electronic Hamiltonian, and $\mathrm{T_N}$ the nuclear kinetic-energy operator defined by:

$$\mathrm{T_N} = -\Sigma_k (\hbar^2/2\mathrm{M}_k)\partial^2/\partial \mathrm{Q}_k^2 .$$

The system is assumed to be initially prepared in a vibronic state in which the donor center is reduced and the acceptor center oxidized, and we intend to find the transition probability to a vibronic state in which the donor is oxidized and the acceptor reduced. These two states, which of course are not stationary states of $\mathscr{H}$, are written as $\chi_{av}(Q)\,\psi_a(r, Q)$ and $\chi_{bw}(Q)\,\psi_b(r, Q)$ respectively, where ψ_a and ψ_b are normalized with respect to r for any value of Q:

$$\langle\psi_a|\psi_a\rangle = \langle\psi_b|\psi_b\rangle = 1.$$

It is convenient to seek a solution $\psi(r, Q, t)$ of the time-dependent Schrödinger equation:

$$\mathscr{H}\psi = i\hbar\,\partial\psi/\partial t, \tag{2}$$

in the form:

$$\psi(r, Q, t) = \chi_a(Q, t)\psi_a(r, Q) + \chi_b(Q, t)\psi_b(r, Q), \tag{3}$$

where the functions χ_a and χ_b are to be determined. By substituting expression (3) into Eq. (2), one finds that χ_a and χ_b must satisfy the following equation [7, 25, 26]:

$$(T_N + T''_{aa} + H_{aa} - i\hbar\,\partial/\partial t)\chi_a = -V_{ba}\chi_b \tag{4}$$

and Eq. 4′ obtained by permutation of a and b in Eq. 4, with:

$$H_{ij} = \langle\psi_i|H|\psi_j\rangle \qquad S_{ij} = \langle\psi_i|\psi_j\rangle$$

$$T_{ij} = (H_{ij} - S_{ij}H_{ii})/(1 - S_{ij}^2) \tag{5}$$

$$T'_{ij} = -\Sigma_k(\hbar^2/M_k)\langle\psi_j|\partial\psi_i/\partial Q_k\rangle\partial/\partial Q_k$$

$$T''_{ij} = -\Sigma_k(\hbar^2/2M_k)\langle\psi_j|\partial^2\psi_i/\partial Q_k^2\rangle$$

$$V_{ij} = T_{ij} + T'_{ij} + T''_{ij} - S_{ij}T''_{ii} \tag{6}$$

In the left-hand side of Eq. (4), second-order cross terms have been neglected.

The determination of the functions χ_a and χ_b that satisfy Eqs. (4) and (4′) and the boundary conditions:

$$\chi_a(Q, 0) = \chi_{av}(Q) \qquad \chi_b(Q, 0) = 0 \tag{7}$$

is straightforward when ψ_a and ψ_b are Born-Oppenheimer type wavefunctions, which depend only parametrically on the nuclear coordinates. This is the case for example if ψ_a and ψ_b are defined as eigenfunctions of two different zero-order Hamiltonians, $H^a(r, Q)$ and $H^b(r, Q)$. In this case, the T_{ij} given by Eq. (5) reduces to:

$$T_{ij} = (\langle\psi_i|H_{int}^i|\psi_j\rangle - S_{ij}\langle\psi_i|H_{int}^i|\psi_i\rangle)/(1 - S_{ij}^2) \tag{8}$$

with

$$H_{int}^i = H - H^i.$$

This formulation emphasizes the importance of the residual interactions H_{int}^i in the electron transfer process.

Another way of defining Born-Oppenheimer wavefunctions is to assume realistic forms for ψ_a and ψ_b and to optimize them variationally, by minimizing the energies H_{aa} and H_{bb} for any value of Q. We shall find later several examples of this procedure. The representation of the initial and final states by Born-Oppenheimer wavefunctions is of course not the most general, and we shall see in Sect. 2.3 that its validity has been questioned.

The functions χ_{av} and χ_{bw} which describe the nuclear motions in the initial and final states, are defined more precisely as belonging to the basis sets $\{\chi_{av'}\}$ and $\{\chi_{bw'}\}$ constituted by the solutions of the eigenvalue equations:

$$(T_N + T''_{aa} + H_{aa} - E_{av'})\chi_{av'} = 0$$
$$(T_N + T''_{bb} + H_{bb} - E_{bw'})\chi_{bw'} = 0 \quad (9)$$

These equations express that $(T''_{aa} + H_{aa})$ (Q) and $(T''_{bb} + H_{bb})$ (Q) constitute potential energy surfaces for the nuclear motions represented by $\chi_{av'}$ and $\chi_{bw'}$. The solution to our problem is readily found by substituting in Eqs. (4) and (4′) the following expansions:

$$\chi_a(Q, t) = \Sigma_{v'} C_{av'}(t)\, \chi_{av'}(Q) \exp(-iE_{av'}\, t/\hbar)$$

$$\chi_b(Q, t) = \Sigma_{w'} C_{bw'}(t)\, \chi_{bw'}(Q) \exp(-iE_{bw'}\, t/\hbar)$$

The coefficient $C_{bw}(t)$ is then obtained by a standard first-order perturbation calculation which takes into account the initial conditions defined by Eq. (7). This gives the transition probability per unit time from the initial state $\chi_{av}\psi_a$ to the isoenergetic continuum of states $\chi_{bw}\psi_b$ in the form:

$$W(av, bw) = (2\pi/\hbar) \left| \int_Q \chi^*_{bw}(Q) V_{ab}(Q) \chi_{av}(Q) dQ \right|^2 \rho(E_{bw}) \quad (10)$$

where $\rho(E_{bw})$ represents the density of final states.

It must be emphasized that this "golden rule" formulation is correct to the extent that V_{ab} is weak enough for the perturbation method to be valid, and is thus appropriate for non-adiabatic processes. We recall that when the interaction is strong enough for the reaction to move into the adiabatic regime, the rate becomes independent of these interactions [1, 2, 3]. It is also interesting to note that expression (10) was obtained by using the sets $\{\chi_{av'}\}$ and $\{\chi_{bw'}\}$ defined by Eqs. (9) through the expectation values H_{aa} and H_{bb} of the whole electronic Hamiltonian H. An alternative choice would be to use the zero-order Hamiltonians H^a and H^b that were introduced previously. It is easy to see that such a procedure would also lead to expression (10) at the first order of the perturbation calculation. The effective calculation of expression (10) requires several simplifying assumptions, which rest on a weak dependence of ψ_a (r, Q) and ψ_b (r, Q) on the nuclear coordinates Q. Usually it is postulated that this dependence is weak enough to ensure that:

i) the different contributions to $V_{ab}(Q)$ vary slowly with Q in the transition region $Q \sim Q^*$ where the nuclear functions χ_{av} and χ_{bw} overlap significantly. This assumption enables the factorization of $V_{ab}(Q^*)$ out of the integral in expression (10) (Frank-Condon factorization)

ii) the contribution T_{ab} (Q^*) is dominant in V_{ab} (Q^*). The validity of this hypothesis is discussed in particular in Ref. [22].
Both assumptions seem reasonable in the case of Born-Oppenheimer wavefunctions. Accordingly, the transition probability can be written:

$$W(av, bw) = (2\pi/\hbar)|T_{ab}(Q^*)|^2 |\int_Q \chi^*_{bw}(Q)\chi_{av}(Q)dQ|^2 \rho(E_{bw}). \tag{11}$$

This expression constitutes the basis of current interpretations of electron transfer processes in biological systems. From Eq. (9), the functions H_{aa} (Q) and H_{bb} (Q) represent potential energy surfaces for the nuclear motion described by χ_{av} and χ_{bw} respectively, if the weak "diagonal corrections" T''_{aa} and T''_{bb} are neglected. Then, the region $Q \sim Q^*$ where χ_{av} and χ_{bw} overlap significantly corresponds to the minimum of the intersection hypersurface between H_{aa} (Q) and H_{bb} (Q). Referring to definition (5), this implies:

$$T_{ab}(Q^*) = T_{ba}(Q^*)$$

This property ensures that expression (11) satisfies the micro-reversibility principle [27]. Note that this result is maintained if the T''_{ii} are kept in (6) and (9). Expression (5) of T_{ab} deserves a few comments. First, T_{ab} appears independent of the zero of energy of H due to the second term of the numerator [27]. Secondly, according to the Wolfsberg—Helmholz approximation [28]:

$$H_{ab} \sim K S_{ab}(H_{aa} + H_{bb})/2;$$

the two terms are expected to be of the same magnitude, and to vary roughly, as does the overlap integral S_{ab}, with orientation and distance.

2.1.2 Expression of the Electron Transfer Rate

The electron transfer rate is obtained by weighing the transition probabilities W(av, bw) given by expression (11) by the Boltzmann factors of all possible initial states. Actually, only one electronic state ψ_a is involved in the process, and the Boltzmann average is performed on the sole nuclear states χ_{av} of energy E_{av}. The rate is, therefore, expressed as the product of an electronic factor $|T_{ab}(Q^*)|^2$ by a nuclear factor which depends explicitly on temperature. In the following, we shall apply the terminology "electronic factor" equally to $|T_{ab}(Q^*)|^2$ and $T_{ab}(Q^*)$. The evaluation of the nuclear factor requires the calculation of thermally averaged squared overlap integrals between the functions χ_{av} and χ_{bw}. It is important to understand the nature of the nuclear motions that have to be considered in this calculation. Generally, electron transfer processes are accompanied by structural rearrangements, like bending or stretching of some bonds, and reorientations of the dipoles in the protein and in the surrounding solvent. The nuclear coordinates of the atoms or groups of atoms affected by these rearrangements then differ in the equilibrium configurations corresponding to the initial and final states. Some nuclear motions of the system lead to fluctuations of these coordinates around their equilibrium values which tend to cancel these differences. Those motions are relevant to the calculation of the nuclear factor, and are said to be coupled to the electron transfer process. However, in the effective calculation, the coupled motions are not represented by solutions of Eqs. (9) corresponding to a particular choice of ψ_a and ψ_b. Instead, they are directly described by simple models which are presented below.

In a first model, these motions are represented by harmonic vibrations, and the functions χ_{av} (Q) and χ_{bw} (Q) are then replaced by products of harmonic oscillator-like wavefunctions. The solutions of Eqs. (9) take this particular form when the T''_{ii} are negligible and when H_{aa} and H_{bb} can be expanded in terms of normal coordinates:

$$H_{aa} = C + \Sigma_n \tfrac{1}{2} k_n (Q_n - Q_{n,a})^2$$

$$H_{bb} = C + \Sigma_n \tfrac{1}{2} k_n (Q_n - Q_{n,b})^2 + \Delta U^0,$$

where $Q_{n,a}$ and $Q_{n,b}$ represent the equilibrium values of Q_n when the system is in state ψ_a and ψ_b respectively, and C is a constant. ΔU^0 is the difference between the minima of the two potential energy surfaces $H_{aa}(Q)$ and H_{bb} (Q). The expressions of the nuclear factor given by this harmonic model are obtained in closed form, and are valid at any temperature. They involve the energy variation ΔU^0, which can be equated to the enthalpy variation ΔH^0, and the characteristics of each oscillator, namely its frequency ω_n and its reorganization energy λ_n. This last quantity represents the energy that has to be supplied to the system assumed to be in state ψ_a, to change the oscillator coordinate from $Q_{n,a}$ to $Q_{n,b}$. It is given by (Fig. 1):

$$\lambda_n = \tfrac{1}{2} k_n (Q_{n,a} - Q_{n,b})^2$$

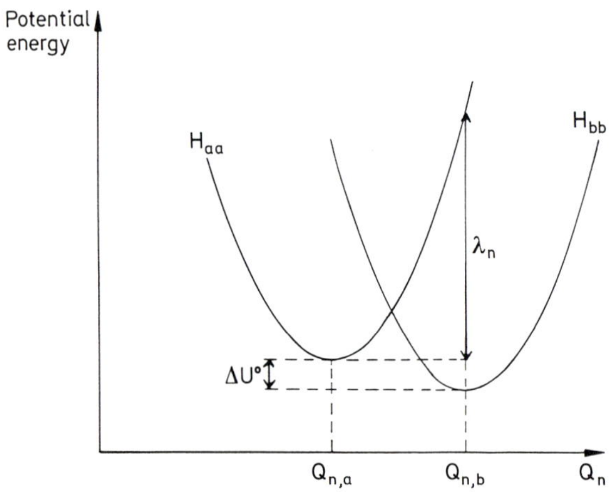

Fig. 1. Definition of the contribution λ_n to the reorganization energy. The figure represents the variations of the potential energy when only Q_n is varied, the other coordinates being kept constant at their equilibrium values

From the expressions given for example in Refs. [4, 9, 29], it can be seen that the nuclear factor, and consequently the electron transfer rate, becomes temperature independent when the temperature is low enough for only the ground level of each oscillator to be populated (nuclear tunneling effect). In the opposite limit where k_BT is greater than all the vibrational quanta $\hbar\omega_n$, the nuclear factor takes an activated form similar to that of Eq. 1 with ΔG^0 replaced by ΔU^0 [4, 9, 29]. The model has been refined to take into account the frequency shifts that may accompany the change of redox state [22].

As pointed out in Ref. [4], no entropy variation appears in the description given by the harmonic model, apart from the weak contribution arising from the frequency shifts of the oscillators. The applications of this model are then a priori restricted to redox reactions in which entropic contributions can be neglected. We shall see in Sect. 3 that the current interpretations of most electron transfer processes which take place in bacterial reaction centers are based on this assumption.

One may wonder whether a purely harmonic model is always realistic in biological systems, since strongly unharmonic motions are expected at room temperature in proteins [30, 31, 32] and in the solvent. Marcus has demonstrated that it is possible to go beyond the harmonic approximation for the nuclear motions if the temperature is high enough so that they can be treated classically. More specifically, he has examined the situation in which the motions coupled to the electron transfer process include quantum modes, as well as classical modes which describe the reorientations of the medium dipoles. Marcus has shown that the rate expression is then identical to that obtained when these reorientations are represented by harmonic oscillators in the high temperature limit, provided that ΔU^0 is replaced by the free energy variation ΔG^0 [33]. In practice, tractable expressions can be derived only in special cases, and we will summarize below the formulae that are more commonly used in the applications.

When the electron transfer process is coupled to classical reorientation modes and to only one harmonic oscillator whose energy quantum $\hbar\omega_v$ is high enough for only the ground vibrational level to be populated, the expression of the electron transfer rate is given by [4, 9]:

$$k = (2\pi/\hbar)|T_{ab}(Q^*)|^2 (4\pi\lambda_s k_B T)^{-1/2} \Sigma_{m=0}^{\infty} e^{-s} S^m/m! \exp[-(\lambda_s + m\hbar\omega_v + \Delta G^0)^2/4\lambda_s k_B T] \tag{12}$$

with

$$S = \lambda_v/\hbar\omega_v,$$

ΔG^0 is the free energy variation, and λ_s and λ_v the contributions to the reorganization energy arising from the dielectric medium and the oscillator, respectively. When the temperature is sufficiently high so that all the nuclear motions can be treated classically, the following expression applies [4, 33]:

$$k = (2\pi/\hbar)|T_{ab}(Q^*)|^2 (4\pi\lambda k_B T)^{-1/2} \exp(-\Delta G^*/k_B T) \tag{13}$$

The activation free energy ΔG^* is given by:

$$\Delta G^* = (\Delta G^0 + \lambda)^2/4\lambda \tag{14}$$

where λ now represents the whole reorganization energy. Usually, λ is analyzed in terms of two additive contributions, λ_i due to the redox centers, and λ_0 due to the medium, protein plus solvent. Expression (13) exhibits an activated type behavior similar to that given by Eq. (1), which was established in the adiabatic limit. Actually, when all the nuclear motions are described classically, it is possible to recover the adiabatic and non-adiabatic expressions as limiting cases of a more general formula [26, 34, 35].

We have seen above that a classical treatment of some nuclear motions allows the electron transfer rate to be expressed in Eqs. (12) and (13) in terms of the free energy variation ΔG^0. Although such a treatment is only valid at temperatures where $k_B T$ is greater than the characterisitic energy of these motions, this result is very important because numerous studies have demonstrated that large entropic factors contribute to the value of the redox potential of proteins in aqueous solution [36–41]. While this phenomenon is generally thought to originate from an important variation of the degrees of freedom of the solvent molecules, it is interesting to note that it is also observed in cytochrome *c* oxidase which is a membrane-bound protein [42]. Thus, in all these systems, the nuclear motions coupled to electron transfer processes cannot be simply described by harmonic vibrations, and more general expressions like (12) or (13) are needed to analyze kinetic data obtained at room temperature. We shall see in Sect. 3.2.1 that the entropy variation ΔS^0 is then explicitly taken into account in the interpretation of experimental results.

As pointed out by Warshel and co-workers, the derivation of the important relation (14) is based on the assumption of non-saturation of the dielectric medium, which does not necessarily applies in the case of a macromolecule in solution [43]. These authors have shown that the validity of relation (14) could be directly tested by simulating the dipole motions through molecular dynamics models [43, 44, 45]. Detailed numerical calculations were carried out for the self-exchange reaction of cytochrome *c* [43], and for the electron transfer between two benzene-like molecules in water [45]. A similar approach was recently developed for the system (Fe^{3+}, Fe^{2+}) in aqueous solution [46]. From these calculations, it was concluded that relation (14) applies provided that λ is evaluated from a microscopic model.

2.2 Calculation of the Electronic Factor

Definition (5) shows that T_{ab}, which is sometimes called the "electronic matrix element", represents the residual interaction resulting from the overlap of the wavefunctions ψ_a and ψ_b. These functions, which describe the initial and final electronic states of the whole system, respectively, depend closely on the nature of the redox centers and of the medium, so that reliable values of T_{ab} are very difficult to obtain from ab initio calculations in complex systems. For that reason, some authors have proposed determining T_{ab} semi-empirically by using the results of spectroscopic measurements. We begin by a brief presentation of these methods, then we review the different theoretical models which have been implemented to effectively calculate the electronic factor.

2.2.1 Semi-empirical Methods

We first recall that the T_{ab} value pertinent in the electron transfer problem is that evaluated for the nuclear configuration $Q \sim Q^*$, where the energy of the intersection surface of $H_{aa}(Q)$ and $H_{bb}(Q)$ is a minimum. In some systems, it may happen that ψ_a and ψ_b are closely related to stationary states of the Hamiltonian H, so that spectroscopic experiments performed on these states may provide useful information about the value of T_{ab} [47, 48]. To clarify this point, we expand the stationary states ψ_i $(i = 1, 2, \ldots)$ of $H(r, Q)$ in the form:

$$\psi_i(r, Q) = C_{ia}(Q)\psi_a(r, Q) + C_{ib}(Q)\psi_b(r, Q) + \ldots \tag{15}$$

Next, we assume that this expansion can be safely limited to the first two terms for the ground state ψ_1 and the first excited state ψ_2. For the nuclear configuration Q_b where the energy difference:

$$U = (H_{aa} - H_{bb})_{Q_b}$$

is much larger than both $|T_{ba}(Q_b)|$ and $|T_{ab}(Q_b)|$ (Fig. 2), the mixing coefficients and energies are given by:

$$C_{1a}/C_{1b} \sim -T_{ba}(Q_b)/U \qquad E_1(Q_b) \sim H_{bb}(Q_b) - T_{ba}^2(Q_b)/U$$

$$C_{2b}/C_{2a} \sim T_{ab}(Q_b)/U \qquad E_2(Q_b) \sim H_{aa}(Q_b) + T_{ab}^2(Q_b)/U$$

When the system occupies the stationary state ψ_1 for $Q \sim Q_b$, the weak mixing that appears in Eq. (15) is responsible for a charge-transfer absorption band corresponding to the transition from ψ_1 to ψ_2, whose energy is practically equal to U (Fig. 2). According to expression (5), $T_{ab}(Q_b)$ and $T_{ba}(Q_b)$ are not equal and their difference originates from the non-orthogonality of ψ_a and ψ_b, which plays an essential role in the formalism developed in Sect. 2.1.1. When the difference between $T_{ab}(Q_b)$ and $T_{ba}(Q_b)$ can be neglected, it is readily shown that the charge-transfer band intensity is proportional to $T_{ab}^2(Q_b)/U^2$ [47, 48]. Under these conditions, the value of $T_{ab}(Q_b)$ may be deduced from the measurement of the intensity of this band. In mixed-valence systems, this is called the intervalence band. Although $T_{ab}(Q_b)$ and $T_{ab}(Q^*)$ are a priori different [26, 49], they are expected to vary similarly as a function of the nature of the redox centers and of the medium.

When the electronic configuration of state ψ_b for instance is such that there is one unpaired electron on each redox site, it gives singlet ψ_b^s and triplet ψ_b^T states which are degenerate, whereas ψ_a is necessarily a singlet. By using Eq. (15) limited to ψ_a and ψ_b, we obtain two singlets ψ_1^s and ψ_2^s and one triplet ψ_1^T identically equal to ψ_b^T. The stabilization of ψ_1^s relative to ψ_1^T is then given by:

$$J = T_{ba}^2(Q_b)/U \tag{16}$$

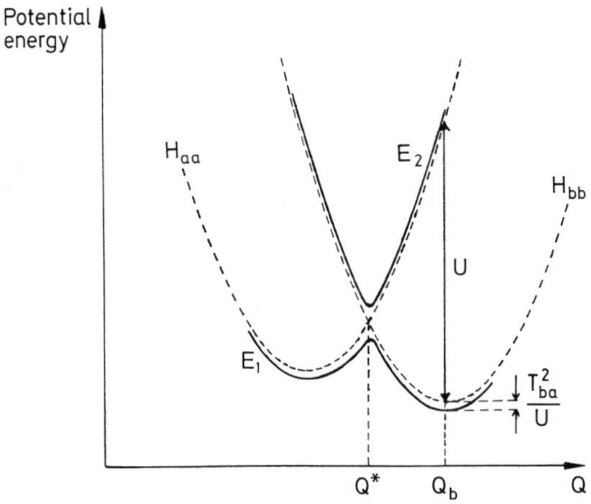

Fig. 2. Energies and nuclear configurations relevant to the semi-empirical methods discussed in the text

This simple model allows the evaluation of $|T_{ba}(Q_b)|$ from the measurement of the exchange parameter J if the energy U is known.

Returning to the general case, we find that for the nuclear configuration $Q \sim Q^*$ where H_{aa} and H_{bb} are equal, the mixing coefficients and the energies satisfy the following relationships:

$$|C_{1b}/C_{1a}| = |C_{2b}/C_{2a}| = 1 \tag{17}$$

$$E_2 - E_1 = 2|T_{ab}(Q^*)| \tag{18}$$

Thus, relation (18) gives directly the value of $|T_{ab}(Q^*)|$ as half the difference between the energies of the two stationary states ψ_1 and ψ_2, calculated at the nuclear configuration $Q = Q^*$. We shall see in Sect. 2.2.3 some examples of theoretical calculations of the electronic factor which are based on this property.

The validity of the preceding methods of evaluation of T_{ab} rest on the assumption that expansion (15) can be limited to ψ_a and ψ_b. This is not always warranted, since while ψ_1 and ψ_2 are uniquely defined from the Hamiltonian H, ψ_a and ψ_b are not and do not constitute a complete basis set [26]. Therefore, these semi-empirical methods are expected to give reliable results only when ψ_a and ψ_b are energetically well separated from other charge transfer-states. This condition is generally fulfilled in mixed-valence molecules, but not necessarily in biological systems where several other low-energy charge-transfer states often exist. This should be kept in mind in the calculation of the weak exchange parameter J, whose value is determined by configuration interactions between ψ_b^s and all singlet states, and between ψ_b^T and all triplet states [50]: while the simple relation (16) may lead to consistent interpretations in some systems [51], we shall see in Sect. 3.5 that more elaborate models are needed in others. Likewise, it will be found in Sect. 2.2.3 that the difference between the values of T_{ab} (Q^*) calculated from relations (5) and (18) may be significant.

2.2.2 One-Electron Theoretical Models

The first model used to evaluate T_{ab} in a biological system was the one-dimension tunneling barrier model initially proposed by Hopfield [8], and subsequently applied by different authors to a number of redox reactions. Although this model has proven conceptually useful, it is now recognized that it cannot be used for quantitative predictions, which require a full orbital description of the system. Now, it is important to realize that the orbitals which have to be explicitly considered in the calculation are essentially those whose nature and/or occupancy are different in states ψ_a and ψ_b. These orbitals are of course centered on the redox sites. In one-electron approximations, only one orbital is considered on each site and the full electronic Hamiltonian H is replaced by an effective monoelectronic Hamiltonian h_{eff} (r, Q) for the "transferable electron". Beside a kinetic energy term, h_{eff} includes a potential energy term due to the interactions with a core representing the nuclei and all the other electrons of the system which are considered as passive. One-electron models give relatively simple expressions, which very often contain the main physical effects that may be confirmed by more elaborate methods. However, they are not expected to give accurate quantitative descriptions, although useful information may be obtained by semi-empirical methods.

In the simplest one-electron model, ψ_a is replaced by φ_D and ψ_b by φ_A, φ_D and φ_A being the centers' orbitals occupied by the electron when it is on the donor and on the acceptor, respectively. Thus, expression (5) reduces to:

$$T^d_{ab} = (\langle\varphi_D|h_{eff}|\varphi_A\rangle - s_{DA}\langle\varphi_D|h_{eff}|\varphi_D\rangle)/(1 - s^2_{DA}) \tag{19}$$

where s_{DA} is the overlap integral between φ_D and φ_A. The suffix d implies that we are dealing with a direct interaction between φ_D and φ_A. The basic properties of T^d_{ab} may be deduced from simple calculations based on the representation of φ_D and φ_A by Slater atomic orbitals, and the use of a coulombic potential. Such calculations demonstrate that T^d_{ab} decreases very rapidly when the intercenter distance increases, and that it is very sensitive to the relative orientation of φ_D and φ_A in the case of *p* orbitals. These findings are conserved if φ_D, φ_A and h_{eff} are described by more elaborate models [52, 53].

In fact, since the early developments of electron transfer theories, it has been recognized that the magnitude of T_{ab} is greatly enhanced if ψ_a and ψ_b are delocalized through the intercenter medium [54, 55]. In the framework of one-electron models, this delocalization can be described by a mixing of φ_D and φ_A with the medium orbitals which leads to the so-called superexchange contribution. The origin of this contribution may be introduced as follows. If only one medium orbital φ_M interacts with both φ_D and φ_A, the initial and final states may be written:

$$\psi_a = N_a(\varphi_D + \alpha_a\varphi_M) \qquad \psi_b = N_b(\varphi_A + \alpha_b\varphi_M),$$

where N_a and N_b represent normalization factors which will be approximated by unity. The mixing coefficients α_a and α_b are determined variationally by minimizing the energies h_{aa} and h_{bb} defined by:

$$h_{aa} = \langle \psi_a | h_{eff} | \psi_a \rangle \qquad h_{bb} = \langle \psi_b | h_{eff} | \psi_b \rangle$$

Using expression (5), one finds for T_{ab}:

$$T_{ab} = T^d_{ab} + T^s_{ab}$$

where T^d_{ab} is the direct contribution given by expression (19), and the superexchange contribution T^s_{ab} is equal to:

$$T^s_{ab} = -t_{DM} t_{MA} / (h_{MM} - h_{aa}) \tag{20}$$

with

$$t_{ij} = h_{ij} - s_{ij} h_{aa} \qquad h_{ij} = \langle \varphi_i | h_{eff} | \varphi_j \rangle$$

The electronic factor of the reaction rate is obtained by evaluating (19) and (20) at the nuclear configuration $Q = Q^*$ where h_{aa} and h_{bb} are equal. Expression (20) is readily generalized to a sequence of *n* orbitals bridging φ_D and φ_A. For $Q = Q^*$, the main term can be written:

$$T^s_{ab}(Q^*) = [(-1)^n t_{D1} t_{12} \ldots t_{nA} / (h_{11} - h_{aa})(h_{22} - h_{aa}) \ldots (h_{nn} - h_{aa})]_{Q*} \tag{21}$$

The terminology "superexchange" was originally introduced to designate the contribution of bridging ligands to the exchange interactions between two paramagnetic centers [56]. We have already discussed in Sect. 2.2.1 the relation between the exchange parameter J and T_{ab} in the simple case of two centers characterized by $S = 1/2$.

From expression (21), it appears that the magnitude of the superexchange contribution depends on the interactions between adjacent orbitals along the bridge and on their energies. This leads to the concept of a favorable "path" for the electron transfer process, constituted by a sequence of orbitals which maximizes expression (21). This path may be composed of segments of different strength, ranging in the more extreme cases from "through bond" interactions involving covalent bonds, to simple "through space" interactions similar to those which determine the direct contribution T^d_{ab} [57]. This point will be discussed later in the context of biological systems. When all the bridge orbitals are identical, expression (21) predicts an exponential decrease of T^s_{ab} as a function of the number of bridging units.

Other approaches to the superexchange problem give general expressions essentially equivalent to Eqs. (20) and (21). They were extensively used to correlate T_{ab} values with intervalence band intensities in mixed-valence systems [49, 58, 59, 60], and to study theoretically the influence of the nature of the bridge on T_{ab} in different models for biological systems [61, 62, 63]. Actually, the very useful concept of bridge-assisted electron transfer has been mainly developed through one-electron models. However, the theoretical expressions given by these models involve quantities like t_{ij} whose physical meaning is not well defined. Usually, they are replaced by resonance integrals of the extended

Hückel method and are evaluated semi-empirically. The reliability of such a procedure cannot be readily assessed, and more elaborate many-electron models are needed to obtain accurate values of T_{ab}.

2.2.3 *Many-Electron Theoretical Models*

Many-electron models give a better description of the change of electronic state induced by the electron transfer process, because they are able to account for effects involving a large set of valence electrons. However, the functions ψ_a and ψ_b are then necessarily antisymmetric with respect to these electrons, so that the calculations become much more complicated than in one-electron treatments.

To illustrate this antisymmetrizing procedure, let us derive the equivalent to expression (21) for the system represented in Fig. 3, where the donor and the acceptor are represented by the molecular orbitals φ_D and φ_A, respectively, and each bridging unit by a doubly occupied molecular orbital φ_i and a vacant orbital φ_i'. The Hamiltonian of the system is written:

$$H = \Sigma_{1=1}^{2n+1} h(1) + \Sigma_{1<m} \frac{e^2}{r_{1m}} + H_c,$$

where h(1) is the mono-electronic Hamiltonian relative to electron 1, and H_c is the core Hamiltonian. Note that these two terms depend on the nuclear coordinates Q. The ground configurations corresponding to the initial and final states are represented by Slater determinants:

$$\Phi_D = N_D |\varphi_D \varphi_1 \bar{\varphi}_1 \; \varphi_2 \bar{\varphi}_2 \ldots \varphi_n \bar{\varphi}_n|$$

$$\Phi_A = N_A |\varphi_A \varphi_1 \bar{\varphi}_1 \; \varphi_2 \bar{\varphi}_2 \ldots \varphi_n \bar{\varphi}_n|$$

N_D and N_A being two normalization factors. The delocalization of ψ_a and ψ_b through the bridge is conveniently described by a mixing of Φ_D, Φ_A with excited charge-transfer configurations implying the bridge orbitals. This method was originally proposed by Halpern and Orgel [54], and was later used in mixed-valence systems [58, 59, 60]. From the orbitals represented in Fig. 3, two kinds of charge-transfer states can be constructed [64]:

i) electron transfer state Φ_i' corresponding to the transfer of an electron from φ_D to φ_i'
ii) hole transfer states Φ_i corresponding to the transfer of an electron from φ_i to φ_A or equivalently of a hole from φ_A to φ_i.

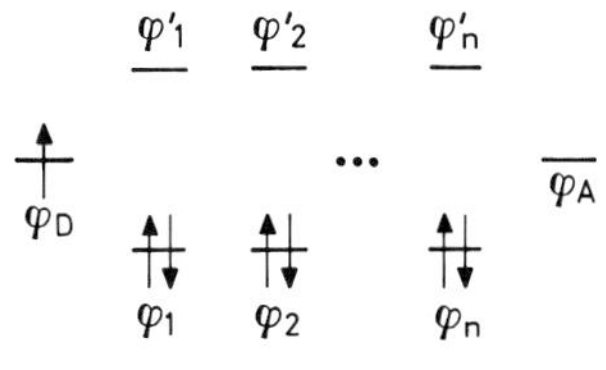

Fig. 3. Molecular orbitals used in the many-electron formulation of bridge-assisted electron transfer

Owing to the interactions between adjacent orbitals along the bridge, these charge-transfer states are mixed with Φ_D and Φ_A. The initial and final states are then taken in the form:

$$\psi_a = N_a[\Phi_D + \Sigma_{i=1}^{n} (a_i\Phi_i + a_i'\Phi_i')]$$

$$\psi_b = N_b[\Phi_A + \Sigma_{i=1}^{n} (b_i\Phi_i + b_i'\Phi_i')]$$

and the mixing coefficients are determined variationally for every value of Q. Using definition (5), T_{ab} is then calculated at the nuclear configuration $Q \sim Q^*$ where the surfaces H_{aa} (Q) and H_{bb}(Q) cross. The final result can be written [64]:

$$T_{ab}(Q^*) = T_{ab}^{m}(Q^*) + T_{ab}^{se}(Q^*) + T_{ab}^{sh}(Q^*),$$

where

$$T_{se}^{m}(Q^*) = (H_{DA} - S_{DA}\, H_{aa})_{Q*}$$

$$T_{ab}^{se}(Q^*) = [(-1)^n T'_{D1}\, T'_{12} \ldots T'_{nA}/(H'_{11} - H_{aa}) \ldots (H'_{nn} - H_{aa})]_{Q*}$$

$$T_{ab}^{sh}(Q^*) = [(-1)^n T_{A1}\, T_{12} \ldots T_{nD}/(H_{11} - H_{aa}) \ldots (H_{nn} - H_{aa})]_{Q*}$$

$$T'_{ij} = H'_{ij} - S'_{ij}\, H_{aa} \qquad H'_{ij} = \langle\Phi_i'|H|\Phi_j'\rangle \qquad S'_{ij} = \langle\Phi_i'|\Phi_j'\rangle$$

$$T_{ij} = H_{ij} - S_{ij}\, H_{aa} \qquad H_{ij} = \langle\Phi_i|H|\Phi_j\rangle \qquad S_{ij} = \langle\Phi_i|\Phi_j\rangle,$$

The expressions of T_{ab}^{se} and T_{ab}^{sh} represent a many-electron formulation of the superexchange contributions, and become equivalent to (21) if bi-electronic terms are omitted from the Hamiltonian H. The "multiple exchange" contribution T_{ab}^{m} results from the antisymmetrical character of ψ_a and ψ_b, and generalizes the "double exchange" term considered by Halpern and Orgel [54]. Among the numerous terms that appear in the expansion of T_{ab}^{m}, one may mention a term equivalent to the direct contribution T_{ab}^{d}, and a main term which results from the overlap between neighboring orbitals along the bridge [64]. The electronic factor T_{ab} then includes three different bridge-assisted contributions. Although their relative magnitude depends on the nature of the system, simple considerations based on the expressions given in Ref. [64] together with the Wolfsberg–Helmholz approximation cited in Sec. 2.1, suggest that the superexchange contributions are predominant in biological systems in which relatively low-energy charge-transfer states are available.

The preceding many-electron formulation is useful to separate the different contributions to T_{ab}, and could be easily improved to include the whole set of valence electrons of the system. However, it is not certain that this method is the most convenient for the effective calculation of the electronic factor. Actually, in the very few many-electron calculations that have been reported in the literature, T_{ab} was evaluated globally. These studies are summarized below.

Larsson and co-workers have used relation (18) to calculate T_{ab} for organic molecules in which two centers are bridged by saturated groups [65, 66], and for mixed valence systems [67]. The stationary states ψ_1 and ψ_2 are determined by a CNDO/S method, with extensive configuration interaction and use of semi-empirical parameters. The nuclear configuration Q* where relation (18) is valid is adjusted so as to satisfy the delocalization property expressed by (17). These calculations yield values of electronic factor in satisfactory agreement with the experimental data.

Newton developed a many-electron model to calculate the electronic factor of the self-exchange reaction between Fe^{2+} and Fe^{3+} in aqueous solution [25, 26, 68, 69]. In this model, the ψ_a and ψ_b states corresponding to the charge-localized electronic configurations [Fe^{2+} $(H_2O)_6$, Fe^{3+} $(H_2O)_6$] and [Fe^{3+} $(H_2O)_6$, Fe^{2+} $(H_2O)_6$] respectively, are represented by SCF-HF functions, which are determined variationally by minimizing H_{aa} and H_{bb}. All the valence electrons are explicitly considered in the Hamiltonian, and inner-shell electron contributions are included through ab initio effective core potentials. T_{ab} was calculated from expression (5) for different relative geometries of the two octahedral complexes, which were supposed in Van der Waals contact through hydrogen atoms. Extensive studies have demonstrated that, for a given geometry, T_{ab} is weakly sensitive to the value of the nuclear coordinates, which are identified to the Fe–O distances in this model. This result justified the Franck–Condon factorization in this system [69]. Newton also compared the values of T_{ab} obtained from expressions (5) and (18), and found that they may differ by about 50% [69]. The origin of this discrepancy is discussed in Ref. [69]. These calculations of the electronic factor were recently extended to other [ML_6^{2+}, ML_6^{3+}] systems (M = Fe, Co, Ru and L = H_2O, NH_3) in a study of the influence of some important parameters, like the nature and the symmetry of the orbitals occupied by the valence electrons [70, 71]. It was shown that the results could be satisfactorily reproduced by effective one-electron expressions in which ligand-mediated contributions appear explicitly [70]. One noteworthy result of all these studies is that, contrary to what had been believed previously, the value of T_{ab} for electron transfers between inorganic ions in solution may be weak enough for these reactions to occur in the non-adiabatic regime.

While it could be thought that detailed numerical calculations are hardly conceivable in biological systems, we shall see in section 3.5 that T_{ab} values have already been calculated from the recently determined structure of bacterial photosynthetic reaction centers. Obviously, in biological systems elaborate calculations are faced with particular difficulties arising from the complex structure of the protein medium. A new approach to this problem was recently proposed, in which the protein regions that mainly contribute to the electronic factor are identified [72]. The model is based on the evaluation of path integrals by a Monte Carlo method, and the use of semi-empirical pseudo-potentials to define the potential energy of an extra electron for each residue of the protein. This method is described in the chapter by Kuki. It would be interesting to know if this very promising method can yield numerical values which may be compared with experimental data.

2.2.4 Experimental Test of Bridge-assisted Electron Transfer Models

For the time being, qualitative or semi-quantitative analysis of the influence of the medium on the electronic factor value rest on the description given by the bridge-assisted electron transfer models which were previously presented. It is then important to examine to what extent this physical description is supported by experiments performed in model systems.

In a first type of systems, redox centers are randomly distributed in a rigid matrix, glass or polymer [73, 74, 75]. Donor or acceptor centers are initially created photochemically or by pulse radiolysis, and the study of the return to equilibrium of the system allows the determination of the law k(R) giving the rate variation as a function of the intercenter distance R. The experimental data are well described by an exponential law, which is considered as reflecting an exponential variation of the electronic factor:

$$T_{ab} \propto \exp(-\alpha R) \tag{22}$$

where α is found to be weakly dependent on the nature of the medium and falls in the narrow range 0.5 to 0.7 $Å^{-1}$. These observations were interpreted in terms of a superexchange model implying the solvent molecules [73, 74, 75]. One may notice that possible variations of the nuclear factor with R, which may be important in some cases [76], were neglected in these analysis.

A second type of model system is provided by organic molecules in which intramolecular electron transfers can be photochemically triggered between a donor and an acceptor linked by a rigid bridge. This bridge is usually built from saturated groups whose number and relative geometry may be modified (see Ref. [77] for a recent review). It is expected that such modifications do not notably affect the redox potential of the centers, so that the nuclear factor variations essentially originate from the contribution λ_0 of the reorganization energy. The magnitude of this contribution can be estimated, thus allowing the determination of the electronic factor variations from the experimentally measured rates [77, 78, 79]. The results of a thorough analysis of the T_{ab} dependence on the number of bridging groups, on their relative geometry and on the attachment stereochemistry of the redox centers are in complete agreement with the qualitative predictions of bridge-assisted electron transfer models [77, 78, 79, 80, 81]. In particular, for a given geometry of the bridge, the dependence on the number of bridging groups may be expressed as a distance dependence similar to relation (22), with α equal to 0.4–0.5 $Å^{-1}$. Moreover, the calculations that have been performed on these systems by ab initio many-electron models account satisfactorily for the experimental results [65, 66, 82].

Similar experimental studies were carried out on molecules in which the bridge is constituted by conjugated groups [83, 84]. General references on photoinduced intramolecular electron transfers may also be found in two recent reviews [85, 86].

2.3 Further Development of the Theory

In Sect. 2.1, the electron transfer rate was defined as the Boltzmann average of transition probabilities, which were calculated through time-dependent perturbation theory by using the Born-Oppenheimer and Frank-Condon approximations. This allowed the rate to be expressed in a compact form, in terms of well-defined parameters whose physical significance is clear. In practice, this formalism does currently constitutes the framework that is usually referred to for the interpretation of experimental data obtained in biological or model systems. However, this "standard" theory involves several assumptions which are not easy to justify from general considerations, and numerous studies have recently been devoted to its improvement. These studies are summarized below, to emphasize the directions in which significant advances are to be expected.

One basic feature of the standard theory that has been criticized is the use of the Born-Oppenheimer approximation to describe initial and final states. While this approximation seems in most cases to be sufficient to calculate the stationary states of molecules and to determine their energetic and spectral properties, its validity to represent initial and final states for long-range electron transfers has been questioned [87, 88, 89]. The argument is that Born-Oppenheimer wavefunctions might fail to reproduce the tails of the true wavefunctions, whose asymptotic behavior determines the electronic factor of the process. As a matter of fact, it was shown that the use of non Born-Oppenheimer wavefunctions in the reaction rate expression could bring about new effects [89]. However, to our knowledge these expressions were always explicitly or implicitly derived within the Born-Oppenheimer approximation, so that a general treatment of these effects remains to be done. A recent study by Deumens and Ohrn [90] constitutes an approach in this direction. The assumption of a weak dependence of ψ_a and ψ_b on the nuclear coordinates, which is invoked to derive expression (11), is no longer warranted in the case of long-range electron transfers. This is because the wavefunctions' tails are then expected to be very sensitive to the relative geometry of the different groups which constitute the effective "bridge", even if the Born-Oppenheimer approximation is used. In this situation, some peculiar nuclear motions that would induce large deformations of the bridge might preclude the Frank-Condon factorization [64]. Different models have been proposed to treat such a situation [57, 89, 91, 92]. It may be noticed that an extreme case of "non Condon" effect could lead to the "conformational gating" phenomenon which is described in Sect. 3.4.3.

The formulation of the electron transfer rate as a Boltzmann average of transition probabilities is valid if two conditions are fulfilled: i) the initial vibronic states are in thermal equilibrium and ii) their energies are well defined. Thermal equilibrium is achieved through relaxation processes that ensure the coupling of the nuclear motions with the bath. Condition i) is then satisfied if the relaxation processes are sufficiently efficient so that the Boltzmann populations are not perturbated by electron transfer transitions. Jortner has examined the consequences of a failure of condition i) and has envisaged the occurrence of such a situation in the case of the ultrafast primary electron transfer in bacterial photosynthetic reaction centers [93]. The current interpretation of this process is discussed in section 3.5. Now, the other limiting case is reached when the efficiency of relaxation processes is so high that the homogeneous linewidth of the vibronic levels becomes comparable to their spacing and condition ii) is no longer valid. This situation has been treated theoretically, and new effects have been predicted [89, 94, 95]. In particular, it may happen that the rate becomes dependent on the relaxation properties of the medium but no longer on T_{ab}.

Some authors have described the time evolution of the system by more general methods than time-dependent perturbation theory. For example, Warshel and co-workers have attempted to calculate the evolution of the function $\psi(r, Q, t)$ defined by Eq. (3) by a semi-classical method [44, 96]: the probability for the system to occupy state ψ_b is obtained by considering the fluctuations of the energy gap between H_{aa} and H_{bb}, which are induced by the trajectories of all the atoms of the system. These trajectories are generated through molecular dynamics models based on classical equations of motion. This method was in particular applied to simulate the kinetics of the primary electron transfer process in the bacterial reaction center [97]. Mikkelsen and Ratner have recently proposed a very different approach to the electron transfer problem, in which the time evolution of the system is described by a time-dependent statistical density operator [98, 99].

Although their conceptual basis is now firmly established, non-adiabatic electron transfer processes are still the subject of intensive theoretical studies. Nevertheless, the framework provided by the standard formalism presented in this section seems sufficiently general to be used for the interpretation of kinetic data obtained in biological systems. Owing to the great number of parameters involved in the theoretical expressions, attainment of useful information requires obtaining numerous data by elaborate experiments. The next section is devoted to a review of the different approaches that have been developed over the last few years.

3 Application to Biological Systems

In the preceding section, we have seen that the expressions given by electron transfer theories may depend on numerous variables. This is particularly true in biological systems which are characterized by a great number of degrees of freedom, and we first examine in this section the physical nature of the different parameters involved by the theory in these systems. The experimental determination of these parameters is the subject of intensive studies which are well represented in the various topics treated in the present volume. The following are some typical approaches that have been implemented:

— temperature dependence of the electron transfer rate,
— variation of the rate as a function of the driving force,
— modification of the intercenter medium.

These are introduced and illustrated by a few examples, and the nature of the information likely to be obtained is discussed. We end this section by a presentation of some recent theoretical works intended to elucidate the mechanism of the primary electron transfer in bacterial photosynthetic reaction centers. These studies illustrate remarkably well the high level currently achieved in the applications of electron transfer theories to biological systems.

3.1 Nature of the Parameters Involved by the Theory in Biological Systems

We recall that two factors of different nature may be distinguished in the rate expression:

1) a nuclear factor determined by the motions coupled to the process, which involves energetic quantities characteristic of the reaction. This factor is in particular responsible for the temperature dependence.
2) an electronic factor determined by the residual interactions between the donor and acceptor centers, which are mediated by the admixture of their orbitals with medium orbitals.

The possible contributions to these factors in biological systems are briefly reviewed.

3.1.1 Contributions to the Nuclear Factor

In the high temperature limit where all the nuclear motions coupled to the process can be described classically, the nuclear factor is expressed in terms of only two parameters: the driving force of the reaction ΔG^0, and the whole reorganization energy λ (expressions (13) and (14)). Detailed calculations carried out in the case of cytochrome *c* have demonstrated that ΔG^0 is a complex quantity, which depends not only on the electronic properties of the redox centers but also on those of the protein and of the surrounding solvent [100]. Usually, ΔG^0 can be evaluated from measurements of redox potentials and of eventual interaction energies between the different parts of the systems (Appendix).

The magnitude of the reorganization energy reflects the importance of the conformational variations that accompany the redox state changes of the system. Such rearrangements may first occur at the redox centers themselves. Although they are expected to be weak for chlorophylls, flavins and quinones in which the valence electrons are largely delocalized, and also in the case of copper centers [101] and low-spin hemes [102, 103], more important effects may be observed in centers built from high-spin transition metal ions, like iron-sulfur clusters [104] and some heme centers [105]. The redox reaction may also induce conformational changes within the protein [102, 106] and reorientations of the solvent dipoles [43].

To these possible contributions to the reorganization energy correspond different types of nuclear motions that may be coupled to electron transfer processes. Owing to their low energy, the motions of the solvent molecules are usually treated classically, even at low temperature. This is also allowed at room temperature for the low-frequency protein modes. Such modes have a collective character, or at least involve large parts of the molecule [31, 107, 108]. On the other hand, the energy of internal vibrations of the redox centers, and of localized protein modes is high enough for their quantified character to appear at room temperature. In this case, we have seen that the nuclear factor depends explicitly on their frequency and on their contribution to the reorganization energy (expression (12)).

3.1.2 Contributions to the Electronic Factor

The electronic factor also involves numerous quantities, and we first examine those originating from the redox centers themselves. Of prime importance is the nature of the orbitals effectively occupied by the valence electrons. According to expressions (20) and (21), the energy, the symmetry and the degree of delocalization of these orbitals determine their interactions with the medium orbitals. These properties are for instance different for a low-spin heme, whose valence electrons are largely localized in the iron *d*-orbitals [109], and for a porphyrin radical whose valence electrons are completely delocalized over the ring. In some electron transfer processes, the spin state of the centers also plays an important role: when the redox reaction involves a net variation of the total spin of the system, T_{ab} differs from zero only through the weak mixing brought about in ψ_a and ψ_b by the spin-orbit coupling terms of the two centers, and the electronic factor is expected to be greatly reduced.

The concept of a favorable path for the transfer, which results naturally from bridge-assisted electron transfer models, is much more difficult to apply in biological systems than in the organic molecules mentioned in Sect. 2.2.4, in which the two centers are linked by a covalent bridge. Although such a bridge could be built from the protein peptide bonds, it could only connect redox centers belonging to the same unit through a complex folding of the polypeptide chain. We therefore infer that efficient paths generally involve a combination of interactions through bonds of different strength like covalent bonds, hydrogen bonds, salt bridges, and also through space interactions [57]. In such a situation, the heterogeneous character of the bridge would preclude the reduction of expression (21) to an exponential law, so that the description of the distance dependence of the electronic factor by relation (22) would lack theoretical support. However, this heterogeneous character could be somewhat attenuated by an averaging effect, if the centers were interacting through a set of parallel paths whose contributions would then add algebraically in first approximation [61]. This hypothesis is suggested by the experimental results obtained in model systems, which were reported in Sect. 2.2.4: the value of the parameter α defined by relation (22) appears to be nearly the same in systems where donor and acceptor centers are randomly distributed in a rigid solvent, and in others where the centers are linked by a covalent bridge. A possible interpretation of this striking result could be that numerous paths implying weak non-bonded interactions may be as efficient as a single bridge built from covalent bonds. Although this important question clearly deserves further experimental and theoretical studies, one may observe that the overall general structure of proteins seems favorable to the formation of such parallel paths.

3.2 Study of the Temperature Dependence of the Electron Transfer Rate

A meaningful comparison of kinetic data obtained in different biological systems should be based on the determination of the respective contributions of the nuclear and electronic factors. The most direct method of separating these contributions consists in the measurement of the temperature dependence of the rate over the widest available range. In the following, we distinguish between experiments performed at room temperature, which are usually interpreted by assuming that all the nuclear motions coupled to the transfer may be described classically, and experiments performed at lower temperature, in which the quantified character of particular vibrational modes may appear.

3.2.1 Classical Treatment of the Nuclear Factor

If the temperature is high enough for all the pertinent nuclear motions to be treated classically, the electron transfer rate is given by expression (13). The activation energy E_a defined by:

$$E_a = -R\partial L_n k/\partial(1/T)$$

is then equal to:

$$E_a = -RT/2 + (\lambda + \Delta G^0)^2 (1 + 2T\Delta S^0/(\lambda + \Delta G^0))/4\lambda, \qquad (23)$$

where

$$\Delta S^0 = -\partial \Delta G^0/\partial T$$

is the entropic contribution to ΔG^0. Expression (23), which is equivalent to expression (35) of Ref. [4], is valid if the temperature dependence of λ can be neglected in the experimental range of temperature. This expression allows the determination of λ from the measurement of E_a, if ΔG^0 and ΔS^0 are known. In the case of redox proteins in solution, we have pointed out in Sect. 2.1.2 that large entropic contributions have been measured, so that ΔS^0 cannot be neglected in Eq. (23). This equation leads to two solutions, λ_1 and λ_2. For $\Delta G^0 < 0$, one solution places the reaction in the normal region ($-\Delta G^0 \leqslant \lambda$), and the other in the inverted region ($-\Delta G^0 \geqslant \lambda$). In particular, the situation $E_a \sim 0$, may occur for two distinct values of λ:

$$\lambda \sim -\Delta G^0 \qquad \lambda \sim -\Delta G^0 - 2\,T\,\Delta S^0$$

that may differ appreciably when entropic effects are important. From expression (23), it is readily seen that when the λ value corresponding to the normal region is such that:

$$|\Delta G^0/\lambda| \ll 1,$$

the relation between E_a and λ takes the simplified form:

$$E_a \sim \lambda/4 + \Delta H^0/2$$

We may illustrate this approach to the determination of the nuclear factor by the elegant studies performed by Gray and co-workers, who have determined the thermodynamic properties and the rate temperature dependence for the electron transfer between $Ru(NH_3)_5^{2+}$ covalently bound to the histidine residues of some proteins, and the redox center of these proteins [110, 111, 112, 113]. The experimental results obtained for cytochrome *c* [110] and azurin [111, 112] are very similar. Using the thermodynamic data and the value or the upper limit of E_a reported in these studies, we deduce from Eq. (23):

cytochrome *c*: 0.9 eV $< \lambda_1 <$ 1.2 eV (normal) 0.15 eV $< \lambda_2 <$ 0.19 eV (inverted)

azurin: 0.9 eV $< \lambda_1 <$ 1.1 eV (normal) 0.18 eV $< \lambda_2 <$ 0.23 eV (inverted)

In the case of myoglobin [113], the only possible solution for λ is equal to 2.3 eV. We now postulate an additive property for the reorganization energy, which seems reasonable for these systems:

$$\lambda(Ru^{2+} - \text{protein ox}/Ru^{3+} - \text{protein red}) \sim \lambda(Ru^{2+}/Ru^{3+}) + \lambda\,(\text{protein ox/protein red})$$

The value of λ (Ru^{2+}/Ru^{3+}) may be estimated to be about 0.6 eV from the measured activation free energy of the $[Ru(NH_3)_5\text{ pyridine}]^{2+/3+}$ electron exchange reaction [114]. We then deduce that all the processes occur in the normal region, and that the protein reorganization energy is about 0.3 to 0.6 eV for cytochrome *c*, 0.3 to 0.5 eV for azurin, and 1.7 eV for myoglobin. The value estimated for cytochrome *c* is somewhat larger than the value 0.14 eV obtained from a microscopic electrostatic calculation using the X-ray crystallographic structure of the oxidized and reduced forms of the protein [43]. The large λ value found for myoglobin may be attributed to the high-spin character of the heme [113], and to the coordination change that accompanies its reduction [105]. From these values of λ, expression (13) could be used to compare the electronic factors of the three electron transfers, which take place over similar distances. However, these values were obtained by assuming that the relevant nuclear motions can be treated classically at room temperature, and should be confirmed by experiments performed in a wider range of temperature. This is especially true in the case of myoglobin, since the important rearrangements that occur at the heme suggest a coupling with relatively high energy internal modes.

Kinetic experiments were performed in other ruthenium-derivative proteins [115, 116, 117, 118], but it is difficult to compare their results with those previously reported as long as the temperature dependence of the rate has not been measured.

3.2.2 Quantum Treatment of the Nuclear Factor

In order to estimate the nature of the nuclear modes coupled to the process, it is necessary to determine the temperature dependence over a wide range of temperature and to describe it using expressions similar to Eq. (12), or more general expressions in which a few frequencies replace the whole spectrum of relevant nuclear modes [9, 22]. When entropic contributions are important, a complication may arise from the temperature dependence of the driving force ΔG^0. This effect has been disregarded in the interpretations of the primary steps of charge separation in photosynthetic reaction centers, which have been reported to be characterized by very weak values of ΔS^0 [119]. In the following, we summarize the main results that have been obtained in this way on these systems, concerning the nature of the nuclear modes coupled to the transfer, and the magnitude of the contributions to the reorganization energy. However, these interpretations could be revised, since large entropic contributions have been measured in other studies [202, 203]. A recent presentation of the reaction center structure and function may be found in Ref. [21], and a schematic representation of the system is given in Fig. 4 in Sect. 3.5.

The first biological electron transfer process that was studied theoretically is the famous reduction of $(\mathrm{Bchl})_2^+$ by a cytochrome in *Chromatium* reaction centers. Since Hopfield's original paper [8], several successive interpretations have been proposed. In the early models, the remarkable rate temperature dependence was ascribed to the coupling with a high-frequency mode of the redox centers [8, 9, 48, 120]. Such a description implied large values for both the reorganization energy λ and the electronic factor, which were considered unrealistic after the ab-initio calculation of λ in cytochrome *c* [43], and the determination of an intercenter distance $R \sim 21$ Å in the *Pseudomonas viridis* reaction center [15, 16]. A new interpretation was then formulated by Jortner and Bixon, in which the data are accounted for by two parallel processes implying the two low potential hemes of the cytochrome. One reaction is activationless with:

$$\lambda \sim -\Delta G^0 = 0.43 \text{ eV}$$

and the other in the inverted region with λ equal to 0.14 eV [121, 122]. In both reactions, the nuclear modes coupled to the process are low-frequency protein modes of energy $\hbar\omega \sim 100$ cm^{-1}. This model leads to electronic factors that were judged compatible with the expected intercenter distances [121, 122]. Another very different interpretation, which is certainly not the last, has been proposed recently to account for the remarkable temperature dependence of this reaction rate [123].

Since the hypothesis of relatively low-frequency coupled modes seems to account for the experimental data for several electron transfers in the reaction center [122], it is interesting to seek their physical origin. Low-frequency modes were predicted by normal mode calculations performed on different proteins [31, 107, 108], which were recently confirmed by neutron diffraction experiments [108]. Their existence is also supported by measurements of the electronic spin-lattice relaxation time of several redox proteins [124]. Moreover, recent experiments carried out on cytochrome *c* provide some evidence for the coupling of such modes to electron transfer processes in proteins. For example, Eden and co-workers have shown that the oxidized form of this protein undergoes more low-frequency large amplitude motion than the reduced form [125]. Similarly, X-ray scattering data collected in solution at room temperature indicate that, at physiological ionic strength, the overall conformations of the oxidized and reduced form differ appreciably [106]. This suggests that low-frequency nuclear motions involving large parts of the protein may be coupled to the electron transfer process.

Now, the contribution of these low-frequency modes could be preponderant in the case of the reaction center for two reasons. First, the nature of the redox centers is such that their valence orbitals are largely delocalized, so that the contribution λ_i is likely to be weak, and therefore high-frequency internal modes are expected to be weakly coupled (see, however, Sect. 3.3.3). Secondly, the contribution to the reorganization energy arising from the polarization of the protein and of the solvent is also very weak, due to the apolar character of the medium [97]. These properties could explain the relatively low value of the total reorganization energy λ, whose main contribution would therefore arise from internal rearrangements of the protein.

It is evident that the preceding considerations do not apply to all biological electron transfer systems. Even in the bacterial reaction center, the transfer between the two quinones $Q_A^- \rightarrow Q_B$, which takes place over 18 Å [18], is characterized in *Rhodobacter sphaeroides* by a large entropic contribution, which has been attributed to the high solvent exposure of Q_B [126]. By using the activation energy value reported in Ref. [126], two very different λ values may be deduced from Eq. (23): $\lambda_1 = 0.1$ eV and $\lambda_2 = 2.5$ eV. The previous considerations are in favor of the higher one, which could largely originate from reorientations of the protein and solvent dipoles [21]. This result could be confirmed by experiments performed over a wider range of temperature. Hoffman and co-workers have measured the rate of the electron transfer between the triplet state of a Zn porphyrin and a high-spin ferric heme between 77K and 300K in a [Zn, Fe] hybrid hemoglobin [127]. Although this transfer is characterized by a variation of the total spin ($\Delta S = 1$), it takes place over 25 Å. The temperature dependence is well described by assuming the coupling to an internal mode of energy $\hbar\omega = 200\ \text{cm}^{-1}$. If the entropic contribution ΔS^0 can be neglected, the experimental activation energy together with Eq. (23) lead to $\lambda_1 = 0.27$ eV, $\lambda_2 = 2.3$ eV. As in myoglobin, the reaction is accompanied by a coordination change of the high-spin heme, so that the highest value seems the most probable [127]. At room temperature, variations of the transfer rate are observed when the axially coordinated water molecule is replaced by other ligands, but more experiments are needed to determine the respective contributions of the electronic and nuclear factors in these variations [128].

3.3 *Study of the Driving Force Dependence of the Electron Transfer Rate*

3.3.1 *Introduction*

Another method of determining the parameters that contribute to the nuclear factor consists in measuring the rate at a fixed temperature, as a function of the driving force of the reaction ΔG^0. The basic difficulty encountered in this method is to keep the other parameters constant while ΔG^0 is varied. Since free energy variations are achieved through chemical modifications of the redox centers, they are expected to be accompanied by changes in their electronic states and therefore of the electronic factor of the reaction rate. As pointed out by Beratan and Hopfield [49], this effect is explictly included in the superexchange model, which predicts that the electronic factor depends on the difference between the energies of the redox centers' orbitals and those of the medium orbitals (Eq. (21)). Chemical modification of the centers may also alter appreciably the reorganization energy, particularly when the contribution λ_i of these centers is large. In fact, the great sensitivity of the kinetics with respect to modifications of the system will be illustrated in Sect. 3.4.3 by many examples showing that simple conformational changes are sufficient to induce substantial rate variations in systems whose chemical composition is otherwise constant.

The preceding considerations suggest that the results given by this method should be interpreted with caution. We first briefly examine how it was implemented in model systems. Closs and co-workers have studied the rate variations, within a series of organic molecules mentioned in Sect. 2.2.4 in which the donor and the bridge are fixed, and the acceptor is varied [77, 78]. The data are reasonably well described by Eq. (12) with a quantum internal mode of energy $\hbar\omega = 1500\ \mathrm{cm}^{-1}$, and the reorganization energy of the solvent is found to vary in the expected direction when its composition is changed. The implicit assumptions and limitations of such a treatment are carefully analyzed in Ref. [78]. Nearly identical results were obtained in glasses [129]. A similar study was performed by Joran and co-workers on a series of photosynthetic molecular models comprised of a Zn porphyrin bridged to a quinone [130]. In spite of the homogeneous character of the series of substituted quinones, the appreciable scattering of the experimental data does not allow a unique interpretation of the results. These studies illustrate the difficulties that may be encountered in the application of this method. We now examine the results obtained in biological systems, and distinguish the studies performed at room temperature which are interpreted through a classical treatment of the nuclear motion, and others performed at lower temperature in which quantified motions are apparent.

3.3.2 Classical Treatment of the Nuclear Factor

The electron transfer processes that take place in the noncovalent complexes constituted on the one hand by cytochrome *c*–cytochrome b_5 and on the other hand by cytochrome *c*–cytochrome *c* peroxidase were studied at room temperature by Miller and McLendon groups [131, 132, 133]. According to the static models built from crystallographic structures, iron-to-iron distances are respectively equal to 18 Å [134, 135] and 24 Å [136] in these complexes. In both systems, ΔG^0 variations are achieved by chemical modifications of the donor or acceptor hemes. The role of the donor and acceptor are sometimes exchanged within the series but this does not affect the interpretation of the results since the electronic factors and the reorganization energies are identical for two reverse reactions. More serious limitations would have been expected from fluctuations of T_{ab} and λ within these series. This is especially true for the study of cytochrome *c*–cytochrome b_5 complexes, in which reactions with $\Delta S=0$ and $\Delta S=1$ are compared. Nevertheless, the experimental data are rather well described by Eq. (13), with $\lambda=0.8$ eV for cytochrome *c*–cytochrome b_5 [131], and $\lambda=1.5$ eV for cytochrome *c*–cytochrome *c* peroxidase [132, 133]. Moreover, these values appear quite reasonable, since in the first system λ is found equal to about twice the value estimated in Sect. 3.2.1 for cytochrome *c* alone, whereas the relatively large value found in the second system may be attributed to the high-spin character of the cytochrome *c* peroxidase heme [132]. This large value has been confirmed by temperature dependent measurements [133].

A similar study was performed on ruthenium-modified myoglobins, in which ΔG^0 variations were obtained by changing the nature of the ruthenium complex covalently bound to the protein, and by substituting a porphyrin to the heme [137]. It is gratifying to observe that, in spite of the rather heterogeneous character of this series, the study leads to an estimation of 1.9 to 2.4 eV for λ which is consistent with the value 2.3 eV derived in section 3.2.1 from temperature dependent experiments. Satisfactory agreement between the results given by the two methods is also observed in the case of ruthenium-modified cytochrome *c* [138].

3.3.3 Quantum Treatment of the Nuclear Factor

Although the previous studies lead to reasonable and consistent values for the reorganization energy, they give no information about the different contributions to λ since they are not sensitive to the nature of the nuclear motions coupled to the process. In order to obtain such information, it is necessary to study the ΔG^0 dependence at different temperatures. This was done by Dutton and co-workers for the electron transfers from Bph^- to Q_A and Q_A^- to $(Bchl)_2^+$ in *Rhodobacter sphaeroides* photosynthetic reaction centers [139, 140], which take place over about 13 Å and 25 Å, respectively [18]. The rates were measured between 10 K and 300 K in series in which quinone substitutions provide ΔG^0 ranges of 0.5 eV and 0.8 eV for the two reactions respectively. The following conclusions were deduced from a thorough analysis of the experimental results:

i) very low-frequency motions like solvent modes are very weakly coupled to the transfers.
ii) intermediate frequency modes with energy $\hbar\omega_M \sim 120\ cm^{-1}$ provide the main contribution $\lambda_M \sim 0.4$ to 0.6 eV to the reorganization energy.
iii) high frequency modes with energy $\hbar\omega_L \sim 1500\ cm^{-1}$ are significantly coupled to the recombination reaction from Q_A^- to $(Bchl)_2^+$, with $\lambda_L \sim 0.2$ eV.

While points i) and ii) are in agreement with the results discussed in section 3.2.2, the information introduced by point iii) is new and suggests the involvement of quinone C=C or C=0 stretching modes [139, 140].

Another method used to vary the ΔG^0 of the recombination reaction without chemical modification of the centers, consists of placing the system in an electric field whose orientation and intensity are well defined [141]. However, the energy level shifts induced by the field also change the electronic factors, so that the interpretation of the experimental results is not straightforward. Bixon and Jortner have proposed using electric field effects to elucidate the nature of the primary electron step in bacterial photosystems [142], a problem that will be discussed in Sect. 3.5. One basic difficulty encountered in this method is the evaluation of the internal field effectively seen by the redox centers in the membrane.

3.4 Variations of the Electron Transfer Rate Due to Modification of the Medium

The theoretical concept of bridge-assisted electron transfer leads naturally to experimental investigations intended to determine the role of the medium in a given biological system. A first line of experiments is provided by site-directed mutagenesis techniques which have been widely developed during the last few years. In the case of bimolecular reactions, another possibility is to change the linking site of the two molecules, which of course modifies the nature of the intercenter medium. Finally, it is sometimes possible to study the influence of the system geometry upon the kinetics, when rate measurements can be performed in different conformations. Numerous experiments have been recently devoted to these methods, and some of them are described in detail in the present volume. In the following section we will just mention some typical examples showing that a definitive interpretation of these experiments often requires complementary studies.

3.4.1 Replacement of Specific Residues By Site-directed Mutagenesis

A Chapter of this volume is devoted to these techniques, which are merely illustrated in this section by one particular example. The electron transfer system that is the most intensively submitted to genetic manipulations is certainly the physiological complex between yeast cytochrome *c* and peroxide-oxidized cytochrome *c* peroxidase, which presents many advantages [143]. Among the modifications performed on cytochrome *c* peroxidase, one may mention the substitution of Trp 191 which interacts directly with His 175 of the heme [144], and of His 181 [145] which was proposed as a bridging unit in a superexchange path involving Phe 87 of cytochrome *c* [136, 146]. On the cytochrome *c* side, Phe 87 has been substituted [147], as well as other residues expected to play an important role in the stabilization of the noncovalent complex [143].

All these modifications induce changes in the activity of the system. The observed variations of the intracomplex transfer rate range from a decrease by a factor greater than 1,000 [144] to an increase by a factor 4 [143]. This last result is conceptually important, since it demonstrates that the chemical composition of a physiological complex must not be a priori considered as that leading to the highest intracomplex electron transfer rate [143]. Generally speaking, the interpretation of these experiments is not straightforward. For example, an X-ray analysis performed on yeast cytochrome *c* has demonstrated that substitution of a serine residue for Phe 87 is accompanied by important structural modifications in different parts of the protein [148]. These modifications are in particular manifested by an increase of the heme accessibility to the solvent, which may be correlated with the observed redox potential lowering of about 50 mV [147]. In addition, the overall structure of the mutant in the reduced state seems close to that of the oxidized wild-type protein, which suggests that the reorganization energies could differ appreciably in the native and modified systems. All these findings emphasize the need for the determination of the respective contributions of the electronic and nuclear factors to the kinetic changes induced by modifications of the medium.

This was attempted by Hoffman and co-workers, who studied the influence of a porphyrin vinyl group on the electron transfer rate in a [Fe, Zn] hybrid hemoglobin [149]. In this study, the chemical modification was simply achieved by preparing the hybrid with Zn deuterioporphyrin. However, entropic contributions are unknown in this system, and activation energies could not be accurately measured. Both effects preclude an accurate determination of the nuclear factor and a definitive attribution of the observed variations to the sole electronic factor.

3.4.2 Change of the Linking Site of Two Molecules

This method is currently very difficult to implement in the case of two proteins, although some of the cross-linking experiments mentioned in the next paragraph may be considered to pertain to it. Therefore, we have illustrated this method by a study in which kinetic data obtained in four ruthenium derivatives of a Zn modified myoglobin were compared [150, 151]. The electron transfer between the Zn porphyrin in the triplet state and a covalently bound Ru^{III} complex is measured to be about a thousand times faster when the complex is linked to His 48 than when it is to His 81, His 116 or His 12. Differences between intercenter distances, which are estimated to about 17 Å for the first system and 25 to 29 Å for the others, were emphasized in the interpretation of these results [151]. However, it must be noted that the experimental activation energy is also lower in the first complex, so that the observed rate differences could partly originate from the nuclear factor.

In a sense, the comparative studies that were performed in the ruthenium derivatives of plastocyanin and azurin are also related to this kind of approach [152].

3.4.3 Conformational Effects

For a given electron transfer process, it is sometimes possible to compare the kinetics corresponding to different conformations of the system. These conformations may be artificially created or naturally present in the studied sample. A first example is given by the comparison of some kinetic data obtained in covalent and noncovalent bimolecular protein complexes. In the system horse cytochrome *c*–cytochrome *c* peroxidase, the intracomplex transfer rate is equal to 1600 s^{-1} in the covalent complex, whereas it amounts to 750 s^{-1} at low ionic strength and 3300 s^{-1} at high ionic strength in the noncovalent complex [153]. This ionic strength dependence is also observed in the noncovalent complex between yeast cytochrome *c* and cytochrome *c* peroxidase [154]. Although it is tempting to simply attribute these rate variations to modifications of the electronic factor brought about by conformational changes, it is not possible presently to rule out an eventual contribution of the nuclear factor [153].

For the cytochrome *c*-plastocyanin complex, the kinetic effects of cross-linking are much more drastic: while the rate of the intracomplex transfer is equal to 1000 s^{-1} in the noncovalent complex where the iron-to-copper distance is expected to be about 18 Å, it is estimated to be lower than 0.2 s^{-1} in the corresponding covalent complex [155]. This result is all the more remarkable in that the spectroscopic and thermodynamic properties of the two redox centers appear weakly affected by the cross-linking process, and suggests that an essential segment of the "electron transfer path" has been lost in the covalent complex. Another system in which such conformational effects could be studied is the physiological complex between tetraheme cytochrome c_3 and ferredoxin I from *Desulfovibrio desulfuricans* Norway: the spectral and redox properties of the hemes and of the iron-sulfur cluster are found essentially identical in the covalent and noncovalent complexes and an intracomplex transfer, whose rate has not yet been measured, takes place in the covalent species [156].

In photosynthetic systems, some electron transfer processes exhibit non-exponential kinetics at low temperature, which are generally attributed to the existence of different conformations of the system. While the differences between the reaction rates corresponding to these conformations do not exceed a factor of four in some cases [157, 158, 159], they are sufficient to lead to different quantum yields in others [160, 161]. Sometimes, the heterogeneous character of the kinetics disappears at room temperature, which probably reflects a fast exchange between the conformations that are frozen at low temperature [157, 158]. A systematic study of all these effects, similar to that performed in Ref. [159], could give useful information about the nature of the conformational differences.

Now, it may happen that the electron transfer rate is so sensitive to the system conformation that a conformational change is needed before any efficient transfer may occur (When only the electronic factor is conformationally dependent, this situation may be considered as an extreme case of "non-Condon" effect). The kinetics of the process then depends on both the rate of the conformational change and the electron transfer rate, which of course greatly complicates the analysis [162, 163, 164]. Examples of the two limiting situations that may be envisaged seem to be encountered in biological systems. Thus, molecular dynamics simulations indicate that a cytochrome *c*–cytochrome b_5 complex whose conformation is initially energy-minimized, evolves in about 50 ps to a conformation in which the two hemes are bridged by Phe 82, and the iron-to-iron distance is 15 Å instead of 18 Å [135]. It is clear that such a fast conformational change does not contribute to the kinetics, which remains limited by the electron transfer step. It has been proposed that the other limiting situation, called "conformational gating", occurs in the cytochrome *c*–flavocytochrome b_2 system in which the electron transfer rate appears experimentally independent of the driving force ΔG^0 [165], contrarily to what is observed in the related system cytochrome *c*–cytochrome b_5 [131]. In fact, this different dynamic behavior is not surprising if one considers the important structural differences that exist between the two complexes [166].

To end this section, it may be added that a temperature-dependent electronic factor reflecting temperature-induced conformational changes has been invoked to explain the significant deviation that is sometimes observed between calculated and measured rate temperature dependences of activationless processes [122, 167, 168].

3.5 Interpretation of the Primary Electron Transfer in Bacterial Photosynthetic Reaction Centers

The recent X-ray crystallographic determination of the structure of *Rhodopseudomonas viridis* [15–17] and *Rhodobacter sphaeroides* [18–20] reaction centers has been the starting point of many theoretical studies. This structure, which is represented in Fig. 4 has confirmed the arrangement of the co-factors that had

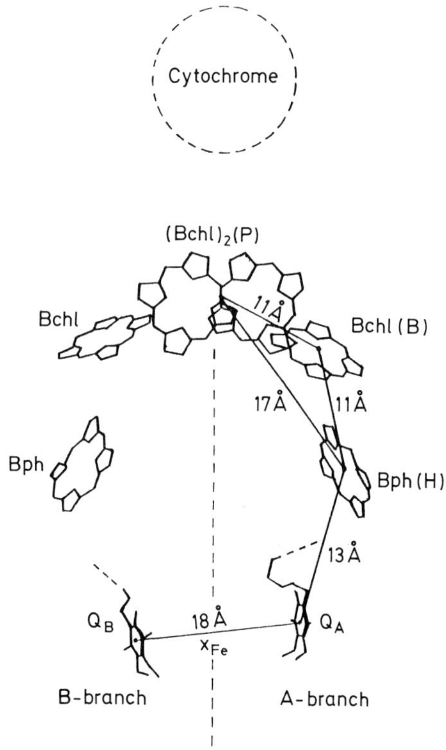

Fig. 4. Schematic representation of the bacterial reaction center (*R. sphaeroides*). The two branches are noted A, B in studies on *R. viridis*. Center-to-center distances are reported from Refs. [18, 21]. The approximate position of the cytochrome is indicated. The simplified notations P, B, H are used in the text

been predicted from spectroscopic and kinetic data, and has brought a lot of new information regarding the distances between the redox centers and relative orientations, as well as the structure of the peptide chains. Two fundamental questions pertaining to the primary electron transfer have immediately constituted a major challenge for theoreticians:

1) What is the role of the "accessory bacteriochlorophyll" B?
2) Is the apparent C_2 symmetry between the *A* and *B* branches compatible with the numerous experiments revealing that charge separation proceeds essentially along the *A* branch [169]?

From the many theoretical studies that have been already devoted to these two questions, it appears that definitive answers will probably require elaborate calculations based on very accurate structural data not yet available at the present level of resolution. The aim of this section is to give the main outlines of these studies, which illustrate well the exactness of the description that may be presently expected from theoretical calculations, and which also demonstrate the high sensitivity of the electron transfer rate to structural details.

We first examine the different alternatives to question 1 by using the experimental data obtained for *R. sphaeroides*. Let us consider the following electronic configurations of the system:

$$1 = P^* B H \qquad 2 = P^+ B^- H \qquad 3 = P^+ B H^-,$$

where P* represents the photoexcited state of the bacteriochlorophyll dimer, and other symbols are defined in Fig. 4. Each configuration gives a singlet and a triplet state. The electron transfer between $^1P^*$ and H is characterized by a very high rate $k_{exp} = 3.6 \times 10^{11}\ s^{-1}$ which is weakly temperature dependent between 10 K and 300 K [167, 170]. Among the different interpretations that were advanced for this process, we consider the two following which were particularly discussed in Refs. [171–174]:

i) two-step sequential transfer:

$$^1(P^* B H) \xrightarrow{k^s_{12}} {^1(P^+ B^- H)} \xrightarrow{k^s_{23}} {^1(P^+ B H^-)}$$

with:

$$k^s_{12} = k_{exp} \qquad k^s_{23} \gg k^s_{12}$$

ii) one-step transfer:

$$^1(P^* B H) \xrightarrow{k^s_{13}} {^1(P^+ B H^-)}$$

with:

$$k^s_{13} = k_{exp}$$

In this last interpretation, B could promote the electron transfer through a superexchange mechanism involving the $^1(P^+ B^- H)$ charge-transfer state. A variant of this model involves an internal charge-transfer state of the Bchl dimer [172].

According to low-temperature kinetic studies, the non-detection of the intermediate $^1(P^+ B^- H)$ implies $k^s_{23} \geqslant 50\ k^s_{12}$ in interpretation i) [167, 170]. This would require an extremely large value for k^s_{23} which, together with other arguments, leads Jortner and co-workers to rule out the two-step mechanism [174]. However, this argument could be revised, since recent experiments support a ratio k^s_{23}/k^s_{12} of about 4 at room temperature [176].

Let us now consider interpretation ii). Its pertinence was essentially discussed on the basis of a relation between the value of the rate k^s_{13} and that of an exchange parameter J. This parameter characterizes the exchange interactions that take place between the P^+ and H^- radicals when electron transfer is blocked beyond H. |J| was estimated to be about $10^{-3}\ cm^{-1}$ from different experiments performed in *R. sphaeroides* [177, 178]. In Sect. 2.2.1, we have described a simplified model leading to relation (16) between the exchange parameter J and the electronic factor T_{ab}, but we shall see that this relation cannot be applied to the primary electron transfer, due to the existence of numerous low-energy charge-transfer states.

The potential energy surfaces of the different states of the system are represented in Fig. 5 according to interpretation ii). The energy splitting between

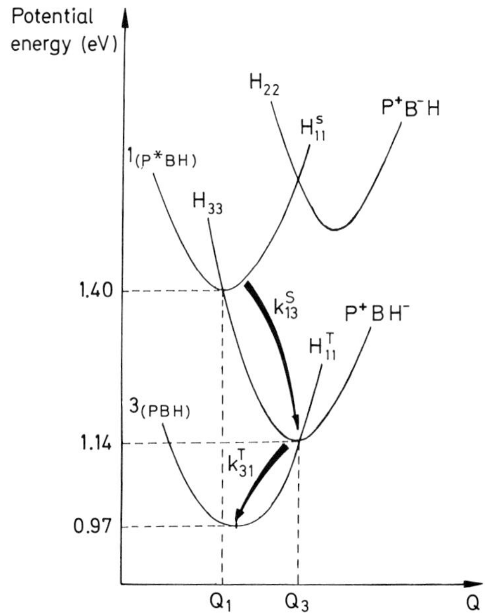

Fig. 5. Potential energy surfaces for different states involved in the interpretation of the primary electron transfer. Energies relative to the ground state PBH are reported from Ref [119]. The arrows indicate electron transfers whose rate have been experimentally measured

singlet and triplet states arising from the same configuration is negligible on the scale of the figure, except for configuration P*BH whose splitting is determined by intramolecular interactions within the Bchl dimer. The energies quoted in the figure are those reported in Ref. [119]. The exchange parameter J is measured at the equilibrium nuclear configuration Q_3 for $P^+ BH^-$ (Fig. 5), and by definition is equal to:

$$J = \Delta E_T - \Delta E_S, \tag{24}$$

where ΔE_S and ΔE_T represent the energy shifts due to the configuration interactions between $^1(P^+ BH^-)$ and all the singlet states, and $^3(P^+ BH^-)$ and all the triplet states of the system, respectively. Among the many configuration interactions that may be envisaged [174], those implying $^1(P^* BH)$ and $^3(PBH)$ involve quantities related to kinetic data. These are the rates k^s_{13} and k^T_{31} of the two electron transfer processes depicted in Fig. 5. Experimentally, these two processes are found activationless and characterized by [167, 179]:

$$k^s_{13} = 3.6 \times 10^{11}\ s^{-1} \qquad \lambda^s_{13} \sim \Delta G^s_{13} = 0.26\ eV$$

$$k^T_{31} = 5 \times 10^8\ s^{-1} \qquad \lambda^T_{31} \sim \Delta G^T_{31} = 0.17\ eV$$

We have seen in Sect. 3.2.2 that in the reaction center the frequency of the nuclear modes coupled to electron transfer processes are sufficiently low for expression

(13) to be used to analyze kinetic data at room temperature. Application of this equation to the preceding numerical values yields:

$$|T^{s}_{13}|_{Q_1} \sim 25\ \mathrm{cm}^{-1} \qquad |T^{T}_{31}|_{Q_3} \sim 1\ \mathrm{cm}^{-1}, \tag{25}$$

where the nuclear configuration relevant to each electronic factor is indicated. Although these numbers are weakly dependent on the exact values of ΔG^{s}_{13} and ΔG^{T}_{31}, it should be noted that the evaluation of $|T^{s}_{13}|$ rests on interpretation ii) whereas that of $|T^{T}_{31}|$ does not. If we assume that the J value is determined primarily by the configuration interactions between $^1(P^+BH^-)$ and $^1(P^*BH)$ on the one hand, and $^3(P^+BH^-)$ and $^3(PBH)$ on the other hand, the calculation leading to Eq. (16) gives:

$$\Delta E^{31}_{S} = -[|T^{s}_{31}|^2/(H^{s}_{11} - H_{33})]_{Q_3} \tag{26}$$

$$\Delta E^{31}_{T} = -[|T^{T}_{31}|^2/(H^{T}_{11} - H_{33})]_{Q_3} \tag{27}$$

In these expressions, the denominator of (26) and the numerator of (27) are known. In order to effectively compare the difference $(\Delta E^{31}_{S} - \Delta E^{31}_{T})$ to the experimental value of J, it is then necessary to evaluate $|T^{S}_{31}|_{Q_3}$ and $(H^{T}_{11} - H_{33})_{Q_3}$. These evaluations cannot be based on kinetic data, and rest on estimations of energetic quantities which are presently largely unknown. This explains in part the different interpretations given particularly in Refs. [173] and [174].

Regarding the difference $(H^{T}_{11} - H_{33})_{Q_3}$, it is certainly weak since the transfer characterized by k^{T}_{31} is known to be activationless. By using the $|T^{T}_{31}|_{Q_3}$ value given by Eq. (25) and reasonable values for λ^{T}_{13}, Michel–Beyerle and co-workers deduce from Eq. (27) that $|\Delta E^{31}_{T}|$ is greater than $2.5 \times 10^{-3}\ \mathrm{cm}^{-1}$ [174]. Since the contributions from configuration interactions implying other triplet states are expected to be negative, the authors conclude that ΔE_T is more negative than $-2.5 \times 10^{-3}\ \mathrm{cm}^{-1}$. This result expresses a relatively large stabilization of the triplet state $^3(P^+BH^-)$, which was obtained independently from the kinetic interpretation ii). The comparison of ΔE_T with the experimental value $|J| \sim 10^{-3}\ \mathrm{cm}^{-1}$ shows that this stabilization of the triplet state is necessarily cancelled by an important stabilization ΔE_S of the $^1(P^+BH^-)$ singlet state.

Such a stabilization is precisely provided by a superexchange mechanism implying the charge-transfer excited state $^1(P^+B^-H)$. When this mechanism prevails, $T^{s}_{31}(Q)$ is obtained from Eq. (20) or the equivalent expression derived in Sect. 2.2.3:

$$T^{S}_{31}(Q) = -(H_{32} - S_{32}H_{33})_Q(H_{21} - S_{21}H_{33})_Q/(H_{22} - H_{33})_Q \tag{28}$$

To evaluate $T^{S}_{31}(Q_3)$ from $T^{S}_{31}(Q_1)$ given by Eq. (25), Michel-Beyerle and co-workers take into account the Q dependence of both the numerator and the denominator of (28). Using Eq. (26), they propose an evaluation of ΔE^{31}_{S} which is compatible with the preceding estimation of ΔE_T and the experimental value of $|J|$ [174].

However, it may be noticed that if ΔE_T^{31} is positive, the negative contributions of other triplet states may lead to a relatively weak overall ΔE_T value. In this situation, the large stabilization ΔE_S given by the superexchange model would not be needed to account for the experimentally measured exchange parameter. Thus, it seems that a definitive conclusion on this point must await a full ab initio calculation of the parameter J, based on configuration interactions with all the important charge-transfer states of the system. Owing to the very weak value of J, it is clear that such a calculation will require very accurate wavefunctions. It should also be noted that expression (28) giving the superexchange contribution would no longer be valid if the difference $(H_{22}–H_{33})_Q$ were much smaller than previously assumed [180, 181, 182]. Thus, it is likely that the last word has yet to be said about this fascinating ultra-fast electron transfer for which interpretations very different from i) and ii) have also been proposed [183, 184].

Let us now consider point 2 concerning the functional asymmetry of the reaction center. Several theoretical studies based on the structures determined recently have been devoted to calculations of wavefunctions describing the electronic states of the pigments, and of the residual interactions between these pigments [169, 184, 185, 186]. From these studies, it was concluded that the unidirectionality of charge separation originates mainly from differences between the electronic factors for the transfers from $(Bchl)_2^*$ to H_A and from $(Bchl)_2^*$ to H_B. These differences arise, on the one hand, from an asymmetric charge distribution of $(Bchl)_2^*$, and, on the other hand, from more favorable contacts between respectively $(Bchl_2)$ and Bchl, and Bchl and Bph in the *A* branch [169, 184, 185]. However, these calculations also reveal that the electronic factors are very sensitive to distances between the nearest atoms of adjacent pigments, which determine primarily the magnitude of "through space" interactions. Moreover, superexchange paths mediated by protein residues or the cofactors phytyl tails could eventually enhance the electronic factors [187, 188]. Lastly, it should not be forgotten that all these calculations are based on the ground state structure of the reaction center, and that conformational relaxation effects induced by electron transfer processes could change appreciably the distances between the different pigment groups. It is possible that these effects have to be taken into account to achieve a good agreement between the calculated values of the electronic factors and those which may be derived from kinetic studies (Eq. (25)).

Another important result that was obtained recently concerns the evaluation of the contribution to the reorganization energy arising from the polarization of the medium, protein and solvent: from a microscopic model including the residual charges and induced dipoles of the protein as well as bound water molecules, a value of about 0.2 eV was calculated for different electron transfer processes [97]. This weak value results from the apolar character of the medium, and is compatible with the kinetic data which indicate that reorganization energies are small in the reaction center (Sect. 3.2.2)

4 Conclusion

It was shown in Sect. 2 that the standard formalism appropriate for non-adiabatic electron transfer processes leads to the definition of an electronic and a nuclear factor in the rate expression. This separation into factors of quite different physical origin is conceptually very useful. As a matter of fact, it is systematically emphasized throughout this presentation to clarify the nature of the different parameters involved in biological electron transfers. It happens also to be very useful when the relation between the kinetics and the biochemical function of these processes is considered. This is illustrated below by a few examples.

We have seen repeatedly that the electronic and nuclear factors corresponding to a given electron transfer are not necessarily optimized, in that the rate measured in the native system may be substantially increased by a suitable chemical modification of the redox centers or of the intercenter medium. In fact, the efficiency of biological electron transfer systems is primarily based first on directionality properties necessary to drive electrons towards specific sites where redox reactions are coupled to enzymatic processes, and secondly on proper energetic yields obtained by minimizing the fraction of driving force ΔG^0 of each intermediate transfer, which is dissipated into heat. The constraints brought about by these two requirements are not expected to optimize simultaneously the electronic and nuclear factors of each transfer. For example, directionality is achieved in bacterial photosynthetic systems by fast charge separation over a large distance, with a quantum yield close to unity. Energetic requirements constrain this process to proceed through a limited number of electron transfer steps. Therefore, redox centers are well separated in the reaction center (Fig. 4) and electronic factors are rather small [140, 188], with the exception of the primary electron transfer (Eq. (25)), which has to compete with very fast relaxation processes. In these conditions, fully optimized nuclear factors are needed in the first steps to attain very fast forward transfers and much slower recombination reactions: the former are activationless with a free energy loss $|\Delta G^0| \sim \lambda$ at each transfer, while the latter occur in the inverted region. Although the reorganization energy seems particularly weak in these systems (Sect. 3.2.2), the overall energetic yield is only about 30% [21]. If we now consider bimolecular reactions in solution, directionality is achieved through the specific recognition of physiological partners. We have already pointed out that the spatial arrangement of the redox centers is not necessarily optimized within the bimolecular complex in which the transfer takes place. On the other hand, the experimental results discussed in Sect. 3 seem to indicate that the reorganization energy is larger for transfers between proteins in solution than for the first steps of charge separation in photosynthetic systems. Optimization of the nuclear factor would then require a free energy loss incompatible with a good energetic yield. These different reasons could explain the much lower rates observed in the case of bimolecular reactions, which are, however, sufficient to ensure the functioning of biological processes.

There is one situation in which there is no particular advantage for the rate to be optimized, namely when the electron transfer is not the limiting step of the process. Recent experiments have demonstrated that some bimolecular reactions which occur in the inner mitochondrial membrane are diffusion-limited or close to that limit [189, 190, 191]. In this case, the kinetics could be described using elaborate models based on molecular structures which have already been proposed to calculate the diffusion rate between two proteins [192]. We have also pointed out in Sect. 3.4.3 the "conformational gating" phenomenon, in which electron transfer must be preceded by a conformational change of the system, which may become the limiting step of the process. This phenomenon provides a simple and efficient means of kinetic control in redox reactions, which may be used in regulation processes. On the other hand, evidence is growing that redox-linked conformational changes are involved in the mechanism of some enzymes, like cytochrome *c* oxidase. This membrane-bound protein is situated at the end of the respiratory electron transport chain, and possess four metallic centers: heme *a*, Cu_a and a heme a_3–Cu_b complex. Four electrons provided by cytochrome *c* are used to reduce O_2 to water at the heme a_3–Cu_b site, and to pump four protons accross the membrane [193]. According to an attractive model proposed by Malmstrom [194, 195], the reduction of both heme *a* and Cu_a combined with a double protonation of the enzyme triggers a conformational change which results in first a large increase of the nuclear factors that enables fast intramolecular electron transfers from heme *a* and Cu_a to the heme a_3–Cu_b site, and secondly the pumping of the two protons. These processes could involve the motion of transmembrane $\alpha-$helix [196]. However, it was shown recently that the four electron transfers from cytochrome *c* to the heme a_3–Cu_b site are non-equivalent with respect to proton pumping [193], and the preceding model should be modified to take this information into account.

We believe that these few examples add to the others cited in this chapter to demonstrate that the standard formalism of electron transfer theories is well adapted to the interpretation of biological processes. The numerous investigations based on this formalism which are presently developed may be divided into two categories:

i) Theoretical calculations of parameters involved in the kinetics: electronic factor, reorganization energy and even driving force ΔG^0, from precise structural and spectroscopic data. Such studies can make allowance for the detailed characteristics of each system, and are expected to give much more significant information than phenomenological treatments based on general expressions like Eq. (22).
ii) Experimental studies intended to measure these parameters, or to elucidate their physical origin in a given system.

These two lines of investigation illustrate the important advances made in this field since the early theoretical interpretations of biological electron transfer processes. Hence, they allow one to envisage the understanding of these processes at the molecular level, and the synthesis of efficient model systems [86, 194].

5 Appendix: Relation Between the Driving Force and the Experimental Redox Potentials

We consider an electron transfer system containing two redox centers D and A, and we are interested by the relation between the driving force ΔG^0 relevant to the electron transfer from D to A, and the redox potentials which are effectively measured. Two different situations may be distinguished:

1 The redox potentials may first be measured directly on the system in which the transfer takes place. This situation corresponds usually to intramolecular processes, but may also be encountered in bimolecular processes when the formation constant K_f is large enough for all the molecules to be complexed in the conditions of the experiment. The four possible redox states of the system are represented in Figure 6, each state being considered in its equilibrium nuclear configuration. The driving force ΔG^0 may be calculated either from Eq. (A1) or from Eq. (A2):

$$\Delta G^0 = \mathscr{F}\,[E^0_{\mathrm{Ared}}(D) - E^0_{\mathrm{Dred}}(A)] \tag{A1}$$

$$\Delta G^0 = \mathscr{F}\,[E^0_{\mathrm{Aox}}(D) - E^0_{\mathrm{Dox}}(A)] \tag{A2}$$

The different potentials involved in these expressions correspond to the redox equilibria indicated in Figure 6. The difference

$$\Delta E^0 = E^0_{\mathrm{Ared}}(D) - E^0_{\mathrm{Aox}}\,(D) \qquad \text{(D)}$$

$$= E^0_{\mathrm{Dred}}(A) - E^0_{\mathrm{Dox}}\,(A) \qquad \text{(A)}$$

represents the interaction potential between the two redox centers. The absolute values of the interaction potentials that have been measured in several multi-center proteins fall in the range 10–50 mV [42, 198, 199, 200]. When ΔE^0 may be neglected, the two redox centers are independent and ΔG^0 is merely given by:

$$\Delta G^0 = \mathscr{F}\,[E^0(D) - E^0(A)]$$

For some intramolecular processes, it may happen that the potentials involved in Eq. (A1) or (A2) cannot be experimentally determined. If $E^0(D)$ and $E^0(A)$ can be

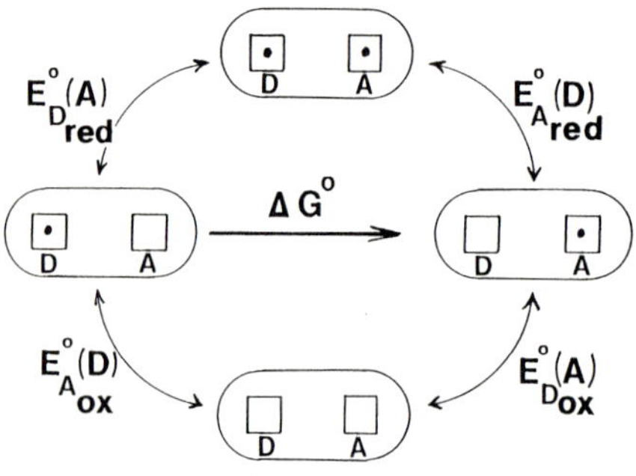

Fig. 6. The four possible redox states of a system comprising two centers. Redox equilibria characterized by the potentials involved in Eqs (A1) and (A2) are indicated

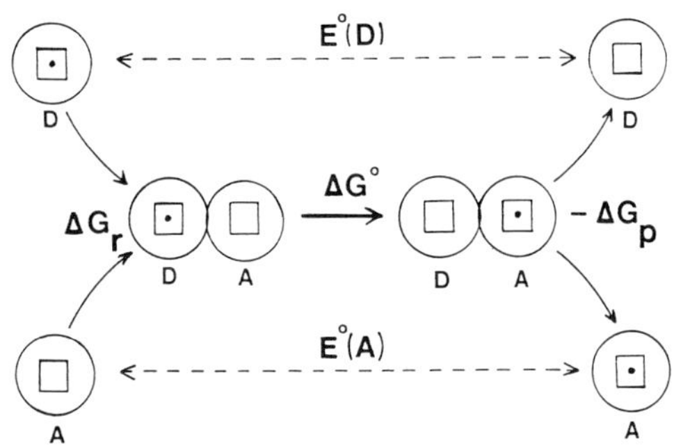

Fig. 7. Free energy scheme for a bimolecular redox reaction

evaluated in closely related systems in which A and D are independent, ΔG^0 may be estimated using the following expression:

$$\Delta G^0 = \mathscr{F}[E^0(D) - E^0(A)] + I(D_{ox}, A_{red}) - I(D_{red}, A_{ox}),$$

where the corrective terms account for eventual interactions between the centers in the studied system.

2 The second situation usually prevails in bimolecular reactions: electron transfer occurs in complexes which represent a negligible fraction of the present molecules, and redox potentials are measured for each molecular species separately. In this situation, the driving force is given by:

$$\Delta G^0 = \mathscr{F}[E^0(D) - E^0(A)] + \Delta G_p - \Delta G_r$$

where $E^0(D)$ and $E^0(A)$ are the redox potentials measured for each species, and ΔG_p, ΔG_r are the free energy of formation of the complexes represented in Fig. 7. These quantities may be determined from measurements of the corresponding formation constants, performed in static or kinetic studies. In the case of physiological partners, these formation constants may depend considerably on the redox state of the centers [154, 201], so that the difference between ΔG_p and ΔG_r cannot be a priori neglected [24].

Note Added in Proof

A recent study (Booth PJ, Crystall B, Giorgi LB, Barber J, Klug DR, Porter G (1990) Biochim. Biophys. Acta 1016: 141) has shown that the free energy difference of the primary electron transfer is dominated by entropic contributions in photosystem II reaction centers as in bacterial reaction centers (Woodbury NWT, Parson WW (1984) Biochim. Biophys. Acta 767: 345), so that the interpretation of the rate temperature dependence should be revised.

Acknowledgements: I would like to acknowledge Dr. J. Breton for helpful discussions.

6 References

1. Marcus RA (1956) J Chem Phys 24: 966
2. Marcus RA (1960) Disc Faraday Soc 29: 21
3. Marcus RA (1965) J Chem Phys 43: 679
4. Marcus RA, Sutin N (1985) Biochim Biophys Acta 811: 265
5. Levich VG (1966) Adv Electrochem Electochem Eng 4: 249
6. Dogonadze RR, Kuznetsov AM, Vorotyntsev MA (1972) Phys Stat Sol B 54: 125
7. Kestner NR, Logan J, Jortner J (1974) J Phys Chem 78: 2148
8. Hopfield JJ (1974) Proc Nat Acad Sci USA 71: 3640
9. Jortner, J (1976) J Chem Phys 64: 4860
10. Pierrot M, Haser R, Frey M, Payan F, Astier JP (1982) J Biol Chem 23: 14341
11. Higushi Y, Kusonoki M, Yasuoka N, Kakudo M, Yagi T (1984) J Mol Biol 172: 109
12. McRee DE, Richardson DC, Richardson JS, Siegel LM (1986) J Biol Chem 261: 10277
13. Verma AL, Kimura K, Nakamura A, Yagi T, Inokuchi H, Kitagawa T (1988) J Am Chem Soc 110: 6617
14. Chance B, Marcus RA, Devault DC, Schrieffer JR, Frauenfelder H, Sutin N (eds) (1979) Tunnelling in biological systems, Academic, New York.
15. Deisenhofer J, Epp O, Miki K, Huber R, Michel H (1984) J Mol Biol 180: 385
16. Deisenhofer J, Epp O, Miki K, Huber R, Michel H (1985) Nature 318: 618
17. Michel H, Epp O, Deisenhofer J (1986) EMBO J 5: 2445
18. Allen JP, Feher G, Yeates TO, Komiya H, Rees DC (1987) Proc Nat Acad Sci USA 84: 5730
19. Allen JP, Feher G, Yeates TO, Komiya H, Rees DC (1987) Proc Nat Acad Sci USA 84: 6162
20. Allen JP, Feher G, Yeates TO, Komiya H, Rees DC (1988) Proc Nat Acad Sci USA 85: 8487
21. Feher G, Allen JP, Okamura MY, Rees DC (1989) Nature 339: 111
22. Devault D (1984) Quantum mechanical tunnelling in biological systems. Cambridge University Press, Cambridge
23. Bertrand P (1986) Biochimie 68: 619
24. McLendon G (1988) Acc Chem Res 21: 160
25. Newton MD (1982) ACS Sym Ser 198: 255
26. Newton MD, Sutin N (1984) Ann Rev Phys Chem 35: 437
27. Efrima S, Bixon M (1976) J Chem Phys 64: 3639
28. Wolfsberg M, Helmholz L (1952) J Chem Phys 20: 837
29. Siders P, Marcus RA (1981) J Am Chem Soc 103: 741
30. McCammon JA, Gelin BR, Karplus M (1977) Nature 267: 585
31. Go N, Noguti T, Nishikawa T (1983) Proc Nat Acad Sci USA 80: 3696
32. Doster W, Cusack S, Petry W (1989) Nature 337: 754
33. Marcus RA (1984) J Chem Phys 81: 4494
34. Zener C (1932) Proc Roy Soc Lond A 137: 696
35. Landau L (1932) Phys Z Sowjetunion 2: 46
36. Sailasuta N, Anson FC, Gray HB (1979) J Am Chem Soc 101: 455
37. Taniguchi VT, Sailasuta-Scott N, Anson FC, Gray HB (1980) Pure and Appl Chem 52: 2275
38. Huang YY, Kimura T (1983) Anal Biochem 133: 385
39. Huang YY, Kimura T (1984) Biochemistry 23: 2231
40. Huang YY, Hara T, Sligar S, Coon, MJ, Kimura T (1986) Biochemistry 25: 1390
41. Reid LS, Taniguchi VT, Gray HB, Mauk AG (1982) J Am Chem Soc 104: 7516
42. Blair DF, Ellis WR, Wang H, Gray HB, Chan SI (1986) J Biol Chem 261: 11524
43. Churg AK, Weiss RM, Warshel A, Takano T (1983) J Phys Chem 87: 1683
44. Warshel A, Hwang JK (1986) J Chem Phys 84: 4938
45. Hwang JK, Warshel A (1987) J Am Chem Soc 109: 715
46. Kuharski RA, Bader JS, Chandler D, Sprik M, Klein ML, Impey RW (1988) J Chem Phys 89: 3248
47. Hopfield JJ (1977) Biophys J 18: 311
48. Hopfield JJ (1977) In: Roux HF (ed) Electrical phenomena at the biological membrane level. Proc 29th Int Congr Chim Phys Elsevier, Amsterdam, p 471
49. Beratan DN, Hopfield JJ (1984) J Am Chem Soc 106: 1584
50. Bertrand, P (1985) Chem Phys Lett 113: 104
51. Okamura MY, Isaacson RA, Feher G (1979) Biophys Biochim Acta 546: 394
52. Siders P, Cave RJ, Marcus RA (1984) J Chem Phys 81: 5613
53. Cave RJ, Siders P, Marcus RA (1986) J Phys Chem 90: 1436

54. Halpern J, Orgel L (1960) Discuss Faraday Soc 29: 32
55. McConnell HM (1961) J Chem Phys 35: 508
56. Kramers (1934) Physica 1: 182
57. Beratan DN, Onuchic JN, Hopfield JJ (1987) J Chem Phys 86: 4488
58. Mayoh B, Day P (1974) J Chem Soc Dalton Trans 846
59. Mayoh B, Day P (1974) Inorg Chem 13: 2273
60. Richardson DE, Taube H (1983) J Am Chem Soc 105: 40
61. Larsson S (1981) J Am Chem Soc 103: 4034
62. Larsson S (1983) J Chem Soc, Faraday Trans 79: 1375
63. Beratan DN (1986) J Am Chem Soc 108: 4321
64. Bertrand P (1987) Chem Phys Lett 140: 57
65. Larsson S, Volosov A (1986) J Chem Phys 85: 2548
66. Larsson S, Volosov A (1987) J Chem Phys 87: 6623
67. Larsson S, Broo A, Kallebring, Volosov A (1988) Int J Quantum Chem Symp 15: 1
68. Newton MD (1980) Int J Quantum Chem Symp 14: 363
69. Logan J, Newton MD (1983) J Chem Phys 78: 4086
70. Newton MD (1986) J Phys Chem 90: 3734
71. Newton MD (1988) J Phys Chem 92: 3049
72. Kuki A, Wolynes PG (1987) Science 236: 1647
73. Miller JR, Beitz J (1981) J Chem Phys 74: 6746
74. Guarr T, McGuire ME, McLendon G (1985) J Am Chem Soc 107: 5104
75. Miller JR (1987) Nouv J Chim 11: 83
76. Isied SS, Vassilian A, Wishart JF, Creutz C, Schwartz HA, Sutin N (1988) J Am Chem Soc 110: 635
77. Closs GL, Miller JJ (1988) Science 240: 440
78. Closs GL, Calcaterra LT, Green NJ, Penfield KW, Miller JR (1986) J Phys Chem 90: 3673
79. Johnson MD, Miller JR, Green NS, Closs GL (1989) J Phys Chem 93: 1173
80. Balaji V, Ng L, Jordan KD, Paddon-Row MN, Patney HK (1987) J Am Chem Soc 109: 6957
81. Oliver AM, Craig DC, Paddon-Row MN, Kroon J, Verhoeven JW (1988) Chem Phys Lett 150: 366
82. Ohta K, Closs GL, Morokuma K, Green NJ (1986) J Am Chem Soc 108: 1319
83. Heitele H, Michel-Beyerle ME (1985) J Am Chem Soc 107: 8286
84. Heitele H, Michel-Beyerle ME, Finckh P (1987) Chem Phys Lett 134: 273
85. Wasielewski MR (1988) In: Fox MA, Chanon M (eds) Photoinduced electron transfer, Elsevier, Amsterdam, Part A p 161
86. Gust D, Moore TA (1989) Science 244: 35
87. Beratan DN, Hopfield JJ, (1984) J Chem Phys 81: 5753
88. Freed KF (1986) J Chem Phys 84: 2108
89. Onuchic JN, Beratan DN, Hopfield JJ (1986) J Phys Chem 90: 3707
90. Deumens E, Ohrn Y (1988) J Phys Chem 92: 3181
91. Kuznetsov AM, Ulstrup J (1982) Phys Stat Sol B 114: 673
92. Mikkelsen KV, Ulstrup J, Zakaraya MG (1989) J Am Chem Soc 111: 1315
93. Jortner, J (1980) J Am Chem Soc 102: 6676
94. Garg A, Onuchic JN (1985) J Am Chem Soc 83: 4491
95. Morillo M, Yang DY, Cukier RI (1989) J Chem Phys 90: 5711
96. Warshel A (1982) J Phys Chem 86: 2218
97. Creighton S, Hwang JK, Warshel A, Parson WW, Norris J (1988) Biochemistry 27: 774
98. Mikkelsen KV, Ratner MA (1989) J Phys Chem 93: 1759
99. Mikkelsen KV, Ratner MA (1989) J Chem Phys 90: 4237
100. Churg AK, Warshel A (1986) Biochemistry 25: 1675
101. Gray HB, Malmstrom BG (1983) Comments Inorg Chem 2: 203
102. Takano T, Dickerson RE (1981) J Mol Biol 153: 95
103. Argos P, Mathews FS (1975) J Biol Chem 250: 747
104. Teo BK, Shulman RG, Brown GS, Meixner AE (1979) J Am Chem Soc 101: 5624
105. Tsukahara K (1989) J Am Chem Soc 111: 2040
106. Trewhella J, Carlson VAP, Curtis EH, Heidorn DB (1988) Biochemistry 27: 1121
107. Levitt M, Sander C, Stern PS (1985) J Mol Biol 181: 423
108. Cusack S, Smith J, Finney J, Tidor B, Karplus M (1988) J Mol Biol 202: 903
109. Mishra KC, Mishra SK, Roy JN, Ahmad S, Das TP (1985) J Am Chem Soc 107: 7898

110. Nocera DG, Winkler JR, Yocom KM, Bordignon E, Gray HB (1984) J Am Chem Soc 106: 5145
111. Kostic NM, Margalit R, Che CM, Gray HB (1983) J Am Chem Soc 105: 7765
112. Margalit R, Kostic NM, Che CM, Blair DF, Chiang HJ, Pecht I, Shelton JB, Shelton JR, Schroeder WA, Gray HB (1984) Proc Nat Acad Sci USA 81: 6554
113. Crutchley RJ, Ellis WR, Gray HB (1985) J Am Chem Soc 107: 5002
114. Brown GM, Sutin N (1979) J Am Chem Soc 101: 883
115. Jackman MP, Mc Ginnis J, Powls R, Salmon GA, Sykes AG (1988) J Am Chem Soc 110: 5880
116. Osvath P, Salmon GA, Sykes AG (1988) J Am Chem Soc 110: 7114
117. Jackman MP, Lim MC, Salmon GA, Sykes AG (1988) J Chem Soc Dalton Trans 11: 2843
118. Farver O, Pecht I (1989) FEBS Lett 244: 379
119. Goldstein RA, Takiff L, Boxer SG (1988) Biochim Biophys Acta 934: 253
120. Buhks E, Bixon M, Jortner J (1981) Chem Phys 55: 41
121. Bixon M, Jortner J (1986) FEBS Lett 200: 303
122. Bixon M, Jortner J (1986) J Phys Chem 90: 3795
123. Knapp EW, Fischer SF (1987) J Chem Phys 87: 3880
124. Bertrand P, Gayda JP, Rao K (1982) J Chem Phys 76: 4715
125. Eden D, Matthew JB, Rosa JJ, Richards FM (1982) Proc Nat Acad Sci USA 79: 815
126. Mancino LJ, Dean DP, Blankenship RE (1984) Biochim Biophys Acta 764: 46
127. Peterson-Kennedy SE, McGourty JL, Hoffman BM (1984) J Am Chem Soc 106: 5010
128. McGourty JL, Peterson-Kennedy SE, Ruo WY, Hoffman BM (1987) Biochemistry 26: 8302
129. Miller JR, Beitz JV, Huddleston RK (1984) J Am Chem Soc 106: 5057
130. Joran AD, Leland BA, Felker PM, Zewail AH, Hopfield JJ, Dervan, PB (1987) Nature 327: 508
131. McLendon G, Miller JR (1985) J Am Chem Soc 107: 7811
132. Cheung E, Taylor K, Kornblatt JA, English AM, McLendon G, Miller JR (1986) Proc Nat Acad Sci USA 83: 1330
133. Taylor Conklin K, Mc Lendon G (1988) J Am Chem Soc 110: 3345
134. Salemne, FR (1976) J Mol Biol 102: 563
135. Wendolovski JJ, Matthew JB, Weber PC, Salemne FR (1987) Science 238: 719
136. Poulos TL, Kraut J (1980) J Biol Chem 255, 10322
137. Karas JL, Lieber CM, Gray HB (1988) J Am Chem Soc 110: 599
138. Meade TJ, Gray HB, Winkler JR (1989) J Am Chem Soc 111: 4353
139. Gunner MR, Robertson DE, Dutton PL (1986) J Phys Chem 90: 3783
140. Gunner MR, Dutton PL (1989) J Am Chem Soc 111: 3400
141. Popovic ZD, Kovacs GJ, Vincett PS, Alegria G, Dutton PL (1986) Chem Phys 110: 227
142. Bixon M, Jortner J (1988) J Phys Chem 92: 7148
143. Hazzard JT, McLendon G, Cusanovich MA, Das G, Sherman F, Tollin G (1988) Biochemistry 27: 4445
144. Mauro JM, Fishel LA, Hazzard JT, Meyer TE, Tollin G, Cusanovich MA, Kraut J (1988) Biochemistry 27: 6243
145. Miller MA, Hazzard JT, Mauro JM, Edwards SL, Simons PC, Tollin G, Kraut J (1988) Biochemistry 27: 9081
146. Liang N, Mauk AG, Pielak GJ, Johnson JA, Smith M, Hoffman BM (1988) Science 240: 311
147. Pielak GJ, Mauk AG, Smith M (1985) Nâture 313: 152
148. Louie GV, Pielak GJ, Smith M, Brayer GD (1988) Biochemistry 27: 7870
149. Gingrich DJ, Noceck JM, Natan MJ, Hoffman BM (1987) J Am Chem Soc 109: 7533
150. Mayo SL, Ellis WR, Crutchley RJ, Gray HB (1986) Science 233: 948
151. Axup AW, Albin M, Mayo SL, Crutchley RJ, Gray HB (1988) J Am Chem Soc 110: 435
152. Jackman MP, McGinnis J, Powls R, Salmon GA, Sykes AG (1988) J Am Chem Soc 110: 5880
153. Hazzard JT, Moench SJ, Erman JE, Satterle JD, Tollin G (1988) Biochemistry 27: 2002
154. Hazzard JT, McLendon G, Cusanovich MA, Tollin G (1988) Biochem Biophys Res Comm 151: 429
155. Peerey LM, Kostic NM (1989) Biochemistry 28: 1861
156. Dolla A, Guerlesquin F, Bruschi M, Guigliarelli B, Asso M, Bertrand P, Gayda JP (1989) Biochim Biophys Acta 975: 395
157. Parrot P, Thiery J, Vermeglio A (1987) Biochim Biophys Acta 893: 534
158. Kleinfeld D, Okamura MY, Feher G (1984) Biochemistry 23: 5780
159. Sebban P, Wraight CA (1989) Biochim Biophys Acta 974: 54
160. Setif P, Mathis P, Wanngard TP (1984) Biochim Biophys Acta 767: 404
161. Vos MH, Van Gorkom HJ (1988) Biochim Biophys Acta 934: 293

162. Hoffman BM, Ratner MA (1987) J Am Chem Soc 109: 6237
163. Cartling BO (1985) J Chem Phys 83: 5231
164. Brunschwig BS, Sutin N (1989) J Am Chem Soc 111: 7454
165. McLendon G, Pardue K, Bak P (1987) J Am Chem Soc 109: 7540
166. Xia ZX, Shamala N, Bethge PH, Lim LW Bellamy HD, Xuong NH, Lederer F, Mathews FS (1987) Proc Nat Acad Sci USA 84: 2629
167. Fleming GR, Martin JL, Breton J (1988) Nature 333: 190
168. Bixon M, Jortner J (1989) Chem Phys Lett 159: 17
169. Michel-Beyerle ME, Plato M, Deisenhofer J, Michel H, Bixon M, Jortner J (1988) Biochim Biophys Acta 932: 52
170. Breton J, Martin JL, Fleming GR, Lambry JC (1988) Biochemistry 27: 8276
171. Marcus RA (1987) Chem Phys Lett 133: 471
172. Bixon M, Jortner J, Michel-Beyerle ME, Ogrodnik A, Lersch W (1987) Chem Phys Lett 140: 626
173. Marcus RA (1988) Chem Phys Lett 146: 13
174. Michel Beyerle ME, Bixon M, Jortner J (1988) Chem Phys Lett 151: 188
175. Won Y, Friesner RA (1988) Biochim Biophys Acta 935: 9
176. Holzapfel W, Finkele U, Kaiser W, Oesterhelt D, Scheer H, Stilz HU, Zinth W (1989) Chem Phys Lett 160: 1
177. Moehl KW, Lous EJ, Hoff AJ (1985) Chem Phys Lett 121: 22
178. Hunter DA, Hoff AJ, Hore PJ (1987) Chem Phys Lett 134: 6
179. Ogrodnik A, Remy-Richter N, Michel-Beyerle ME, Feick R (1987) Chem Phys Lett 135: 576
180. Joachim C (1987) Chem Phys 116: 339
181. Hu Y, Mukamel S (1989) Chem Phys Lett 160: 410
182. Reimers JR, Hush NS (1989) Chem Phys 134: 323
183. Scherer POJ, Fischer SF (1987) Chem Phys Lett 141: 179
184. Scherer POJ, Fischer SF (1989) Chem Phys 131: 115
185. Plato M, Mobius K, Michel-Beyerle ME, Bixon M, Jortner J (1988) J Am Chem Soc 110: 7279
186. Warshel A, Creighton S, Parson WW (1988) J Phys Chem 92: 2696
187. Yeates TO, Komiya H, Chirino A, Rees DC, Allen JP, Feher G (1988) Proc Natl Acad Sci USA 85: 7993
188. Plato M, Michel-Beyerle ME, Bixon M, Jortner J (1989) FEBS Lett 249: 70
189. Gupte SS, Hackenbrock CR (1988) J Biol Chem 263: 5241
190. Gupte SS, Hackenbrock CR (1988) J Biol Chem 263: 5248
191. Chazotte B, Hackenbrock CR (1989) J Biol Chem 264: 4978
192. Northrup SH, Boles JO, Reynolds JCL (1987) J Phys Chem 91: 5991
193. Wikstrom M (1989) Nature 338: 776
194. Brzezinski P, Malmstrom BG (1987) Biochim Biophys Acta 894: 29
195. Malmstrom BG (1989) FEBS Lett 250: 9
196. Williams RJP (1987) FEBS Lett 226: 1
197. Hopfield JJ, Onuchic JN, Beratan DN (1989) J Phys Chem 93: 6350
198. Gayda JP, Benosman H, Bertrand P, More C, Asso M (1988) Europ J Biochem 177: 199
199. Benosman H, Asso M, Bertrand P, Yagi T, Gayda JP (1989) Eur J Biochem 182: 51
200. Santos H, Moura JJG, Moura I, Legall J, Xavier AV (1984) Eur J Biochem 141: 283
201. Hintz MJ, Mock DM, Peterson LL, Tuttle K, Peterson JA (1982) J Biol Chem 257: 14324
202. Woodbury NWT, Parson WW (1984) Biochim Biophys Acta 767: 345
203. Booth PJ, Crystall B, Giorgi LB, Barber J, Klug DR, Porter G (1990) Biochim Biophys Acta 1016: 141

Electronic Tunneling Paths in Proteins

Atsuo Kuki

Department of Chemistry, Baker Laboratory, Cornell University, Ithaca, NY 14853, USA

The connections between basic electron transfer theory and the methods currently available for analyzing quantum electronic interactions of long-range electron transfer within and between proteins are examined. Both traditional basis set expansion as well as quantum path integral Monte Carlo descriptions of propagation through the intervening medium are described and compared. The development of the concepts and technical aspects of superexchange are traced from simple to complex implementations, with an emphasis on the unifying framework of the sum over tunneling path amplitudes. Directions for the future refinement of quantitative biomolecular computations are identified.

1 Introduction

1.1 Goals: The Electronic Consequences of Protein Structure

The complexity of the structure and reactivity of the folded protein molecule presents a formidable challenge to theoretical analysis. Theoretical research on electron transfer proteins has been motivated by this challenge, and by intriguing experimental results, to develop in two ways. First, new tools have been developed for quantum electronic analysis specifically designed for the complex, three-dimensional, inhomogeneous, and aperiodic nature of protein structures. The nature and applicability of the new tools, as well as old tools adapted to the biomolecular realm, will be discussed here. Secondly, the theories are being refined to achieve increasingly realistic treatments of specific proteins currently under experimental investigation and thus to provide quantitative predictions which biochemists and biophysicists can use. The Marcus theory of electron transfer [1, 2] provides a splendid example of a successful theory with extensive predictive and correlative utility, and it has guided our basic understanding of electron transfer in both small molecules and proteins. But the theoretical approaches addressing the complexities of biological electron transfer reactivity are still at a primitive state with respect to their quantitative implementation. Indeed, though the fundamental equations and tools are well understood, great simplifications have been so far necessary in order to apply them to proteins. Here we will point out current developments being pursued.

In this chapter we examine routes to an understanding of the nature and role of the electronic interactions important in governing the reactivity of electron transfer proteins. Electron transfer proteins are crucial components in many bioenergetic pathways within oxidative phosphorylation [3, 4] and photosynthesis [5, 6] and their function is to participate selectively in electron transfer interactions. In both natural and artificial molecular assemblies, selectivity implies molecular engineering of the kinetics, and the extent to which nature tunes electronic interactions in proteins remains an open question. Extensive experimental and theoretical work has been devoted to understanding both the electronic and nuclear factors that regulate electron transfer reactivity between redox centers in protein environments. A central goal ever since the speculations of Szent-Gyorgi more than forty years ago [7] has been to determine the role of the intervening protein medium in modifying the rate of electron transfer by controlling or directing the motion of the tunneling electron. A theoretical rate expression containing a distance dependence only is a spherically symmetric theory and is inappropriate for a structured inhomogeneous system. We would like to reach a quantitative understanding of the role of various amino acid residues which are found intermediary in space between redox centers in long-range biological electron transfer processes.

1.2 Nuclear Motion

Basic electron transfer theory, summarized compactly by Sutin [8, see also Bertrand, Chapter 1 in this volume], and reviewed in the biological context by Jortner [9], separates the reaction dynamics into nuclear and electronic dynamics. This basic separation is very central to the simplification of a complex dynamical phenomenon, and a few words about the nuclear factors are in order here, before we proceed to the electronic factors.

Nuclear in this context means vibrational ("inner sphere") rearrangements at the redox centers as well as the reorganization of the electrostatic polarization of the surrounding medium ("outer sphere"). In polar fluids outer sphere reorganization means the alteration of solvent dipole orientational probability distributions, initially in equilibrium with the reactant redox state, to those in equilibrium with the product redox state. In such homogeneous media a critical parameter known as the outer sphere reorganization energy, proportional in the linear regime to $(1 - 1/\varepsilon)$, enters into the theoretical activation energy (ε = dielectric constant). In proteins, the corresponding outer sphere reorganization energy cannot be described by a homogeneous dielectric constant, but nevertheless this energy arises similarly from the requirement that fluctuations occur in the positions and orientations of polar groups within the protein and in the nearby solvent in order for the reaction to occur. The polar groups which matter in outer sphere rearrangement are those coupled electrostatically to the change in charge at the redox sites. The free energy of activation is in principle computable by adapting classical protein molecular dynamics, commonly applied to equilibrium states, instead to a treatment of the transition state "dividing hyperplane". However it is important to note that in the crystallographic X-ray structures of ferri- and ferro-cytochrome c the conformational differences are rather small and detectable only by the highest resolution refinement [10, 11], possibly indicating that the outer sphere rearrangements are of very small amplitude and are highly distributed. An experimental estimation of the outer sphere reorganization energy in cytochrome c is available [12]. Multiple minima or large scale conformational dynamics, particularly the motions of two proteins relative to each other in a protein-protein complex, will be viewed in our analysis here as a heterogeneity of reactive complexes, each form or geometry being characterized by a microscopic electron transfer rate constant. This static inhomogeneity viewpoint is an approximation, and formal expressions for examining the breakdown of this Born-Oppenheimer type approximation are available [13]. The nuclear motion can be often be treated classically, insofar as the rate constants exhibit classical activated rate behavior near room temperature (the bacterial photosynthetic reaction center being a prominent exception [6]). But the electronic motion is inherently quantum mechanical and hence absolutely requires a quantum theory.

1.3 Electronic Motion

The general framework of the quantum mechanical rate expression for long-range electron transfer processes in the very weak or non-adiabatic regime will be presented in Sect. 2 with an emphasis on the inclusion of superexchange interactions. The relation between the simplest case of direct donor-acceptor interactions, on the one hand, and long-range electronic interactions important in proteins, on the other, is considered in terms of the elements of electron transfer theory.

Traditional quantum electronic methods are valuable for single amino acids and prosthetic groups but prove intractable for problems involving whole proteins. Methods based on semi-empirical one-electron theories have been advanced by Beratan and Hopfield [14, 15, 16], and by Larsson [17, 18] as models of through-bond interaction in bridged donor-acceptor systems. The relation of such through-bond models, which involve chains of intermediate basis functions, to the general multi-path perturbation expressions for long-range electronic interaction are examined in Sect. 3 below.

One of our principal objectives is to examine tools based on real-space quantum propagators as they avoid excessive dependence upon the tails of basis functions used in the traditional basis set quantum methods. The path integral theory developed by the author in collaboration with Wolynes [19] is an example of such a computational method and this will be examined in Sect. 4. One important distinction of this approach is that the path sampling quantitatively identifies the zone in a protein of maximum importance to the electron tunneling amplitude. No a priori assumptions are required about which pathways or what type of pathway may be important to the long-range electronic interaction. Nevertheless, the method has so far been implemented in a protein at the one-electron Hamiltonian level, and an important objective of current research is its generalization to multi-electron Hamiltonians. These directions are touched upon in the concluding Sect. 5.

In the concluding remarks we also summarize the different patterns of electronic interaction in an carefully defined but uncomplicated classification scheme.

2 The Quantum Mechanics of Long-Range Electron Transfer Kinetics

2.1 The Beginnings of Superexchange Analysis

Superexchange, a term borrowed from its original use in non-nearest neighbor magnetic interactions [20, 21], means in a broad sense non-nearest neighbor electronic interactions intimately involving the electrons of the intervening molecular components.

The concept of electronic interaction by superexchange in electron transfer processes was considered first some time ago, by Halpern and Orgel, in a discussion of electron transfer between transition metal ions [22]. Their basic model described a situation in which the molecular orbitals of a ligand provides overlap with both the donor and acceptor metal-centered orbitals simultaneously, which would lead to a significantly greater coupling than that due to the direct interactions between the metal-centered orbitals. McConnell also presented a seminal study of self-exchange processes between aromatic radical anions tethered together by a polymethylene chain [23], in which he provided a perturbation theory model involving the unoccupied molecular orbitals of $(CH_2)_n$. One of the key features of this analysis was the identification of the self-trapping regime in which the electronic coupling leads to hopping between the two trapped localized states, rather than to electronic delocalization. This remarkable and independent work, in retrospect, can be recognized in our current unified view as the Marcus theory [1, 2] of the double-minimum potential energy surface due to solvent polarization, combined with a superexchange extension of the Levich-Dogonadze theory of non-adiabatic electron transfer [24].

What are we trying to calculate in the superexchange analysis? Much of the time we are interested in using quantum mechanics to compute an electronic interaction energy, also given the apt name "charge resonance energy" [23]. The kinetic significance and indeed the definition of this electronic interaction energy arises through either or both the classical Marcus and the quantum mechanical Levich-Dogonadze rate theories. The measurement of a remarkable electron transfer rate [25, 26, 27, 28] over ten, fifteen, twenty angstroms or more, in or between proteins is thus translated into a tantalizing question of how this electronic interaction energy can be so significant and non-vanishing at this distance. A general and unified framework of the theoretical connection between rates, or the time-dependence, and energies, obtainable through time-independent analysis, has been detailed elsewhere and the salient features are outlined next.

Some familiarity with the elements of the Marcus theory of the activation free energy barrier between solvent trapped localized electronic states will be assumed. The desired mixed quantum/classical rate expressions will be obtained in two ways, in order to set the stage for the superexchange extensions. The Landau-Zener route will lead to an approach we elaborate into the quantum thermodynamic path integration method (Sect. 4). The time-dependent perturbation theory route naturally leads to the superexchange expressions involving intermediary basis functions (Sect. 3). Thus while it will be obvious to experts that the two parallel routes contain exactly the same physics, we will trace them both in order to gain a strong appreciation and a unified view of superexchange. This preparation is strictly an overview of the role of electronic factors in electron transfer rate theory with an eye towards superexchange, and the reader is referred to the original references [2, 29, 30] for the derivations of the equations, as well as to the comprehensive reviews by Marcus, Sutin, Newton, Jortner, and Schmidt [9, 31, 32, 33, 34].

2.2 The Landau-Zener Transmission Coefficient and the Meaning of Δ

We start from classical transition state theory which provides the basic factorization of the electron transfer rate constant:

$$k_{ET} = \kappa_{EL}\,A\,\exp(-\Delta G^{\neq}/k_B T). \tag{1}$$

The Marcus classical free energy of activation is $\Delta G^{\neq}$, the adiabatic pre-exponential factor A may be taken from Eyring's Transition State Theory as $(k_B T/h)$, and κ_{EL} is a dimensionless transmission coefficient $(0 < \kappa_{EL} < 1)$ which includes the entire effect of electronic interactions between the donor and acceptor, and which becomes crucial at long range. With κ_{EL} set to unity the rate expression has only nuclear factors and in particular the inner sphere and outer sphere reorganization energies mentioned in the introduction are dominant parameters controlling $\Delta G^{\neq}$ and hence the rate. It is assumed here that the rate constant may be taken as a unimolecular rate constant, and if needed the associated bimolecular rate constant may be constructed by incorporation of diffusional processes as:

$$D + A \underset{k_d}{\overset{k_a}{\rightleftharpoons}} DA \xrightarrow{k_{ET}} D^+ + A^- \tag{2}$$

$$k_{\text{bimolecular}} = \frac{k_a k_{ET}}{k_d + k_{ET}} \approx \left(\frac{k_a}{k_d}\right) k_{ET} \tag{3}$$

by the standard steady state analysis or pre-equilibrium approximation, respectively. The factor (k_a/k_d) in the pre-equilibrium approximation has a value of close to $0.5\ M^{-1}$ in a solvent with a viscosity like that of water.

This factorization of the rate of the elementary process (Eq. 1) leads (with a few approximations) to the compartmentalization of the experimental parameters in the following way: the dependence of the rate upon reaction exothermicity and upon environmental polarity controls and is reflected in the activation energy and the temperature dependence, whereas the dependence of the rate upon distance, orientation, and electronic interactions between the donor and the acceptor controls and is reflected in κ_{EL}. We refer to this electronic interaction energy as "Δ" rather than the common matrix element symbol H'_{if}, since we require that Δ include contributions from high-order perturbations and in particular superexchange processes. Experimentally, the y-intercept of the Arrhenius plot of the electron transfer rate yields the prefactor $[\kappa_{EL} A \exp(+\Delta S^{\neq}/k_B)]$, and hence the true activation entropy must be known in order to extract κ_{EL}. An interesting example of the extraction of the temperature independent prefactor has been presented in Isied's polyproline work [35].

The compartmentalization of experimental parameters is not rigorously exact, but provides a useful and practical approach to which further refinements such as the distance dependence of the exothermicity [36] and reorganization energy [35, 37, 38] or the temperature dependence of Δ can be appended if necessary.

The transmission coefficient κ_{EL} in Eq. 1 above becomes less than unity when the quantum mechanical coupling of the reactant and product electronic states becomes very weak such that only a fraction (κ_{EL}) of the reactant systems successfully reaching the transition state actually cross into the product state. Depicted in Fig. 1 is this branching process which occurs just at the transition state and whose branching ratio is given by $\kappa_{EL}/(1 - \kappa_{EL})$. At close range or direct contact the electronic interaction is sufficiently strong (a few hundred cm^{-1} or greater) such that κ_{EL} becomes unity, and the particular value of Δ is no longer important to the rate. On the other hand at longer range and weaker Δ, the κ_{EL} factor starts to become significantly less than unity. The essential physics of this crossover from adiabatic ($\kappa_{EL} = 1$) to the non-adiabatic regime appropriate to long-range electron transfer is contained in the crossover formula known as the Landau-Zener transition probability (κ_{LZ}) expression [39]:

$$\kappa_{EL} = \kappa_{LZ} = 1 - \exp\left\{\frac{-\Delta^2 \pi^{3/2}}{\hbar\omega(\lambda k_B T)^{1/2}}\right\}. \qquad (4)$$

The thermally averaged transition probability above is expressed in a form appropriate for harmonic reactant and product potential energy surfaces, in terms of the harmonic frequency ω and reorganization energy λ [40, 41]. It is Eq. 4 which provides us with the required connection between interaction

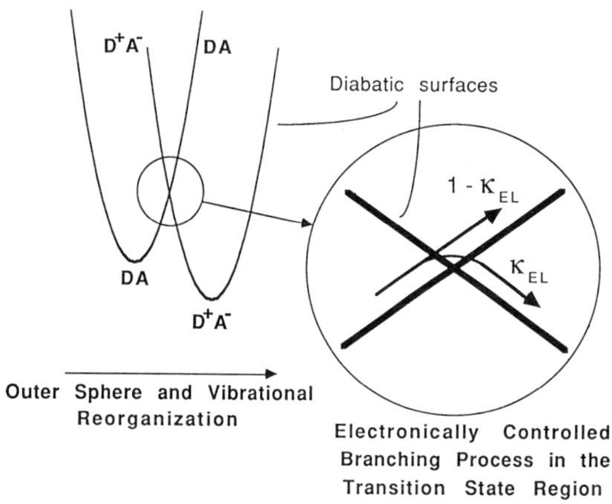

Fig. 1. The Marcus parabolic free energy surfaces corresponding to the reactant electronic state of the system (DA) and to the product electronic state of the system (D^+A^-) cross (become resonant) at the transition state. The curves which cross are computed with zero electronic tunneling interaction and are known as the diabatic curves, and include the Born-Oppenheimer potential energy of the molecular system plus the environmental polarization free energy as a function of the reaction coordinate. Due to the finite electronic coupling between the reactant and charge separated states, a fraction κ_{EL} of the molecular systems passing through the transition state region will cross over onto the product surface; this electronically controlled fraction κ_{EL} thus enters directly as a factor into the electron transfer rate constant

Primordial Two-State Case

$$\mathbf{H} = \begin{bmatrix} E_o & -\Delta_o \\ -\Delta_o & E_o \end{bmatrix} \qquad \Delta, \Delta_o > 0$$

$$\mathbf{H}\,\mathbf{c}_i = E_i\,\mathbf{S}\,\mathbf{c}_i \qquad \Delta = \frac{\Delta_o + sE_o}{1 - s^2}$$

$$\text{where, } \Delta = \frac{h}{2}\nu_{Rabi}$$

Fig. 2. The quantum mechanics of the two-state problem provide a paradigm for the much more extensive electronic state space of a real molecular or macromolecular system. The eigenvectors c_i of the Hamiltonian are symmetric and antisymmetric linear combinations of the localized basis vectors with an eigenvalue splitting of 2Δ, where s is the overlap integral and Δ_o is the direct coupling (the only kind possible in this case)

energies and rates. The relation of a computed off-diagonal Hamiltonian matrix element to the Δ in the primordial 2×2 case is given in Fig. 2. The identification of Δ is facile because it corresponds to a Rabi frequency in the time-dependent quantum mechanics of a system initially in the reactant electronic configuration which then oscillates into one other (product) electronic state;

$$\nu_{Rabi} = 2\Delta/h. \tag{5}$$

The Landau-Zener expression is calculated in a time-dependent semiclassical manner from the diabatic surfaces (those depicted in Fig. 1) exactly because these surfaces, which describe the failure to react, are the appropriate zeroth order description for the long-range electron transfer case. As can be seen, in the very weak coupling limit (small Δ) the κ_{EL} factor and hence the electron transfer rate constant become proportional to the absolute square of Δ:

$$k_{ET} = \left(\frac{k_B T}{h}\right)\left(\frac{\Delta^2 \pi^{3/2}}{\hbar\omega(\lambda k_B T)^{1/2}}\right)\exp(-\Delta G^{\neq}/k_B T). \tag{6}$$

This mixed classical/quantum expression is valid for classical nuclear behavior and, strictly speaking, for the case of direct two-site interaction rather than superexchange, as the Landau-Zener expression was derived from the time-dependent Schrodinger equation assuming a two-state (reactant/product) electronic system with direct coupling. Nevertheless, it becomes clear on physical grounds that the form of Eqs. 4–5 can serve to define an effective Δ in the superexchange case in terms of the Rabi precession frequency characteristic of the two trap sites embedded in the complex system; wherein $2\Delta/h$ would be computed from this net effective Rabi precession frequency.

Without need for mathematical elaboration, it is simple to see that the Landau-Zener approach (and Fig. 1) enables a direct interpretation of Δ even in the superexchange case, where the direct coupling $H'_{if} = 0$. During the (brief) time in which the molecular system is in the transition state region an electronic oscillation occurs between the electronic state characterized by the reactant redox states to the electronic state with product redox state character. In the very weak coupling regime, this oscillation, far from completing one cycle, mixes in only a little of the product state and the transmission coefficient is accordingly very much less than unity. Landau-Zener analysis tells us that, in this limit, κ_{EL} is proportional to the square of this oscillation frequency, known also as a Rabi frequency. The key observation is that the Δ in Eq. 6 above signifies the Rabi frequency (multiplied by h/2) regardless of whether this oscillation arises due to direct electronic coupling H'_{if} or through indirect superexchange mechanisms. Thus we have a definition of Δ valid even in complex and large systems such as proteins, which is independent of any particular approximate forms, perturbational or otherwise, derived from simpler cases. The Landau-Zener perspective also makes it clear that the electronic energetics (diagonal Hamiltonian elements) relevant to the computation of Δ are non-standard redox potentials of the donor and acceptor at those nuclear configurations which are isoenergetic as DA or D^+A^- (see also in this regard [42, 43]). Correspondingly we may define this unique non-standard redox potential as a "transition state redox potential" [44]. Higher order generalizations of the Landau-Zener non-perturbative time-dependent procedure are also possible to characterize additional features of the case where $H'_{if} = 0$ [45].

McConnell's idea of an effective pseudopotential (more properly called "effective off-diagonal matrix element") analyzes the general conditions for the representation of indirect couplings as an effective direct interaction, as illustrated in Fig. 3. The basic condition is that the excitation energy to the mediating electronic states from either the DA or D^+A^- states is large compared to any of the perturbation matrix elements. (This is also essentially the same condition as that of Kuznetsov and Ulstrup's "High energy intermediate states" limit and Lin's Case I [46, 47].) Stated in other terms, a set of numerous indirect couplings may be converted into an equivalent effective Δ in a reduced state space when the mediating molecular components have relatively inaccessible redox potentials, though they may interact with the tunneling centers by polarization and multiple electron exchange interactions. When this conversion to an effective 2×2 Hamiltonian is possible, then this effective off-diagonal coupling Δ divided by h is equal to a Rabi frequency of the full coupled system which yields a generalized superexchange transmission coefficient in exactly the same form (Eq. 4) as in the original two-state case. The path integral Monte Carlo instanton approach (Sect. 4.4) obtains precisely this effective Δ in a three-dimensional and highly structured macromolecular system.

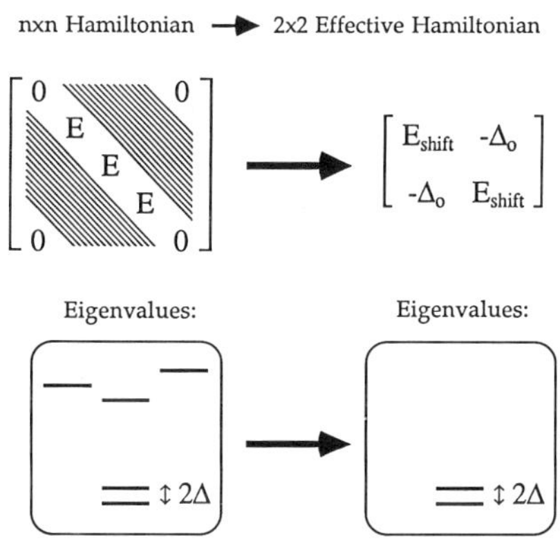

Fig. 3. McConnell's idea of the effective 2×2 Hamiltonian is based on matching the two lowest eigenvalues of the $n \times n$ state space with an effective 2×2 matrix. Note that the true $n \times n$ Hamiltonian can have zero direct off-diagonal coupling between reactant and product, but non-zero values in the shaded off-diagonal region. The reactant and product states may be considered trap states embedded in the higher energy medium, and the resulting lowest eigenvalues (must) correspond principally to linear combinations of the trap states

2.3 Fermi's Golden Rule: Nuclear Factors and Electronic Interaction Energies

Turning to the fully quantum mechanical approach, we find that the lowest order rate theory for general non-radiative relaxation processes also provides a factorized rate expression:

$$k_{ET} = \frac{2\pi}{\hbar} |\Delta|^2 \, \mathrm{FCWD} \tag{7}$$

where $\Delta = H'_{if}$, the direct electronic interaction matrix element of the perturbation Hamiltonian, and FCWD is the Franck-Condon weighted density of vibrational states. This class of rate expressions is important because they serve to incorporate quantum mechanical nuclear vibrations, as is appropriate for high frequency modes at room temperature. Known in general as Fermi's Golden Rule #2, this expression was developed by Levich, Dogonadze, Jortner, Hopfield, Ulstrup and others for the case of long-range electron transfer [24, 29, 30, 48]; it is also the underlying foundation for many other rate theories as well, such as those of intersystem crossing and singlet and triplet excitation transfer, not to mention linear optical absorption and emission. For one harmonic nuclear mode in the low temperature limit, the FCWD can be evaluated to yield:

$$k_{ET} = \left(\frac{2\pi}{\hbar^2 \omega}\right) \Delta^2 \exp(-S) \left(\frac{S^p}{p!}\right). \tag{8}$$

The exothermicity dependence is in p, ($p = -\Delta G^o/\hbar\omega$) whereas the reorganization energy is expressed in s, ($s = \lambda/\hbar\omega$). This limit can be appropriate for the inner sphere reorganization; whereas for the outer sphere reorganization one can assume very low frequencies and take the high temperature (classical) limit, obtaining

$$k_{ET} = \frac{\Delta^2}{\hbar} \left(\frac{\pi}{\lambda k_B T}\right)^{1/2} \exp(-\Delta G^{\neq}/k_B T). \tag{9}$$

This is, not surprisingly, nearly identical to the Landau-Zener corrected transition state theory above (Eq. 6); exact equivalence is obtained if the transition state theory prefactor is taken to be (ω/π) rather than Eyring's $(k_B T/h)$, and the numerical difference between these two conceptually different prefactors [49] is often small enough as to be of little experimental consequence.

A very practical comprehensive rate expression appropriate for long-range electron transfer in proteins and other large molecules yet which retains ease of computation by anyone and on the smallest computer is obtained by assuming one high frequency harmonic mode (inner sphere reorganization) and one very low frequency harmonic mode (outer sphere reorganization) [30, 46]:

$$k_{ET} = \frac{\Delta^2}{\hbar}\left(\frac{\pi}{\lambda_s k_B T}\right)^{1/2} \exp(-S) \sum_{p=0} \exp\left\{-\frac{(\Delta G^\circ + p\hbar\omega + \lambda_s)^2}{4\lambda_s k_B T}\right\} \frac{S^p}{p!} \tag{10}$$

One advantage of this mixed quantum/classical rate expression is that, by use of the Marcus outer sphere activation free energy $\Delta G^{\neq}$, we recover the inclusion of a realistic activation entropy which is entirely missing from the low temperature quantum expression (Eq. 8). The activation entropy as an independent factor can only appear within a rate expression if the model includes degrees of freedom orthogonal to the reaction coordinate (i.e. more modes than one). Two modes may still seem a bit meager for a description of a protein but the idea is that we let the classical outer sphere mode represent the free energy associated with many degrees of freedom. This two mode model has been parameterized by revealing and important experiments on organic donor/acceptor molecules [50, 51], non-covalent donor/acceptor systems [52], and photosynthetic reaction centers [53].

Table 1 provides values of the prefactor valid for either Eq. 9 or Eq. 10 above; in addition the rate constant is equal to this prefactor in the special case of $\Delta G^{\neq} = 0$ (Eq. 9) or the summation = 1 in Eq. 10, which is the maximum possible rate constant. The fact that one should be able to experimentally locate the $\Delta G^{\neq} = 0$ maximal speed point by varying the donor/acceptor reaction energetics has been successfully exploited by Miller and Closs [50] to obtain a direct measurement of this non-adiabatic pre-factor.

Now Fermi's golden rule, Eq. 7, arises from standard lowest order time-dependent perturbation theory, and in particular is derived by assuming direct coupling in a two dimensional (reactant/product) electronic state space, as was the original Landau-Zener expression. To extend the Landau-Zener theory we seek a more general route to the Rabi oscillation frequency; to extend Fermi's golden rule here we naturally pursue higher order perturbation theory. When Eq. 7 is rederived in the more general case of indirect or superexchange

Table 1. The non-adiabatic prefactor in terms of Δ

Reorganization energy, temperature		Rate expression prefactor (s^{-1})
$\lambda = 0.2$ eV	298 K	$\Delta^2 \times 5.78 \times 10^8$
$\lambda = 0.5$ eV	298 K	$\Delta^2 \times 3.65 \times 10^8$
$\lambda = 1.0$ eV	298 K	$\Delta^2 \times 2.58 \times 10^8$
$\lambda = 1.0$ eV	233 K	$\Delta^2 \times 2.92 \times 10^8$
$\lambda = 1.5$ eV	298 K	$\Delta^2 \times 2.11 \times 10^8$

The electronic interaction energy Δ is assumed to be in units of cm^{-1}, and the prefactor given is the pre-exponential of Eq. 9, or the pre-exponential of Eq. 6 (at 298 K) if $\omega/2\pi$ is equal to 104 cm^{-1}. The prefactor expression is valid only when the exponential in Eq. 4 can be linearized; for this range of parameters, Δ must be less than 100–200 cm^{-1}.

interactions (Sect. 3.1) the exact same form results, and the explicit expressions for Δ in terms of the specific combination of the various non-zero electronic matrix elements involving the intervening medium are obtained. This factorization of the rate into a superexchange electronic factor and a nuclear factor, where the nuclear factor is the same as that obtained in lowest order perturbation theory has been derived in the case of "high energy intermediate states"; see Ulstrup, Kuznetsov, and Lin [46, 47, 54]. Hence the same ultimate conversion from Δ to rates through Eqs. 7–10 and Table 1 are performed in the end to compute rates.

Looking ahead, the extension to the long range case can be accomplished either through infinite or high order perturbation theory (Sect. 3), or through quantum thermodynamic path integration (Sect. 4). Perturbation theory is an old and trusty tool which is best suited for the situation in which the identification of a single dominant chain of intermediate states does not present a problem. A key strength of the path integral approach, which is a new and developing tool in biomolecular theory, is that the method serves to identify the regions of intervening space important in mediating and controlling the long-range electronic interaction, and indeed the method is particularly well-suited if there is a multiplicity of such electronically important paths.

3 Superexchange and Perturbation Theory

3.1 The General Transition Matrix Element of Arbitrarily High Order

Intermediate states corresponding to oxidized or reduced states of the intervening medium appear explicitly in the high order perturbation theory for electron transfer, with the understanding that we are still examining a single elementary chemical reaction process. In sharp contrast to actual chemical intermediates within a classical kinetic scheme (with corresponding concentrations), intermediacy in this sense means quantum mechanical superposition with very low coefficients (coherent probability amplitudes) on these intermediary sites. Nevertheless the explicit formulas for the perturbation sum are readily interpreted as a type of sum over paths, and this sum over paths perspective can provide a general framework which embraces practically all treatments of electronic tunneling interactions in proteins including those based upon covalent and non-covalent contacts through selected bridge orbitals [16, 17]. We focus only on the path aspects here; the formulas are venerable and the aim is to raise broader questions.

The rate expression described in Sect. 2, Fermi's golden rule, is commonly derived from a minimal model of two interacting quantum states plus environmental (bath) states which serve to provide the localizing fluctuations. As the resulting rate expression is based upon lowest order time-dependent perturbation theory [55, Eq. 35.14], it thereby lacks intermediate states altogether and hence is not useful for a treatment of superexchange phenomena.

An alternative and more general time-dependent procedure is available from scattering theory, and this has been recognized by many theoreticians as providing the foundation for a perturbation theory of superexchange [46, 54]. The common feature (as far as theory is concerned) between quantum particle scattering and long-range electron transfer is that there is a final state which is degenerate or resonant with the initial state, and that the net transition is mediated by many other non-resonant states mixed in by the perturbing interactions which are considered time-independent. In addition, there are other requisite conditions such as a high density of final states, which is broadly relevant to condensed phase electron transfer as it is to particle scattering into continuum states. Known as the reaction or transition matrix method, a clear exposition may be found in Schiff's textbook [55, Sect. 37], which illuminates the general connection between the time-dependent transition probability and the time-independent high order transition matrix elements of the form $\langle\psi_f^o|V|\chi^+\rangle$, where V is the coupling and χ^+ is defined below. Under rather general conditions, the electron transfer rate constant is given by a perturbation-analysis-free treatment [55, pp. 312–4] of the time-dependent Schrodinger equation to be proportional to $|\langle\psi_f^o|V|\chi^+\rangle|^2$. Perturbation theory then enters in only at the time-independent level to provide approximations to this transition matrix element, which is an explicit representation of Δ, $\Delta = \langle\psi_f^o|V|\chi^+\rangle$.

Omitting all discussion of the mathematical properties and the subtleties with regard to the continuum of final state energies, we will hop to the perturbation expansion in three short equations. The initial state describing the unperturbed reactant (DA) system describes the solution to the zeroth order Schrodinger equation:

$$(H_o - E_i^o)\psi_i^o = 0 \tag{11}$$

At long range the direct interaction with the final state (D^+A^-), similarly a zeroth order state, may be negligibly small relative to the contribution from superexchange interactions at that distance. The (unknown) solution for the Hamiltonian including interactions ($H = H_o + V$), is χ^+,

$$(H - E_i^o)\chi^+ = 0 \quad \text{or} \quad (E_i^o - H_o)\chi^+ = V\chi^+ \tag{12}$$

and is defined under the boundary condition that $\chi^+ \rightarrow \psi_i^o$ as $V \rightarrow 0$. The energy E_i is that of the electronic state DA (which equals that of D^+A^-) at the Marcus transition state [44]. Considering the inverse operator to $(E_i^o - H_o)$, and including the small additive constant $i\varepsilon$ to condition its singularity, χ^+ can be expressed as the solution to an inhomogeneous equation known as the Lippmann-Schwinger equation,

$$\chi^+ = \psi_i^o + (E_i^o - H_o + i\varepsilon)^{-1}V\chi^+ \tag{13}$$

which includes the known homogeneous solution ψ_i^o according to the boundary condition. Now expanding the χ^+ on the right hand side recursively, we can generate the desired infinite series in terms of the zeroth order states,

$$\chi^+ = \psi_i^o + \sum_{j \neq i, f} \psi_j^o [(E_i^o - E_j^o)^{-1} \langle \psi_j^o | V | \psi_i^o \rangle]$$

$$+ \sum_{j,k \neq i,f} \psi_j^o \left[\frac{\langle \psi_j^o | V | \psi_k^o \rangle \langle \psi_k^o | V | \psi_i^o \rangle}{(E_i^o - E_j^o)(E_i^o - E_k^o)} \right] + \sum_{j,k,l,\dots}$$

$$\times \psi_j^o \left[\frac{\langle \psi_j^o | V | \psi_k^o \rangle \langle \psi_k^o | V | \psi_l^o \rangle \langle \psi_l^o | V | \psi_{\dots}^o \rangle}{(E_i^o - E_j^o)(E_i^o - E_k^o)(E_i^o - E_l^o)} \times \dots \right] \tag{14}$$

The transition matrix describing the overall coupling including superexchange is obtained by operating from the left by $\langle \psi_f^o | V$ upon the χ^+ above to yield $\langle \psi_f^o | V | \chi^+ \rangle$, and we find the expected lowest order first term $H'_{if} = \langle \psi_f^o | V | \psi_i^o \rangle$, from the first term on the right hand side of Eq. 14, plus the second order term from the $\sum_{j \neq i, f}$, plus an infinite series of higher order terms. The Nth order term in $\langle \psi_f^o | V | \chi^+ \rangle$ thus consists of a sum over all possible chains consisting of $N - 1$ intermediate states of the product of $\langle \psi_f^o | V | \psi_j^o \rangle$ times $N - 1$ coupling matrix elements each divided by a corresponding energy denominator [56]. Each chain within the sum for each order contributes an amplitude to χ^+ and an energy to $\Delta = \langle \psi_f^o | V | \chi^+ \rangle$ from a particular path.

3.2 Many Approximations and Single Paths

As formal expressions Eq. 14 and the corresponding expansion of $\Delta = \langle \psi_f^o | V | \chi^+ \rangle$ provide a clear definition of the target for theoretical computation, but when implemented various stages of simplification are applied, which must be understood in interpreting the results. A hierarchy of simplifications may include:

i) The states are generally vibronic states, but, as mentioned before, in the "high energy intermediate states" limit [46] the quantum mechanical vibrational states of the intermediates can be dropped from the expression. (Treatments including quantum vibrations on bridging groups have also been presented [16].) Note that this refers to the vibrations in the virtual states ψ_j^o as being distinct from those of the system in the initial (DA) and final (D^+A^-) states. Protein molecular dynamics which describe the conformational states of the protein system in the (DA) and (D^+A^-) states are entirely another matter and have been touched upon in the introduction.
ii) The electronic states consist in general of multi-electron Slater determinants, where each such state involves wavefunctions on all sites. This is the approach, at the CI level, taken by Morokuma [57], and at the Hartree-Fock level, taken by Newton [58]. The alternative simplification is to drop

the antisymmetric many-electron picture and refer to single electron molecular orbitals localized on distinct sites as the initial, final, and intermediate states in place of the corresponding multi-electron states of the system. This is also known as the one-electron hopping model.

iii) The great majority of the electrons may be considered "frozen" and the calculation restricted to the orbitals for a few valence electrons.

iv) Approximate (semi-empirical) rules may be used to estimate the integrals in each matrix element. These rules are calibrated on small molecules, which do not provide calibrations of non-bonded interactions. When extended Huckel is used [15, 16], each intermediate state becomes an atomic orbital, and the energy of an electron in one of the resultant molecular orbitals is independent of the population of any of the other molecular orbitals. Interference effects between different bonded paths are, however, retained [15].

v) Only a single path from within the summation corresponding to a single order in the perturbation series (Eq. 14) is evaluated [23].

The approximation in item (ii) above, abandonning all correlation including between electrons of like spin, may have very significant consequences in long-range biomolecular electron transfer interactions, but the correlated alternatives are also the hardest to test and assess in such large systems; very little is known or established. In our laboratory we are currently pursuing a many-electron computation based on taking approximation (iv) (at the MNDO level) without approximation (ii) in order to investigate the role of correlation and non-local exchange [59]. The uncorrelated single particle Hamiltonian framework is common to the extended Huckel and band theory studies by Beratan and by Larsson [14, 15, 16, 17], and to the existing path integral study by Kuki and Wolynes [19]. New path integral methods are being designed, however, and in particular several electron path integral computations for tunneling including correlation are under active exploration in the author's laboratory.

Simplifying approximations (i)–(iv) are elements of a well-understood hierarchy in electronic eigenstate computations in general; but in our superexchange problem in particular approximation (v) stands out as the prominent issue the treatment of which, either implicitly or explicitly, must influence the entire framework of a theoretical approach. How can quantitative studies shed light on the question of the relative role of different types of paths, particularly the role of the through-bond vs through space classes? Put another way, assume that various levels of numerical electronic structure technology are brought to bear in analyzing the orbitals on various prosthetic groups and amino acids in a protein, and the level of sophistication (items (i)–(iv)) is selected appropriately in recognition of the large scale of the macromolecular problem (sophistication inverse to some power of the number of electrons). Now if direct Hamiltonian matrix diagonalization is out of the question, the selection of a framework into which these computed numerical integrals are installed becomes a question of the choice of paths, a restriction of the set of paths to be digested by the computer. To choose likely-looking paths on the basis of logic built upon small molecules which are covalently-bridged appears to be not much fundamental progress from the musings of Szent-Gyorgi; for while the semiconductor analogy he proposed [7] is probably off the mark, there is still very limited rigorous basis for recognizing paths of macromolecular electronic superexchange in largely insulating and very three-dimensional proteins.

The single path approximation, in the context of Eq. 14, can be readily identified in a model in which a single chain of intermediate sites is extracted from the summation over all intermediate sites {j, k, l . . .}, so that

$$\chi^+ \cong \psi_i^o + \psi_j^o \left[\frac{\langle \psi_j^o | V | \psi_k^o \rangle \langle \psi_k^o | V | \psi_l^o \rangle \langle \psi_l^o | V | \psi_{\ldots}^o \rangle}{(E_i^o - E_j^o)(E_i^o - E_k^o)(E_i^o - E_l^o)} \times \ldots \right] \tag{15}$$

is evaluated for a particular j, k, l. . . only, and the product over hopping amplitudes between intermediate sites in this single chain is evaluated. The single path approximation (Eq. 15) is to be distinguished from fully retaining the first two terms on the right hand side of Eq. 14, which is indeed a low order approximation, but not a single path approximation [34, Eq. 428]. Even more extreme, though not uncommon is the truncation to

$$\Delta = \langle \psi_f^o | V | \psi_i^o \rangle + \frac{\langle \psi_f^o | V | \psi_j^o \rangle \langle \psi_j^o | V | \psi_i^o \rangle}{(E_i^o - E_j^o)}; \tag{16}$$

this may be the dominant correction at shorter donor-acceptor separation but even there it is clear that distinct choices of the nature of the particular ψ_j^o retained commonly gives rise to "new" models of electron transfer mechanisms.

Important electron transfer kinetic measurements have been carried out on a number of lovely organic molecules in which the donor and acceptor are directly covalently bridged [36, 60, 61, 62, 63, 64], and a theoretical analysis of these results by use of the single path approximation is rather natural given the conspicuous covalent link. Nevertheless, the domination of a single through-bond pathway in these smaller molecules [61] certainly does not imply a similar domination in electron transfer proteins, and the sufficiency or accuracy of the single path approximation in folded proteins is a theoretical as well as conceptual question which merits fresh analysis. The perturbation expansion for the general transition matrix element $\langle \psi_f^o | V | \chi^+ \rangle$ clearly shows that the one path approximation is a severe truncation of the exact formula. Even an approach based upon Hamiltonian diagonalization, while formally non-perturbative in the Hamiltonian, can still be a one or few path approximation simply by severely truncating the space of intermediate states to a limited basis set confined to within a narrow geometric sub-structure, thus in effect assuming that many off-diagonal couplings which are components of the chain of amplitudes for other paths are zero. Beratan and Hopfield's original band theory approach for linear bridges with translational symmetry [14, 15], for example, is a one-electron non-perturbative treatment in which an important approach to the direct diagonalization of major sub-blocks of the Hamiltonian matrix is developed, and yet if applied to a peptide backbone in a protein [16], assumes the importance of one-dimensional molecular structures.

3.3 Many Approximations but Many Paths

The transition matrix expression for Δ at long range may alternatively be evaluated by including in the Hamiltonian matrix large numbers of intermediate states corresponding to electron configurations involving orbital population shifts on many sites in a generous three-dimensional zone encompassing the donor and acceptor. The Hamiltonian or large sub-blocks thereof is then diagonalized to obtain χ^+. This basic type of procedure automatically incorporates many or all of the paths constructed from permutations among the network of intermediate wavefunctions, and is followed by evaluation of $\langle\psi_f^o|V|\chi^+\rangle$. This research with large path numbers is in progress in the author's laboratory in a multi-electron version [59]. Once the multi-path versions for inhomogeneous and aperiodic macromolecular structures are in hand, we hope to be able to assess, by comparative methodological studies on the same model for a protein, the relative importance of different classes of chains in this manner. The path integral method described next is intrinsically a many path method as well. In the quantum path integral theory of electronic tunneling the relative importance of paths of different geometries is not restricted at the outset but are rather determined by the course of the Monte Carlo simulation itself. Many approximations remain; items (i), (ii), and (iii) are approximated but (v) is not. The method has semi-empirical elements in very different ways from molecular orbital calculations, and is more closely related to solvated electron calculations with the use of effective potentials. Most particularly, the method is built and tailored from the outset for the problem of tunneling between sites in complex three-dimensional media.

Before leaving the perturbation theory of superexchange it is instructive to examine a simple cubic lattice model for the explosive growth in the number of paths connecting two sites as one considers paths of less than the maximal coupling. The existence of very large numbers of alternative paths of smaller amplitudes than the optimal becomes more and more pronounced as the sites are pulled apart in three dimensions. Figure 4 and Table 2 presents the number of paths according to path length in a model one-particle three-dimensional lattice hopping problem where the minimum number of hops is ten. Each hop in each path, of course, corresponds to an additional factor of the form $\langle\psi_k^o|V|\psi_i^o\rangle/(E_i^o - E_k^o)$ in the tunneling amplitude for that path, and it is assumed that the intervening medium is sufficiently insulating such that this factor is less than unity. The table displays the exact formula for the number of paths of length N given the minimum number of steps N_o (the number of kinks is defined as $K \equiv (N - N_o)/2)$ [65]:

$$N(N, N_o) = \sum_{k_x=0}^{K} \sum_{k_y=0}^{K-k_x} \frac{N!}{(N_0 + k_x)!k_x!(k_y!)^2(k_z!)^2}, \quad k_z = K - k_x - k_y \quad (17)$$

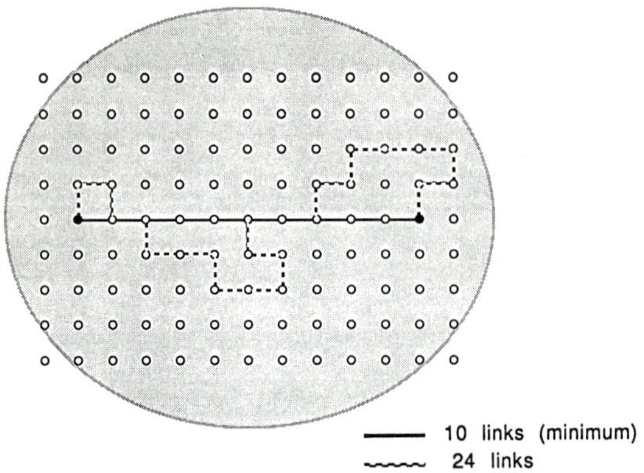

Fig. 4. Illustrative model of paths between two trap sites embedded in a three-dimensional cubic lattice. The dashed 24-link line has 7 "unnecessary" kinks which reduce its contribution to the path sum, but there are many of them (Table 2); note that the kinks in the figure are two-dimensional but the count in Eq. 17 is three-dimensional. The paths corresponding to terms in Eq. 14 may in general cross over themselves and backtrack, but may not visit the initial or final sites twice. The latter condition does not arise directly from Eq. 13 but rather from the irreversibility concept underlying the theory of the rate constant

Table 2. Path counts in a cubic lattice as a function of path length between two points ten links apart

Path Length	Non-termini Avoiding (Eq. 17)	Corrected (j, k, . . . $\neq$ i, f)
10	1	1
12	276	264
14	40,495	37,111
16	4,033,680	3,526,524
18	313,891,740	262,170,396
20	20,650,661,328	16,529,889,600
22	1,204,985,102,013	927,681,664,005
24	$.6432854208660 \times 10^{14}$	$.4779880277166 \times 10^{14}$
26	$.3209004307141 \times 10^{16}$	$.2308786026698 \times 10^{16}$
28	$.1518313483814 \times 10^{18}$	$.1060829285522 \times 10^{18}$
30	$.6888055826133 \times 10^{19}$	$.4685847654680 \times 10^{19}$
32	$.3020640456480 \times 10^{21}$	$.2005440997327 \times 10^{21}$
34	$.1288497474190 \times 10^{23}$	$.8365321110427 \times 10^{22}$
36	$.5370826083568 \times 10^{24}$	$.3416553285062 \times 10^{24}$
38	$.2196349778560 \times 10^{26}$	$.1371126902448 \times 10^{26}$
40	$.8837752481937 \times 10^{27}$	$.5422273190439 \times 10^{27}$

as well as the number of paths in which those that visit the initial or final site more than once are discarded. This latter correction, which embodies the restriction on the summation over intermediate states in Eq. 14, can be seen to discard about a third of the paths at long path length, and hence does not alter the essential pattern of growth given by the formula (Eq. 17), which is slightly less than exponential in N. From Table 2 it follows that the single N = 10 path of maximum amplitude must, in the simple analysis, contribute an amplitude more than 200 times that of one of the N = 12 (one kink) paths or else there will be 265 important paths; the number of paths of yet longer length continues to grow further by a factor of 36–100 for each additional increase in N. Put another way, if the square of $\langle\psi_k^o|V|\psi_i^o\rangle/(E_i^o - E_k^o)$ is not less than 1/36 then convergence in path numbers will be extremely slow.

This 3D lattice Hamiltonian with two traps can be elaborated further by adding filled lattice orbitals (3D bridge orbitals), which will then exhibit interference effects, and which could be computed by an extension of the Hopfield-Beratan method. Nevertheless the path counting given in Table 2 serves to illustrate the dramatic character of the difference between the single path model (Eq. 15) and the many path result (Eq. 14) in three dimensions.

4 Monte Carlo Sampling of Tunneling Paths: The Path Integral Instanton Method

4.1 Tunneling and Wavefunction Tails

Consider path integrals to arise from a foundation built from the solutions of two elementary (meaning simple yet profound) quantum mechanical problems, a one-dimensional tunneling problem [66], and a propagation in two or more dimensions by more than one alternative path [67].

A quantum particle in a one-dimensional potential V(x) has energy eigenvalues and energy eigenstates obtained by integrating the Schrodinger equation from minus infinity to plus infinity in x, with necessary boundary conditions, and the particle is defined to be tunneling in any region in which the total particle energy is less than the potential energy. For example in case of the double well potential in Fig. 5, we assume that the Schrodinger equation has been solved, and the ground state energy eigenvalue is found to be E_+, and so it is evident that the three regions marked by the heavy bars are the "classically forbidden regions" for the ground state, the regions in which the particle is tunneling. When, as illustrated, the two wells are fairly isolated (separated by a substantial classically forbidden region) the two lowest exact eigenvalues of the global system are predictably found just below (E_+) and just above (E_-) the single well eigenvalue E_0; this is the tunneling splitting. The straightforward inspection of the kinetic energy second derivative operator will indeed reveal that this tiny splitting $E_- - E_+ = 2\Delta$ reflects the higher curvature of the antisymmetric solution due to the presence of the node (which then pushes slightly more probability amplitude outward into the costly left and right classically forbidden regions); but this is dissection post mortem in terms of the solution (the delocalized global wavefunction) rather than the input.

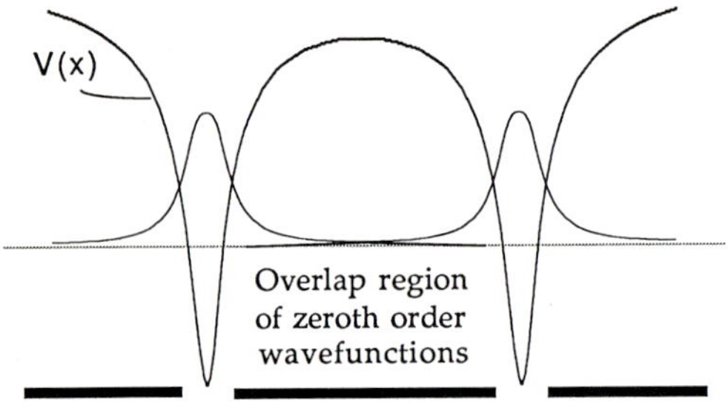

Fig. 5. Tunneling describes the quantum mechanical delocalization of a particle into the region where the total energy (dotted line) is less than the potential energy V(x). The tunneling barrier $V(x) - E_o$ enters directly into the WKB formula for the wavefunction tail, whereas the zeroth order wavefunctions are typically expressed in a basis set expansion which may or may not be optimized for its tunneling characteristics

To obtain a more penetrating view on Δ as an interaction between the original single wells, we compare two analytical routes to the evaluation of the basic two-center charge resonance integrals in this limit; WKB and the low-dimensional perturbation formula.

The WKB analysis is a semiclassical theory which describes the tunneling behavior of the particle directly in terms of the local structure of the central tunneling region. Referring to the two deep regions of high particle probability as the redox sites, we take the wavefunctions within each local site to be unaltered by the presence of the other site, and determine E_o. Recall that at the transition state the two vertical redox potentials are necessarily equal to each other so the energies are both equal to E_o. Now the tail of one of the wavefunctions (either one) is extended from its classical turning point ($x_{t.p.}$) into the tunneling region by use of the simple and explicit WKB formula:

$$\psi^{WKB}(x) \approx \exp\left\{ - \int_{x_{t.p.}}^{x} \sqrt{\frac{2m(V(x') - E_o)}{\hbar^2}}\, dx' \right\} \tag{18}$$

which becomes for a flat barrier potential, even simpler:

$$\psi^{WKB}(x) \approx \exp\left\{ - \sqrt{\frac{2m(V - E_o)}{\hbar^2}}\,(x - x_{t.p.}) \right\} \tag{19}$$

Since the particle (electron) in one redox site can respond to the presence of a second deep well out in the distance only to the extent that this quantum tail is non-zero, the tunneling splitting Δ decays essentially exponentially just as the wavefunction tail in Eqs. 18–19 above. More precisely, we write:

$$\Delta = \int \psi_f^o(x)\, V(x)\, \psi_i^{WKB}(x)\, dx \tag{20}$$

The mechanics of single particle tunneling captured in these expressions (Eqs. 18 and 20) shows Δ to be directly controlled by the potential actually visited by the tunneling particle; this is the essence of the WKB idea which we wish to preserve and translate into the fully three dimensional and much more complex, and many electron, world of protein molecules.

By contrast the alternative approach of perturbation theory formulates an approximation for Δ in terms of the original single well wavefunctions as

$$\Delta = \int \psi_f^o(x)\, V(x)\, \psi_i^o(x)\, dx \tag{21}$$

The problem here is that, while the zeroth order single well wavefunctions may be highly optimized with regard to the site energy and may describe superbly the electronic properties of the redox site itself, the Δ computed from them is wholly dependent upon the details of the tails of the wavefunctions, which depends on the atomic orbital basis set chosen, as well as upon the environment used in the single well calculation. Adding extra basis orbitals in the tunneling region is an important procedural improvement, particularly at "medium range" [58, 68, 69]. If the extra basis functions are on atomic sites in the tunneling region then higher order perturbation theory provides elaborations of Eq. 21 as we have seen, but the fact remains that the optimization of the basis set is a special and distinct problem in tunneling calculations since the variational principle applied to any selected basis set will optimize the molecular orbital coefficients only insofar as they lower the total energy, which is dominated by interactions close to the bound site.

4.2 Path Integral Quantum Mechanics and Statistical Thermodynamics

Path integrals in either their time-dependent or thermodynamic garb can be considered to be a way of extending WKB into three-dimensions, where now there are many different possible routes, as we have seen above, connecting donor and acceptor. Path integrals are real space quantum propagators [67], and, in a certain sense convert a three-dimensional tunneling problem into a sum over many one-dimensional tunneling problems (sum over paths). Consider Fig. 6, where quantum propagation along two paths gives rise to the well-known interference pattern of, say, electron diffraction. The two paths depicted are the dominant paths, each characterized by a path amplitude which is stationary with regard to small deformations of the path around that dominant path. For any space-time path $\mathbf{r}(t)$ with boundary conditions $\mathbf{r}(0) = a$, $\mathbf{r}(\tau) = b$, Feynman showed that the path amplitude

$$\exp\left(\left(\frac{i}{\hbar}\right) S[\mathbf{r}(t)]\right) = \exp\left\{\left(\frac{i}{\hbar}\right)\int_0^{\tau} L\left(\mathbf{r}(t'), \frac{d\mathbf{r}(t')}{dt'}, t'\right)dt'\right\} \tag{22}$$

is the contribution which that path makes to the overall path integral. The action S is the time integral of the Lagrangian L exactly as in classical mechanics, but the complex exponential and Planck's constant are distinctly quantum. The paths to be summed are not just the dominant paths but all

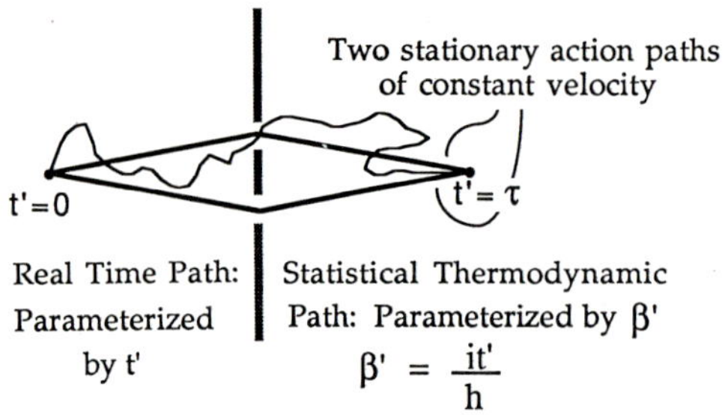

Fig. 6. All paths leading from the initial to the final points in time τ contribute an interfering amplitude to the path sum describing the resultant probability amplitude for the quantum propagation. In this double slit free particle case, two paths of constant speed are local functional stationary points of the action, and these two dominant paths provide the basis for a (semiclassical) classification of subsets of paths which contribute to the path integral. In the statistical thermodynamic path expression, the path sum is equal to the off-diagonal electronic thermal density matrix element, and the analogous dominant paths are local functional minima of the integral $\int_0^\beta H(\mathbf{r}(\beta'))\, d\beta'$ in Eq. 23

possible paths, though the dominant paths provide the basis for approximate (semiclassical) analytic evaluations of the path integral. In order to develop the quantum simulation technology needed for long range tunneling, computational imperatives require us to shift to the quantum thermodynamic or imaginary time theory parallel to this suggestive real time path integral theory. Yet it can already be perceived that Feynman's path integral formulation of quantum mechanics (and of quantum statistical thermodynamics) can be developed to establish the meaning of electron transfer paths in proteins based on a sum over action-weighted paths.

Tunneling gives rise to a splitting 2Δ which can be determined by examination of the thermal off-diagonal electronic density matrix at the transition state nuclear configuration. As a feature of the energy level spectrum the tunneling splitting has thermodynamic consequences which in principle can be detected (computationally) several ways, but the off-diagonal thermal propagator is particularly sensitive and increasingly optimal particularly as Δ becomes very small. The off-diagonal electronic density matrix element between two states, 1 and 2, is determined by the extent to which they are coherently coupled as measured by the thermal density operator $\exp(-\beta H)$, where β is the inverse thermal energy $[(k_B T)^{-1}]$. Let us describe the principal features of the path integral Monte Carlo instanton method with one-particle paths $\mathbf{r}(\beta')$; the extension to N electrons becomes important and is discussed later. This one-particle electronic density matrix element, $\langle 1|e^{-\beta H}|2\rangle$, can be represented in the form of a real-space propagator as a path integral:

$$\langle 1|e^{-\beta H}|2\rangle = \int d\mathbf{r}_i \psi_1(\mathbf{r}_i) \int d\mathbf{r}_f \psi_2(\mathbf{r}_f) \int \mathcal{D}\, \mathbf{r}(\beta) \exp\left\{ -\int_0^\beta H(\mathbf{r}(\beta'))d\beta' \right\} \quad (23)$$

where

$$H(\mathbf{r}(\beta)) = \frac{m}{2\hbar^2}\left|\frac{d\mathbf{r}(\beta)}{d\beta}\right|^2 + V(\mathbf{r}(\beta)) \tag{24}$$

and $V(\mathbf{r}(\beta))$ is the potential on which the electron moves with path $\mathbf{r}(\beta')$.

The term $\exp\left\{-\int_0^\beta H(\mathbf{r}(\beta'))d\beta'\right\}$ is the exponential of the action for imaginary time motion, and is the key path-dependent weight which depends upon the nature of the protein environment visited by the specific path within the tunneling region. In the computation on ruthenated myoglobin, for example, $V(\mathbf{r})$ consisted of the sum of atom-centered pseudopotentials that describe an electron moving around each of the 1217 non-hydrogen atoms in the protein. The off-diagonal nature of the density matrix element is achieved by choosing ψ_1 to be a wave function that is localized on the donor and ψ_2 to be a wave function localized on the acceptor. The main requirement of ψ_1 and ψ_2 is that they pin down the ends of the tunneling path so that the path is forced to cross the tunneling region; only in this way can this density matrix element become and remain an effective tool for weaker couplings and very small Δ.

The exponentials in Eqs. 18, 22, and 23 are all distinct yet may be related by various transformations; note in particular that in the WKB expression it is the energy which is the fixed and known parameter, whereas in the quantum real time expression, Eq. 22, it is the total propagation time τ, and in the quantum statistical thermodynamic Eq. 23 it is the inverse simulation temperature β. The variable β' plays the role of a time index in parameterizing the path trajectories, and will occassionally be referred to in our discussion as "time". The connection between β and the real time variable is purely a mathematical relationship; the analytic continuation of the path integrand to the imaginary time axis ($\beta = i\tau/\hbar$) relates the real time to quantum statistical thermodynamic path integrals. The reader is referred to the seminal texts by Feynman and Hibbs [67, Sect. 10.2] and by Feynman [70, Chap. 3] for the development of the quantum path integral expression for canonical ensemble density matrices; the former treatment makes the connection from the real time propagator whereas the latter starts directly from the statistical thermodynamics. The WKB expression is an approximation to the full quantum mechanics, whereas the path integral expressions above are exact.

4.3 Path Integral Monte Carlo Simulation

To determine Δ we need to implement the path integration as a numerical path summation. The path integral in Eqs. 23–24 is in fact isomorphic to the configuration integral of a flexible polymer which interacts with the external potential $V(\mathbf{r}(\beta))$. This analogy can be made more explicit by considering a discrete approximation to the path integral [71]. If the path is cut up into P segments we obtain the discretized path $\{\mathbf{r}_1 \ldots \mathbf{r}_P\}$ whose path weight can be recognized as equal to the statistical weight of a harmonic polymer of P links with a particular Hamiltonian. The discretization replaces the integral $d\beta'$ by the sum over links of β/P times the Hamiltonian operator. The characteristic quadratic form of the harmonic polymer arises as each link now corresponds to a small inverse temperature of β/P, hence a high-temperature approximation can be made for the exponential integrand describing each of the links,

$$\begin{aligned}\langle \mathbf{r}_j|\exp(-(\beta/P)H)|\mathbf{r}_k\rangle &= \exp(-\beta V(\mathbf{r}_j)/2P)\langle \mathbf{r}_j|\exp(-(\beta/P)\hat{T})|\mathbf{r}_k\rangle \\ &\times \exp(-\beta V(\mathbf{r}_k)/2P) = \exp(-\beta V(\mathbf{r}_j)/2P) \\ &\times \exp\left\{-\frac{\pi|\mathbf{r}_j - \mathbf{r}_k|^2}{[\Lambda(\beta/P)]^2}\right\}\exp(-\beta V(\mathbf{r}_k)/2P) \qquad (25)\end{aligned}$$

where the thermal deBroglie wavelength is given by $\Lambda(\beta) = (2\pi\beta\hbar^2/m)^{1/2}$.

Inserting the high-temperature propagator for each short link, and integrating over all possible coordinates of the intermediate points, we can write the path integral as:

$$\begin{aligned}\int \mathscr{D}\,\mathbf{r}(\beta)\exp\left\{-\int_0^\beta H(\mathbf{r}(\beta'))d\beta'\right\} &= \int d\mathbf{r}_1 \int d\mathbf{r}_2 \ldots \int d\mathbf{r}_{P-1}[\Lambda(\beta/P)]^{-3P} \\ &\times \exp\left\{-\sum_{j=1}^{P-1}\left(\frac{\pi|\mathbf{r}_j - \mathbf{r}_{j-1}|^2}{[\Lambda(\beta/P)]^2}\right.\right. \\ &\left.\left.+\frac{\beta[V(\mathbf{r}_j) + V(\mathbf{r}_{j-1})]}{2P}\right)\right\} \qquad (26)\end{aligned}$$

We see that the kinetic energy in Eq. 24 has been translated into the potential energy of harmonic springs that connect neighboring points in the isomorphic polymer chain (albeit with unusual spring constants containing the mass of the electron, Planck's constant, and so on). Equation 26 is evidently of the form of a classical configuration integral $\int \ldots \int \exp\{-\beta\tilde{H}_P\}$, where $\tilde{H}_P$ is the isomorphic harmonic Hamiltonian for the polymer of P links;

$$\tilde{H}_P = \sum_{j=1}^{P-1}\left(\frac{\pi|\mathbf{r}_j - \mathbf{r}_{j-1}|^2}{\beta[\Lambda(\beta/P)]^2} + \frac{[V(\mathbf{r}_j) + V(\mathbf{r}_{j-1})]}{2P}\right) \qquad (27)$$

This particular discrete version of the path sum is particularly suggestive of both physical analogies and computational procedures. We wish to compute a free energy (configuration integral) of this isomorphic classical polymer chain consisting of Hooke's law springs [72]. The discretization according to Eqs. 26–27 is one of a number several possible factorizations whereby the propagation over the full potential is represented as segmental propagations using approximate analytic forms known from simple potentials. Exact results require taking the limit $P \to \infty$ but the finite P approximation can be used for numerical computation, and considerable activity has been recently focussed upon improved factorizations of the discretized path integral so as to achieve superior convergence with regard to P [73, 74]. These appear to be valuable technical improvements, and are of substantial importance in the actual computation, though the harmonic Hamiltonian above presents in a simpler form the essential qualitative features of the forces that govern the relative probabilities of the alternative paths.

Now the computer comes in. Through discretization we have converted the path integral of Eq. 23, which is an infinite-dimensional integral, to the 3P-dimensional multiple integral of Eq. 26. With P a number of typically in the hundreds, depending principally upon β and the complexity of the potential, this is still a challenging integral. In the structured, aperiodic, inhomogeneous molecular environment of a protein there is little hope that the analytical semiclassical route, dependent upon the normal mode approximation for the classical polymer, will be quantitatively or qualitatively useful. Metropolis Monte Carlo [75] is a powerful and highly adaptable path (or polymer configuration) sampling method ideally suited for the simulation and identification of the important sub-volumes within high-dimensional integrals. The paths which contribute the most to the path summation are located by one of several technical implementations of this biased random walk algorithm in configuration space. Equally important is the fact that configurational entropy enters in directly as a determinant of the set of paths sampled, which we may recognize as the role of large numbers of tunneling paths.

There are many non-biological applications of this general quantum simulation method which are described in the papers of Wolynes, Chandler, Rossky, Stratt, and Hall, and others [71, 76, 77, 78], and in the instructive review by Berne and Thirumalai [79]. A treatment of electronic tunneling interactions in ruthenium-modified myoglobin was performed by Kuki and Wolynes [19] using path integral Monte Carlo and the instanton method described below. Algorithmic details such as the Monte Carlo determination of the requisite relative free energy by thermodynamic integration from a reference potential to the full potential in the protein are described there.

4.4 Instantons and Tunneling Paths

The analysis of tunneling in a multistable potential relies on a classification of the paths that contribute to the path integral in Eq. 23 [19, 71, 80]. Particularly in the three-dimensional situation it is important to see that there will be one or more minimum action (maximum weight) paths determined by the vanishing first functional derivative of the path weight. These particular paths whose weights are local minima are stretched from the donor to the acceptor with a delicate balance between deformations into attractive regions of the potential (relatively deep V or regions where the barrier is not so high) and the minimization of the tension (kinetic energy) of the harmonic springs in the polymer. These extremal paths satisfy an Euler-Lagrange equation which is a second order differential equation with respect to β', and which has the form of Newton's equation of motion (see "The Classical Limit", Sect. 2–3, in [67]); in the statistical thermodynamic tunneling context these paths are known as instantons [80]. Around these minimum action paths there are bundles of reasonably probable but highly intricate and tangled paths which follow no law of classical motion, and whose contributions give rise to the heat capacity and entropy of the classical polymer system. When these fluctuations around the extremal paths are treated harmonically we recover the semiclassical approximation to the quantum path integral, which constitutes the dominant analytic route to evaluating path integrals in those special cases in which analytic solution is possible. The total free energy (more precisely, the configuration integral) of this isomorphic polymer in the tunneling region is equal to the free energy (partition function) of the quantum electron in its tunneling segment; this is the target of quantum mechanical interest.

Consider the total imaginary time β to be very long (low simulation temperature). In particular, we will invoke the same limit as in the "high energy intermediary states" mentioned before in related contexts, that we can select the parameter β such that $\beta\Delta$ is of order unity while $\beta\Delta E \gg 1$ where ΔE is the excitation energy from the quasi-degenerate redox levels to the third lowest level. In this case a typical path that contributes to Eq. 23 spends a great deal of time in one redox center before rapidly making a transit to the other redox center, where it again spends a good deal of time. Referring to Fig. 7, we will call such a rapid transit a "kink". An instanton is precisely a kink of the particular shape (trajectory in β') such that its action is a local functional minimum. Other paths that contribute to the path sum will have a larger but necessarily odd number of such kinks. For redox sites separated by substantial classically forbidden regions the kinks will be well separated in time and can be considered to be noninteracting. The excursion represented by such a kink is costly in energy; however, if β is large, multikink configurations are favored because of their large number.

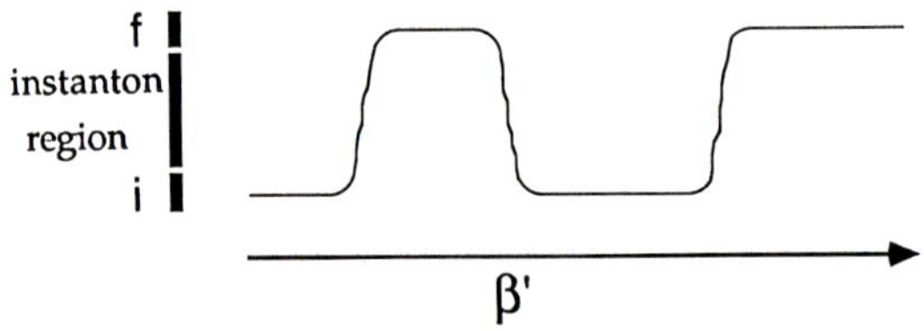

Fig. 7. The tunneling paths in a double minimum potential like that of Fig. 5 may be classified as having one or more (odd) numbers of instantons, or tunneling segments. Three such traverses of the classically forbidden region are shown above. The classification of all paths according to the number N of instantons is the basis for evaluating the path integral as the sum in Eq. 29; note however that the F_K therein is constructed to include non-harmonic (beyond semiclassical) fluctuations around the minimum action instanton paths, which are evaluable by Metropolis Monte Carlo

The free energy cost of introducing a kink is F_k and is the difference between the free energy of a one kink path and that of a path of equivalent β that is confined to one well, E_o (since the wells are isoenergetic at the crossing point it does not matter which one [81]). The path sum can then be written as an expansion in kinks. The contribution of the set of paths with N kinks to the overall partition function consists of a factor from each kink of $Q_k = \exp(-\beta F_k)$, and is

$$\left(\frac{1}{N!}\right) e^{-N\beta F_k} e^{-\beta E_o} \tag{28}$$

if the interaction between kinks can be neglected. In this expression, the free energy F_k includes the integration over the position of the center of the kink, that is, the translational entropy of the kink within a polymer of length β. As a result, $\exp(-\beta F_k)$ grows linearly with β. To be more precise, the translational entropy introduces a factor into the partition function which grows linearly in β; but in addition the ratio Q_k/β or $(1/\beta)e^{-\beta F_k}$ can in fact be shown to be a constant, in the dilute instanton limit, and equal to the number density of kinks. The factor of $1/N!$ arises from the indistinguishability of kinks. In the total path sum there is a set of paths with $N = 1$, a set of paths with $N = 3$, and so on. Thus one finds that the amplitude for a transition from state 1 to state 2 can be represented as the sum over such sets,

$$e^{-\beta E_o} \sum_{N\,\mathrm{odd}} \left(\frac{1}{N!}\right) e^{-N\beta F_k} = e^{-\beta E_o} \sinh(e^{-\beta F_k}). \tag{29}$$

Now we observe from ordinary quantum mechanics (Fig. 2) that the two lowest quasi-degenerate eigenstates of the global system have energies $E_o \pm \Delta$, and hence the same off-diagonal density matrix element can be written as $[e^{-\beta E_o} \sinh(\beta\Delta)]$. This interaction energy Δ, whose value we seek, includes contributions of all orders (Fig. 3). The hyperbolic sine of $\beta\Delta$ is nothing other than the analytic continuation to imaginary time of the oscillatory circular sine function describing the desired Rabi oscillation of the tunneling system. By examining the argument of the sinh in Eq. 29, we therefore see that the charge resonance energy is exponentially related to the free energy of introducing a kink:

$$\Delta = \left(\frac{1}{\beta}\right) e^{-\beta F_k}. \tag{30}$$

This free energy in turn can be obtained by looking at configurations with only a single transit from donor to acceptor well, which is a critical observation, as multikink paths and longer β values cause unnecessary computational difficulties in the simulation. The number of kinks which will form in the polymer chain is approximately $\beta\Delta$, hence the desired one instanton tunneling simulation of the off-diagonal electronic density matrix is arranged by setting β to a value such that $\beta\Delta$ is less than unity. The Δ thus obtained from F_k is independent of the β used in the calculation [82].

4.5 Paths and Potentials

The derivation above of the formula given by Eq. 30 is completely general, with no assumptions about the nature or simplicity of the tunneling barrier, and is not dependent upon any (semiclassical) assumptions that the number of dominant paths contributing to F_k is small. In addition, the general instanton formula for Δ, Eq. 30 holds for a path integral treatment of many-electron tunneling as well as the one tunneling electron model described. In such a case the initial, final, and intermediate states are all N-particle states in 3N-dimensional space (see below). The path analysis is also very revealing as it is clear that it is the collective involvement of a whole set of tunneling paths, and not only the minimum action tunneling path, which contributes to the free energy F_k and hence to Δ. Particularly at long range, the accessibility of a larger cross-section to the bundle of tunneling paths is very important to an enhancement of the coupling; such accessible paths contribute favorably to the entropy of F_k.

The protein controls the paths through the potential term in Eqs. 26–27; and this potential can be divided into longer range and shorter range terms. The longer range terms are the partial electrostatic charges on all the atoms and a sum over atom-centered (or group-centered) polarizabilities of the charge-induced-dipole energy, $-(\alpha/2r^4)$, where α is the polarizability. There are three shorter range terms which are more subtle but very important, (i) the penetration of the electrostatic screening of the closed shell electron clouds, (ii) the exchange energy which reduces electron-electron repulsion between electrons of like spin, and (iii) a repulsive pseudopotential which serves to incorporate the Pauli Exclusion principle. The screening requires a reasonable model of the electron density field in the protein, and improved methods for obtaining such descriptions simultaneous to computation of partial charges are becoming available [83]. Exchange and the repulsive pseudopotential both arise directly as consequences of fermionic antisymmetry, in a multi-particle computation, and hence the adoption of some form of effective potential is necessary in a one-electron model [84]. The local potential adopted in a myoglobin path integral tunneling simulation as well as details of other aspects of the potential may be found in the paper by Kuki and Wolynes [19]; also highly instructive is the solvated electron path integral work of Schnitker and Rossky [77, 85].

The balance between attractive regions of the potential that lie off the straight-line path between the donor and acceptor and the "tension" of the harmonic springs of the polymer, from the kinetic energy term in the propagator, determine the overall straightness of the bundle of paths which appear in the Monte Carlo simulation. The balance between narrowly defined "valleys" of attractive potential and the chain entropy dictate whether a few narrowly defined paths, each analogous to a single chain in high order perturbation theory, will dominate. The implications are that the pattern of paths emerging from the Metropolis Monte Carlo are significant in themselves, in addition to providing the means for the computation of Δ. Wolynes and this author quantitated this information by interpreting the functional derivative of the F_k with respect to the potential as an effective classical density of the tunneling electron. The ratio of the classical partition functions for closely related potentials V_2 and V_1 is given approximately by the standard perturbation expression,

$$\ln(\Delta_2/\Delta_1) \cong -\beta \int \frac{\delta F_k}{\delta V(\mathbf{r})}[V_2(\mathbf{r}) - V_1(\mathbf{r})]d^3\mathbf{r} \tag{31}$$

The required functional derivative, which reveals the sensitivity of the overall configuration sum (the Δ) to the value of the potential at location $\mathbf{r}$, is the particle density,

$$\frac{\delta F_k}{\delta V(\mathbf{r})} = \frac{\int \ldots \int \frac{1}{P}\sum_j \delta(\mathbf{r}_j - \mathbf{r})e^{-\beta \tilde{H}_P}}{e^{-\beta F_k}} = \rho_{kink}(\mathbf{r}) \tag{32}$$

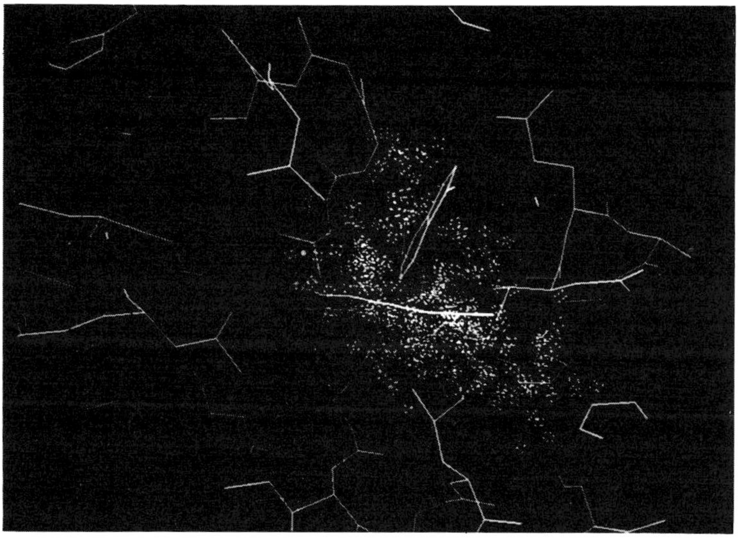

Fig. 8. A view into the interior of a ruthenium modified myoglobin where the amino acids in the vicinity of Trp-14 are shown. The dots correspond to the statistical density $\rho_{kink}(\mathbf{r})$ of (discretized) tunneling path vertices (r_j in Eq. 26) from 500,000 tunneling paths [19]. The $\rho_{kink}(\mathbf{r})$ is clustered in a cylindrical zone centered on the average path, shown as the light line appearing in the center and emerging toward the viewer. The computation modeled paths of electron transfer in Ru(His-12) myoglobin studied experimentally by Gray and coworkers [88]

If a limited region of the protein is slightly changed, for example by site-directed mutagenesis in the intervening region, then the change in the tunneling amplitude may be predicted from the kink or tunneling path density in the path integral Monte Carlo simulation of the unperturbed protein. Thus, a local alteration of the potential, $\Delta V(r)$, will change the logarithm of Δ by an amount approximately given by the product of the protein dependent factor, $\rho_{kink}(\mathbf{r})$, and the amino acid dependent factor $\Delta V(r)$. This perturbation analysis applied to the three-dimensional quantum mechanical tunneling problem may be referred to as a "linear action relationship" by analogy to linear free energy relationships of classical barrier-crossing kinetics; linear here means linear in the $\Delta V(r)$.

Depicted in Fig. 8 is an example of the computed $\rho_{kink}(\mathbf{r})$ in the vicinity of a tryptophan (Trp-14) in ruthenium modified myoglobin. The tryptophan lies partly in the cylindrical zone defined by the tunneling paths between the zinc-substituted protoporphyrin IX and the ruthenium site, and causes a noticeable but small enhancement in the tunneling path density above and below the aromatic plane, which is largely due to the polarizability term, $-(\alpha/2r^4)$, summed over the heavy atoms of the aromatic ring. Yet in the case of the model potential used, the tryptophan failed to seriously distort the majority of the tunneling paths, which generally proceeded without large detours. The number of paths simulated in the path integral Monte Carlo was 500,000 (at each stage of the thermodynamic integration required to evaluate the relative free energy F_k), and the computation and the potential were limited to the one-electron level.

5 Summary and Outlook

5.1 Classification of Paths in Superexchange

We have stressed that paths quantified by their path amplitudes provide a unifying framework for the analysis of indirect mechanisms contributing to the electronic tunneling interaction in the case of long-range electron transfer. The path amplitudes in high order perturbation theory and in path integral Monte Carlo are mathematically different; perturbation theory describes chains of basis states while path integrals describe real space quantum propagation. Yet in both cases the desired Δ is obtained by a path summation over intermediates in the intervening protein medium, and hence in both cases it is very important to understand the classes of tunneling paths.

A useful and uncomplicated classification scheme which is a needed refinement over the simple through-space/through-bond terminology is:

i) **Conjugated or ligated.** The direct bonding between, say, a metal and its ligand, or two functional groups on an aromatic, or groups separated by an olefinic bridge. Here, there is no question that an approach based on molecular orbitals is appropriate. The electrons in such a bonding unit should be treated as completely delocalized.

ii) **Through-bond.** The interaction through a covalent sigma-bonded bridge which connects the donor and acceptor. The relevant atoms in the bridge are on the order of 1.5 Å apart. The experimental results of both the Closs and Miller [61] steroid project and the Verhoeven, Paddon-Row and Hush [86] Diels-Alder project indicates excellent agreement with molecular orbital pictures and predictions of the classical through-bond type [87].

iii) **Non-bonded superexchange.** Donor and acceptor interact through intermediary molecules or residues across non-bonded gaps. This is obviously the case when two separate proteins interact, and may be the case when two groups both within a single folded polypeptide chain yet remote in sequence interact. This is not likely to be found in most small molecules of the donor-bridge-acceptor type, which have been important in other ways as models for proteins. The non-bonded distances involved in the superexchange pathways are likely to be of order of van der Waals contact, even though the full donor-acceptor separation may be considerably greater. These superexchange interactions at non-bonded distances result from electrons which are neither fully delocalized nor (evidently) fully localized, and are the primary focus of theoretical research in the author's group.

iv) **Through-space** (vacuum in the computer). This is the situation where there are literally no other molecular components in the model other than the donor and acceptor. The exponential decay of this interaction with distance is severe, with exponential factors on the order of two times faster (four times the tunneling barrier) than experimentally observed on the average in proteins. Thus the vacuum model is clearly a poor theory for the overall tunneling decay, let alone its modulation by structure.

5.2 Toward Fewer Approximations, Several Electrons, Many Paths

There remains a fundamental puzzle concerning the nature of the tunneling barrier as gauged by experimental measurements of distance dependence collected throughout the past decade. The distance dependence of Δ can be gauged experimentally in both protein systems containing two redox sites on one polypeptide, and also between aromatics randomly arrayed in frozen organic glasses. The overall exponential decay constants for Δ are not that different; 0.44–0.50 $Å^{-1}$ [88, proteins], 0.55–0.6 $Å^{-1}$ [31, proteins], and 0.60–0.65 $Å^{-1}$ [89, organic glasses], and correspond to a barrier height of merely 0.8 to 1.5 eV or so, in WKB terms, or to remarkably large ratios $\langle\psi_k^o|V|\psi_i^o\rangle/(E_i^o - E_k^o)$, in the superexchange expansion. This implies that through-bond coupling which is conceivable through the backbone of polypeptides, or through their hydrogen bonds [16], can not be the definitive explanation for these small decay constants as the same small decay constant is present in frozen organic solvents lacking protic functionalities. This is a fascinating result which deserves considerably more attention [90]; but is clearly relevant to the resolution of the important paths of superexchage as classified above. Here we simply suggest that these results indicate that class (iii), non-bonded superexchange, may be very significant and possibly dominant in electron tunneling interactions within proteins as well as in non-bonded frozen matrices.

Non-bonded superexchange will require the activation of more than one electron in a path integral model. Current directions in our laboratory include multi-electron path integral Monte Carlo studies, which require both fundamental path integral development as well as new protein models. We will be able to use Eq. 30 directly, as a result of its generality, but the effective path Hamiltonian (Eq. 27) becomes altered by the addition of the key (e^2/r_{12}) interelectronic repulsion between correlated electrons. This means removing the approximation labeled as Item (ii) in Sect. 3.2, and re-activating correlation and explicit antisymmetric exchange processes. As a technical matter antisymmetric path integral Monte Carlo is extremely challenging, but we have recently made progress in this area of chemical physics [91] and are preparing to apply the methodology to biomolecular applications. Nevertheless the promise of a theory which actively includes explicit multi-electron exchange processes within long-range electronic tunneling is that the average tunneling barriers in organic media may be correctly understood, as well as the mechanics of its alteration by particular residues within protein structures.

In closing, the theoretical problem of quantitatively understanding electronic interactions in complex aperiodic macromolecular structures such as proteins remains a fascinating technical and scientific challenge, for which we have attempted to outline the available tools and the broader questions. But theoreticians who probe the control of quantum electronic tunneling in proteins still have a long way to go. We will know we have progressed when we can accurately quantify the range of variability in Δ available in protein structures, and pinpoint where the charge resonance interactions of a given folded polypeptide molecule lies within this range.

Acknowledgements. It is a pleasure to acknowledge Peter Wolynes for introducing me to quantum simulations, for his encouragement, and for many stimulating discussions. I and this chapter have also benefitted greatly from many discussions with Jim Gruschus. Bill Newman is responsible for Eq. 17. In addition I thank the National Science Foundation (DMB-8717673) and the Camille and Henry Dreyfus Teacher-Scholar Grant Program for their valuable support.

6 References

1. Marcus RA (1956) J Chem Phys 24:966
2. Marcus RA (1964) Annu Rev Phys Chem 15:155
3. Hatefi Y (1985) Annu Rev Biochem 54:1015
4. Anraku Y (1988) Annu Rev Biochem 57:101
5. Govindjee (ed) (1982) Photosynthesis: Energy conversion by plants and bacteria. Academic Press, New York
6. Michel-Beyerle ME (ed) (1985) Antennas and reaction centers of photosynthetic bacteria. Springer, Berlin Heidelberg New York
7. Szent-Gyorgi A (1941) Science 93:609
8. Sutin N (1981) Acts Chem Res 15:275
9. Jortner J (1980) Biop Bioc Acta 594:193
10. Takano T, Dickerson RE (1981) J Mol Biol 153:95
11. Williams G, Moore GR, Williams RJP (1985) Comments Inorg Chem 4:55
12. Meade TJ, Gray HB, Winkler JR (1989) J Am Chem Soc 111:4353
13. Onuchic JN, Beratan DN, Hopfield JJ (1986) J Phys Chem 90:3707
14. Beratan DN, Hopfield JJ (1984) J Am Chem Soc 106:1584
15. Beratan DN (1986) J Am Chem Soc 108:4321
16. Beratan DN, Onuchic JN, Hopfield JJ (1987) J Chem Phys 86:4488
17. Larsson S (1983) J Chem Soc Faraday Trans. II 74:1375
18. Larsson S, Broo A, Kallebring B, Volosov A (1988) Int J Quantum Chem Quantum Biol Symp 15:1
19. Kuki A, Wolynes PG (1987) Science 236:1647
20. Anderson PW (1950) Phys Rev 79:350
21. Keffer F, Oguchi T (1959) Phys Rev 115:1428
22. Halpern J, Orgel LE (1960) Discuss Faraday Soc 29:32
23. McConnell HM (1961) J Chem Phys 35:508
24. Levich VG (1966) In: Delahay, Tobias (eds) Adv Electrochemistry Electrochem Eng, vol 4, p 249
25. Mayo SL, Ellis WR Jr, Crutchley RJ, Gray HB (1986) Science 233:948
26. Therein MJ, Selman M, Gray HB, Chang I-J, Winkler JR (1990) J Am Chem Soc 112:2420
27. Liang N, Mauk AG, Pielak GL, Johnson JA, Smith M, Hoffman BM (1988) Science 240:311
28. McLendon G (1988) Acts Chem Res 21:160
29. Hopfield JJ (1974) Proc Natl Acad Sci 71:3640
30. Jortner J (1976) J Chem Phys 64:4860
31. Marcus RA, Sutin N (1985) Biophys Biochem Acta 811:265
32. Newton MD, Sutin N (1984) Annu Rev Phys Chem 35:437
33. Schmidt PP (1975) In: Electrochemistry. The Chemical Society, London, p 21 (Specialist Periodical Report, vol 5)
34. Schmidt PP (1978) In: Electrochemistry. The Chemical Society, London, p 128 (Specialist Periodical Report, vol 6)
35. Isied SS, Vassilian A, Wishart JF, Creutz C, Schwarz HA, Sutin N (1988) J Am Chem Soc 110:635
36. Oevering H, Paddon-Row MN, Heppener M, Oliver AM, Cotsaris E, Verhoeven JW, Hush NS (1987) J Am Chem Soc 109:3258
37. Sutin N, Brunschwig BS (1990) In: Johnson MK et al (eds) Electron transfer in biology and the solid state, American Chemical Society, Washington DC, p 65 (Advances in Chemistry)
38. Johnson MD, Miller JR, Green NS, Closs GL (1989) J Phys Chem 93:1173
39. Zener C (1932) Proc Royal Soc A137:696
40. Hush NS (1968) Electrochim Acta 13:1004
41. Hush NS (1982) In: Rorabacher DB, Endicott JF (eds) Mechanistic aspects of inorganic reactions. American Chemical Society, Washington, DC, p 301 (ACS Symposium Series)
42. Kuznetsov AM (1982) Faraday Discuss Chem Soc 74:49
43. Mikkelsen KV, Ulstrup J, Zakaraya MG (1989) J Am Chem Soc 111:1315
44. Consider each of the two Marcus free energy parabolas to consist in turn of the sum of parabolas

for the solvation of the donor and the solvation of the acceptor (this is possible within a linear dielectric model). Then the non-standard redox potentials for the two half-reactions become (linear) functions of the reaction coordinate, with the critical feature that the exoergicity at the transition state is zero. We may thus define a unique Transition State Redox Potential to be that particular non-standard potential common to both half reactions at the transition state. The Transition State Redox Potential can be simply shown to be $E^{\text{transition state}} = [\lambda_D E^o \text{ (acceptor)} + \lambda_A E^o\text{(donor)}]/(\lambda_D + \lambda_A)$ where we have assumed the simple partitioning of $\lambda = \lambda_D + \lambda_A$. This expression was also obtained by Beratan, Beratan DN (1986) J Am Chem Soc 108:4321
45. Kuki A (to be published)
46. Kuznetsov AM, Ulstrup J (1981) J Chem Phys 75:2047
47. Lin SH (1989) J Chem Phys 90:7103
48. Ulstrup J, Jortner J (1975) J Chem Phys 63:4358
49. The conversion of (k_BT/h) to (ω/π) may be derived within the Eyring Transition State Theory as due to the inclusion in the prefactor of the reactant vibrational (harmonic) partition function.
50. Closs GL, Calcaterra LT, Green NJ, Penfield KW, Miller JR (1986) J Phys Chem 90:3673
51. Liang N, Miller JR, Closs GL (1990) J Am Chem Soc 112:5353
52. Kemnitz K (1988) Chem Phys Lett 152:305
53. Gunner MR, Dutton PL (1989) J Am Chem Soc 111:3400
54. Ulstrup J (1979) Charge transfer processes in condensed media, Springer, Berlin Heidelberg New York
55. Schiff LI (1968) Quantum mechanics, 3rd edn, McGraw-Hill, New York
56. As pointed out by Miller in his definitive experiments on superexchange processes in organic glasses, the energy denominators give rise to a dependence of Δ upon the binding energy of the electron in one trap (donor) and of the hole in the other (acceptor) which is characteristically distinct from the square root relation of WKB. See Miller JR, Beitz JV (1981) J Chem Phys 74:6746, and Miller JR (1985) In: Michel-Beyerle ME (eds) Effects of distance, free energy and molecular struct, Springer, Berlin Heidelberg New York, p 234 (Chemical Physics, vol 42)
57. Ohta K, Closs GL, Morokuma K, Green NJ (1986) J A Chem Soc 108:1319
58. Newton MD (1988) J Phys Chem 92:3049
59. Gruschus J, Kuki A, multiconfigurational computation of superexchange including correlation and non-local exchange (in progress)
60. Leland BA, Joran AD, Felker PM, Hopfield JJ, Zewail AH, Dervan PB (1985) J Phys Chem 89:5571
61. Closs GL, Miller JR (1988) Science 240:440
62. Finckh P, Heitele H, Volk M, Michel-Beyerle ME (1988) J Phys Chem 92:6584
63. Gust D, Moore TA, et al (1987) J Am Chem Soc 109:846
64. Gust D, Moore TA (eds) (1989) Covalently linked donor-acceptor species. (Tetrahedron Symposia-in-Print, vol 45, No 15) Pergamon Press, New York)
65. Assume that the two sites are separated by N_o steps in the x-direction. (The formula can be straightforwardly generalized to arbitrary locations of the two trap sites.) The number of extra kinks in the path of length $N > N_o$ is K, which is the number of pairs of non-essential steps. The formula given then follows from the combinatorials of N total steps which may be taken in any order, and which consist of six classes of objects, $N_o + k_x$ forward steps (in the + x direction), k_x backward steps ($-$ x), k_y sideways steps (+ y), k_y sideways steps back ($-$ y), and k_z (+ z), and k_z ($-$ z), where we must also have that $k_x + k_y + k_z = K$. Asymptotically for large $N > N_o$ the number of paths grows as $6^N/N^{(3/2)}$
66. DeVault D (1984) Quantum mechanical tunneling in biological systems, Cambridge University Press, Cambridge
67. Feynman RP, Hibbs AR (1965) Quantum mechanics and path integrals, McGraw-Hill, New York
68. Logan J, Newton MD (1983) J Chem Phys 78:4086
69. Cave RJ, Baxter DV, Goddard WA, Baldeshwieler JD (1987) J Chem Phys 87:926
70. Feynman RP (1972) Statistical Mechanics. Benjamin WA, Reading
71. Chandler D, Wolynes PG (1981) J Chem Phys 74:4078. Quantum Path Integral Monte Carlo can also provide a way to evaluate the rate in non-adiabatic reactive flux correlation theory; Wolynes PG (1987) J Chem Phys 87:6559
72. In understanding quantum simulation work it is important to appreciate that the isomorphic classical Hamiltonian H_p is unusual in that it is β-dependent; the spring constants are proportional to $(1/\beta^2)$

73. Mak CH, Andersen HC (1990) J Chem Phys 92:2953
74. Makri N, Miller WH (1989) J Chem Phys 90:904
75. Metropolis N, Metropolis AW, Rosenbluth MN, Teller AH, Teller E (1953) J Chem Phys 21:1087
76. Schweizer KS, Stratt RM, Chandler D, Wolynes PG (1981) J Chem Phys 75:1374
77. Schnitker J, Rossky PJ (1987) J Chem Phys 86:3471
78. Hall RW, Wolynes PG (1986) Phys Rev B33:7879
79. Berne BJ, Thirumalai D (1986) Ann Rev Phys Chem 37:401
80. Coleman S (1979) In: The whys of subnuclear physics, 1977 international school of subnuclear physics, Erice. Plenum, New York, p 805
81. This E_o site energy in the exponential of Eq. 28 may be replaced with the perturbed site energies in the interacting system ($E_o + E_{shift}$) to include environmentally induced diagonal perturbations, but this has no consequence at all to the instanton analysis and the derivation of Eq. 30. In treatments of low symmetry molecular systems donor and acceptor site energies may be differentially perturbed, which means that one is not properly at the exact transition state with respect to electrostatic interactions in the medium. However the Monte Carlo path simulation will automatically signal the condition of excessive imbalance by a tendency of the instanton to slide to one end of the chain if the product of β and this site energy imbalance is not sufficiently small.
82. The value of β is important only with respect to technical considerations. The presence of finite orbital overlap between the initial and final states slightly alters the β dependence of the relationship between F_k and Δ. The overlap corrections for the instanton analysis arise from the slightly different normalization of the symmetric and antisymmetric eigenstates, which can be obtained from the simple model in Fig. 2. The coefficients of the symmetric and antisymmetric eigenstates in terms of the zeroth order states, nominally $\pm 1/\sqrt{2}$, become instead $c_{SYM}^2 = 1/2(1 + S)$, which is slightly less than $c_{ANTI}^2 = 1/2(1 - S)$. Equation 29 remains unchanged but Eq. 30 can be refined as follows. Recall that F_k is the free energy of introducing a kink, and define $p = c_{SYM}^2 + c_{ANTI}^2$ (nominally one) and $m = c_{SYM}^2 - c_{ANTI}^2$ (nominally zero). Then F_k is related to the ratio $\langle 1|\exp(-\beta H)|2\rangle / \langle 1|\exp(-\beta H)|1\rangle = \tanh(\exp(-\beta F_k))$ by the instanton analysis, whereas expanding the ratio in terms of the tunneling eigenstates yields $\langle 1|\exp(-\beta H)|2\rangle / \langle 1|\exp(-\beta H)|1\rangle = (p\sinh(\beta\Delta) + m\cosh(\beta\Delta))/(p\cosh(\beta\Delta) + m\sinh(\beta\Delta))$, which holds for the exact p and m.
If we insert the 2 × 2 model for p and m (from c_{SYM}^2 and c_{ANTI}^2) then elementary manipulations of power series results in:
$$\beta\Delta = (s + s^3/3) + T + (-2s^3)T^2 + (1/3 - 2s^2)T^3 + (-s - 5s^3)T^4 + O(s^4) + O(T^5)$$
where $T \equiv \tanh(\exp(-\beta F_k))$. The series given here can be seen to reduce to $\tanh^{-1}(T)$ as $S \to 0$, as it must, thereby yielding Eq. 30. In practice the β-dependence of the computed $\exp(-\beta F_k)$ can be monitored to check if the overlap corrections are detectable (principally the leading constant: s) or insignificant.
83. Gruschus JE, Kuki A (1990) J Comput Chem 11:978
84. Szasz L (1985) Pseudopotential theory of atoms and molecules, Wiley, New York
85. Schnitker J, Rossky PJ (1987) J Chem Phys 86:3462
86. Oliver AM, Craig DC, Paddon-Row MN, Kroon J, Verhoeven JW (1988) Chem Phys Lett 150:366
87. Hoffman R (1971) Acts Chem Res 4:1
88. Axup AW, Albin M, Mayo SL, Crutchley RJ, Gray HB (1988) J Am Chem Soc 110:435
89. Miller JR, Beitz JV, Huddleston RK (1984) J Amer Chem Soc 106:5057
90. Closs and Miller have also remarked that the intermolecular interaction across van der Waals gaps in frozen organic glasses does not give rise to large attenuations relative to intramolecular couplings (Science (1988) 240:446).
91. Newman WH, Kuki A (submitted to J Chem Phys)

Long-Range Electron Transfer Within Metal-Substituted Protein Complexes

Brian M. Hoffman, Michael J. Natan, Judith M. Nocek, and Sten A. Wallin

Department of Chemistry, Northwestern University, Evanston, IL 60208-3113, USA

Substitution of a closed-shell metalloporphyrin (MP) within one partner of a protein-protein complex makes it possible to study long-range electron transfer (ET), including direct measurement of both photoinitiated and thermal ET between the MP and heme (FeP) groups. Mixed-metal [M, M′] hemoglobin hybrids have a structure that is fixed and well-characterized, which permits us to explore the factors that control the ET process by judicious selection of both the heme ligands and the substituting metal ion. In contrast, complexes between MP-substituted cytochrome *c* peroxidase (MC*c*P) and cytochrome *c* (FeC*c*) are conformationally mobile and the photoinitiated redox cycle provides a tool with which to explore the interplay between conformational interconversion and ET. Through a combination of metal substitution and site-directed mutagenesis, we probe the interfacial regulation of these processes.

1 Introduction

Intra- and inter-protein electron transfer reactions are central to biology [1] and recently techniques have been developed to study such reactions without the complication of second order processes, through use of modified proteins [2]. One approach, developed by several groups, involves the study of intra-protein electron transfer within proteins modified by covalent attachment of redox-active inorganic complexes to surface amino acid residues [3, 4]. For example, $[(L)_5Ru]^{+2}$, when bound to a histidine residue on the outside of redox proteins such as cytochrome *c* or myoglobin, can exchange an electron with a metal-containing redox center located on the inside of the protein. In parallel, we and others [2a, 6] focused on studies of long-range inter-protein electron transfer within protein-protein complexes. Our approach involves replacing the heme (FeP) of one protein partner with a closed-shell porphyrin MP (M = Zn or Mg; P = protoporphyrin IX), and studying electron transfer between the MP and FeP groups [7, 8]. A reversible electron transfer cycle within such metal-substituted complexes (Scheme **I**) is initiated by laser flash photoproduction of the ^{3}MP triplet state (**A***). The ^{3}MP is a strong reductant and can reduce the ferriheme partner (Fe^{3+}P) by long-range electron transfer with a photo-initiated electron transfer rate constant k_t (Eq. (1)).

$$^{3}MP + Fe^{3+}P \xrightarrow{k_t} (MP)^{+} + Fe^{2+}P \tag{1}$$

The resulting charge-separated intermediate, $[(MP)^{+}, Fe^{2+}P]$ (**I**) returns to the ground state by thermally-activated electron transfer from Fe^{2+}P to the cation radical, $(MP)^{+}$ (Eq. (2)) with

$$(MP)^{+} + Fe^{2+}P \xrightarrow{k_b} MP + Fe^{2+}P \tag{2}$$

rate constant k_b. In our studies we have used transient absorption and emission techniques to monitor **A*** and **I**, thereby allowing us to measure both k_t and k_b.

Two issues may be addressed with this approach. If the electron donor/acceptor redox pair is rigidly held by the protein complex, then the system can be used to explore the factors that control the electron-transfer process itself, for example, the dependence on energetics or the intervening medium. Mixed-metal [M, M′] hemoglobin hybrids provide us with such a protein-protein electron transfer system [7]. Under the conditions of our experiments the hemoglobin tetramer in solution adopts the deoxy-Hb (T-state) structure that has a crystallographically known structure. Thus, electron transfer occurs between redox centers held at a rigidly fixed and crystallographically known distance and orientation. In contrast, physiological complexes such as that between yeast cytochrome *c* peroxidase (C*c*P) and cytochrome *c* (C*c*) are thought to exhibit a preferred binding mode, but are not restricted to a unique, static docking geometry. In this case, the study of long-range electron transfer (ET) as well as energy transfer within the metal-substituted [M*Cc*P, *Cc*] complex can be used to probe the dynamics of docking rearrangements [8].

The standard formalisms for describing ET processes assume that in reactions such as Eqs. (1) and (2) there is but a single stable conformational form for each of the precursor and successor electron-transfer states. However, for a system that displays two (or more) alternative *stable* conformations with different ET rates, dynamic conformational equilibrium can modulate the ET rates. Major protein conformational changes can occur at rates that are competitive with observed rates of ET [9], and such "gating" [10] may occur in non-rigid complexes such as that between zinc cytochrome *c* peroxidase (Zn*Cc*P) and cytochrome *c* (see below) or even within cytochrome *c* [5].

In summary, a rigid complex such as the [M, M′] Hb hybrids can be used to study ET; electron transfer (and energy transfer) can be used to study conformational reorganization within a mobile complex such as [Zn*Cc*P, *Cc*].

2 Mixed-Metal Hemoglobin Hybrids

Preparation of mixed-metal hemoglobin hybrids is achieved by separation of hemoglobin into its constituent α and β chains, followed by demetallation of one of the chains, reconstitution with MP, and chain recombination, yielding the tetrameric [α_2(MP), β_2(Fe^{3+}P)] or [α_2(Fe^{3+}P), β_2(MP)] species [11]. Thus, MP $\rightarrow$ FeP electron transfer might in principle occur between α_1/β_1 or α_1/β_2 subunits. However, the distance between α_1 and β_1 hemes is over 10 Å greater than the α_1 and β_2. This extra distance is expected, and indeed is found, to reduce ET rates by several orders of magnitude. Hence, for all practical purposes we may treat the $\alpha_2\beta_2$ tetramer in terms of two independent [α_1, β_2] electron transfer complexes.

The expectation that the geometry of mixed-metal hemoglobin hybrids is fixed in the known structure of T-state (deoxy) Hb requires that metal replacement does not perturb that structure. The first issue of structure that must be considered is local: does substitution of Zn or Mg cause significant perturbations? The structure of MgHb, in which all four prosthetic groups are MgP, has recently been determined crystallographically at 2.2 Å resolution [12]. Using the atomic model of deoxy Hb as the starting point in a least-squares refinement, only trivially small structural differences were noted, meaning the replacement of Fe^{2+} with Mg^{2+} in hemoglobin does not significantly alter the structure.

The second issue is global: is the quaternary structure of a mixed-metal hybrid significantly different from that of T-state Hb? Here, the answer again is no, as indicated by an X-ray structure of the [α(FeCO), β(Mn)] hybrid [13]. Thus, the distances and geometric relation of the heme groups involved in electron transfer within the [α_1, β_2] ET complex are preserved in the metal-substituted species (Fig. 1).

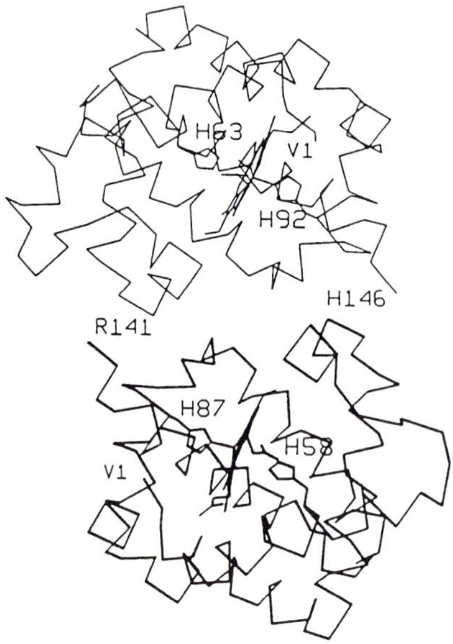

Fig. 1. α-Carbon backbone of an $[\alpha_1 \beta_2]$ dimer within the $[\alpha(FeCO)P, \beta(Mn^{3+}P)]_2$ hybrid. (From Ref. [13])

With this corroborative structural data it is possible to discuss the structure of the $[\alpha_1, \beta_2]$ complex with high precision. We consider as an example the [Fe, M] hybrids, M = Mg, Zn. In the $[\alpha_1, \beta_2]$ ET complex (Fig. 1), the β_2(MP) and α_1(FeP) are roughly parallel, with distances of 25 Å between metals, and about 17 Å edge-to-edge [14]. This structurally defined but chemically manipulable system offers many avenues for study. Recently we have focused on the effects of changing ET energetics. One method for doing this is to vary the closed-shell MP. The MgP^+/MgP reduction potential is about 100 mV lower than the ZnP^+/ZnP reduction potential. Consequently, $-\Delta G$ for the photoinitiated $\mathbf{A}^* \rightarrow \mathbf{I}$ process is 1.0 eV for $[^3(ZnP), Fe^{3+}P]$, and 1.1 eV for $[^3(MgP), Fe^{3+}P]$, while for the $\mathbf{I} \rightarrow \mathbf{A}$ ET, $-\Delta G$ is 0.8 eV for $[(ZnP)^+, Fe^{2+}P]$ and 0.7 eV for $[(MgP)^+, Fe^{2+}P]$.

Variation of the exogenous heme-ligand provides an even more effective means of altering the energetics for electron transfer without changing the structure of the ET complex. At neutral pH, the $Fe^{3+}P$ in met-hemoglobin has H_2O in the distal coordination site, with the remaining five sites occupied by the nitrogens of the porphyrin and of the proximal histidine [15]. The coordinated H_2O can be replaced by other ligands L, both neutral ($L = L^\circ$ = imidazole) and anionic ($L = X^- = CN^-, F^-, N_3^-$). Because the $Fe^{3+}P/Fe^{2+}P$ redox potential depends on L, the driving force for ET changes correspondingly. Importantly, X-ray crystallographic measurements of liganded hemoglobins show negligible changes upon ligand variation, and thus the heme geometry is retained [16]. The ability to vary the ligands in the hybrids can be contrasted with the situation in redox proteins such as cytochromes *c* and b_5, where the two axial endogenous ligands are fixed by the protein and cannot be changed without major disruption of its structure.

The combination of metal/ligand variation in these hybrids not only allows alteration of the ET energetics but also provides a means to study mechanistic questions, such as whether ET is direct or involves a hopping mechanism.

2.1 Triplet-State Decay Kinetics

Reversible electron transfer within Mg and Zn-substituted hemoglobin hybrids is initiated by flash photoproduction of the long-lived triplet state (^{3}MP). According to Scheme **I**, the triplet return to the ground state involves two pathways, intrinsic triplet decay (with rate constant k_D) and electron transfer quenching (with rate constant k_t).

Figure 2 shows the progress curves for triplet decay in the [MgP, Fe^{2+}P], [MgP, $Fe^{3+}(H_2O)$P], [ZnP, Fe^{2+}P], and [ZnP, $Fe^{3+}(H_2O)$P] hybrids, as monitored by ^{3}MP/MP absorbance difference spectra [12]. The data are shown normalized to unit triplet population. The triplet decay for both reduced hybrids, [MgP, Fe^{2+}P] and [ZnP, Fe^{2+}P], is exponential for more than five half-lives. The rate constant for this intrinsic triplet decay, k_D (Scheme **I**), is 20(2) s^{-1} at 25°C for M = Mg, and 53(5) s^{-1} at 25°C for M = Zn; identical rate constants are obtained by following triplet-state emission.

Triplet decay in the [Mg, $Fe^{3+}(H_2O)$] and [Zn, $Fe^{3+}(H_2O)$] hybrids monitored at 415 nm, the $Fe^{3+/2+}$P isosbestic point, or at 475 nm, where contributions from the charge-separated intermediate are minimal, remains exponential, but the decay rate is increased to $k_p = 55(5)$ s^{-1} for M = Mg and $k_p = 138(7)$ s^{-1} for M = Zn. Two quenching processes in addition to the intrinsic decay process (k_D) can contribute to deactivation of ^{3}MP when the iron containing-chain of the hybrid is oxidized to the Fe^{3+}P state: electron transfer quenching as in Eq. (1) (rate constant k_t), and Förster energy transfer (rate constant k_o). The triplet decay in oxidized hybrids thus is characterized by k_p, the net rate of triplet disappearance ($k_p = k_D + k_t + k_o$). The difference in triplet decay rate constants for the oxidized and reduced hybrids gives the quenching rate constant, $k_q = k_p - k_D = k_t + k_o$, which is thus an upper bound to k_t.

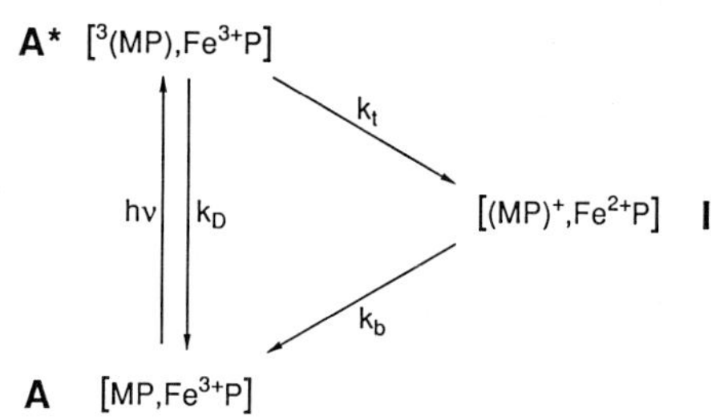

Scheme I

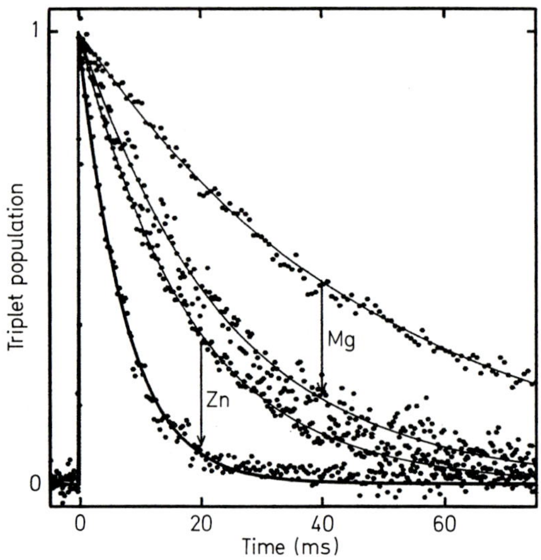

Fig. 2. Normalized triplet decay curves for [MP, FeP] hybrids. For a given M, (Zn, Mg) the arrow is directed from the curve for the Fe^{2+} state toward that for the Fe^{3+} state

Subtraction yields $k_q = 85(10)\ s^{-1}$ for [ZnP, $Fe^{3+}(H_2O)P$] and $k_q = 35(5)\ s^{-1}$ for [MgP, $Fe^{3+}(H_2O)P$].

We have measured kinetic progress curves for triplet decay as a function of heme ligand L (L = H_2O, imidazole, CN^-, F^-, and N_3^-) for [β(MP), α(Fe^{3+}(L)P)] hybrids. If we take the intrinsic triplet decay rate constant, k_D, to be largely independent of heme ligation, the differences displayed reflect inequivalent values of k_q for the various ligands. The data clearly falls into two classes: for M = Zn, hybrids with ferriheme coordinated to the neutral ligands (H_2O and imidazole) give $k_q \approx 80\ s^{-1}$, while those with bound anionic ligands give significantly reduced values [$3(2)\ s^{-1} < k_q < 12(3)\ s^{-1}$]. Similar trends (with smaller values for k_q were found for M = Mg. In our initial studies, which were limited to measurements of triplet quenching, we considered whether energy transfer could be contributing to k_q. Förster energy transfer would be proportional to spectral overlap between the ^{3}(MP) emission spectrum and the Fe^{3+}(L)P absorption spectrum. Thus, a lack of correlation between the ligated heme optical spectra and the observed k_q indicated that for L = L° = H_2O (and imidazole), most, if not all, of triplet quenching is associated with electron transfer [7c]. However, with the smaller rate constants for triplet quenching by the anion-ligated hemes, it was by no means clear whether electron transfer was the predominant quenching mechanism. That is, a small value for k_o would have minimal consequences for L = L°, but could account for much or all of k_q in the case where L = X^-. Thus, direct observation of the charge-separated intermediate **I**, in addition to yielding k_b, the rate constant for the thermal process, is required to confirm the very existence of long-range electron transfer in cases where k_q is small.

2.2 Direct Observation of Charge-Separated ET Intermediates

The time course of the charge-separated intermediate **I** can be measured in a flash photolysis experiment that monitors the (**I** – **A**) transient absorbance difference at a ground state/triplet state isosbestic point (e.g., 432 nm for Mg, and 435 nm for Zn). We have observed this intermediate for the [M, Fe] hybrids with M = Mg, Zn; representative kinetic progress curves are shown in Fig. 3 [7a]. In a kinetic scheme that includes Eqs. (1) and (2) as the only electron-transfer processes, when the **I** → **A** step is slow ($k_b < k_p$) the intermediate builds up (exponentially) during the lifetime of **A*** and exponentially disappears with rate constant k_b (Fig. 4A). This behavior is not observed for the hybrids, where the **I** → **A** process is more rapid than **A*** → **I**, with $k_b > k_p$. In this case, **I** appears exponentially at early times with rate-constant k_b and is expected to disappear completely in synchrony with **A*** in an exponential fall with rate-constant k_p (Fig. 4B).

The occurrence of a persistent absorbance change ($\Delta \mathbf{A}_\infty$) for the [M, Fe] hybrids is a slight complication that requires an extended kinetic model (Scheme **II**) in which $(MP)^+$ is reduced not only by $Fe^{2+}P$ (regenerating the [MP, $Fe^{3+}P$] state), but also by an as yet unidentified amino acid residue X and/or solution impurities (Eq. (3)) leading to [M, Fe^{2+}]:

$$[(MP^+), Fe^{2+}P] + X \xrightarrow{k_m} [M, Fe^{2+}P] + X^+ \tag{3}$$

Solution of the kinetic equations implicit in Scheme **II** indicate that the magnitude of $\Delta \mathbf{A}_\infty$ is proportional to k_m and that **I** appears exponentially with

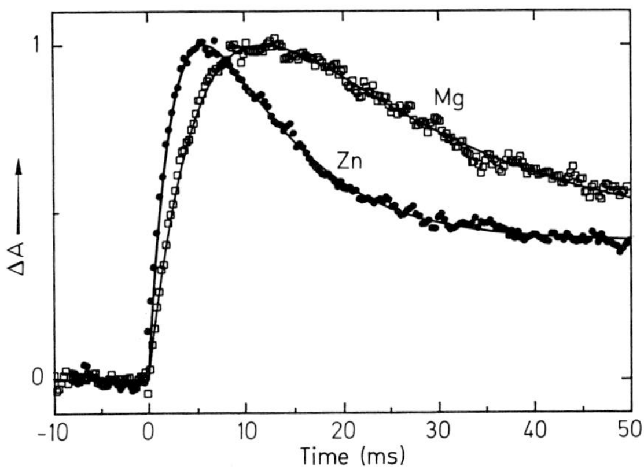

Fig. 3. Normalized kinetic progress curves at 5°C for electron-transfer intermediate (**I**) plus photoproduct (**C**) (See Scheme **II**) formed upon flash photolysis of the mixed-metal Hb hybrids: [β(MgP), α($Fe^{3+}P$)] (λ = 432 nm); [β(ZnP), α($Fe^{3+}P$)] (λ = 435 nm). *Solid lines* are non-linear least squares fits to the equations in Ref. [7a]. For [Mg, Fe], $k_b = 155(15)\ s^{-1}$, $k_p = 47(5)\ s^{-1}$, and $k_m = 20(5)\ s^{-1}$; for [Zn, Fe], $k_b = 350(35)\ s^{-1}$, $k_p = 122(10)\ s^{-1}$, and $k_m = 40(8)\ s^{-1}$. Buffer: 0.01 M KP_i, pH 7.0

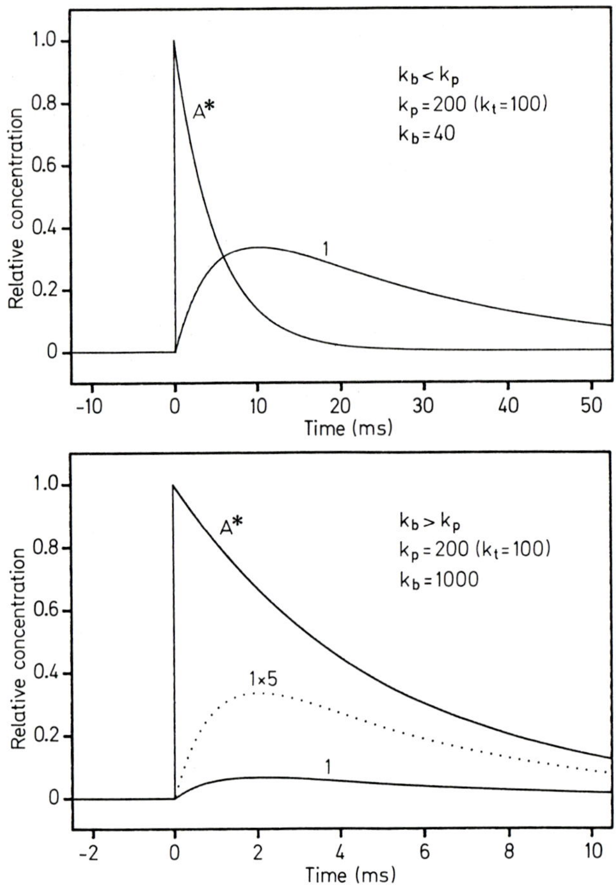

Fig. 4. Simulated kinetic progress curves for [A*(t)] and [I(t)] as predicted from Scheme **I**. Comparison of the relative concentrations of [A*(t)] and [I(t)] when (upper) $k_b < k_p$ and (lower) $k_b > k_p$. The values used for the rate constants of Scheme **I** are given

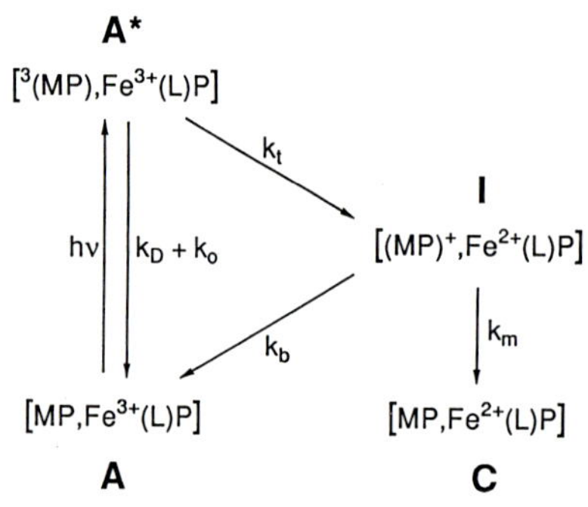

Scheme II

rate constant $k_x = k_b + k_m$. Figure 3 shows that the kinetic traces for the Zn- and Mg-substituted hybrids are well-described by non-linear least squares fits to the solutions of these kinetic equations. Identification of the absorbance transients of Fig. 3 with the charge-separated intermediate, **I**, is confirmed as follows: i) the magnitudes of the transients are proportional to the concentration of $Fe^{3+}P$; ii) reduction of [MP, $Fe^{3+}P$] to [MP, $Fe^{2+}P$] with a stoichiometric addition of $Na_2S_2O_4$ eliminates the transient; iii) the signs and magnitudes of the absorbance changes at ^{3}(MP)/MP isosbestic points are entirely consistent with formation of $Fe^{2+}P$ in accord with Eq. (2) (e.g., for Mg, the transient absorbance is positive at 432 nm and is negative at 542 nm, as is the ([MgP^+, $Fe^{2+}P$] − [MgP, $Fe^{3+}P$]) difference spectrum). The data in Fig. 3 show that the time course of the intermediate [$(MP)^+$, $Fe^{2+}P$] (**I**) strongly depends on M. At 5°C for Mg, $k_b = 155(15)\ s^{-1}$, $k_p = 47(5)\ s^{-1}$, and $k_m = 20(5)\ s^{-1}$; for Zn, $k_b = 350(35)\ s^{-1}$, $k_p = 122(10)\ s^{-1}$, and $k_m = 40(8)\ s^{-1}$.

The existence of long-range electron transfer within [^{3}(MP), Fe^{3+}(L)P] has been verified for both M and for all L by direct observation of **I**, the charge-separated intermediate. Figure 5 shows a comparison of the kinetic progress curves obtained for [β(ZnP), α($Fe^{3+}(H_2O)$P)] and [β(ZnP), α(Fe^{3+} (CN^-)P)] [7b]. The exponential fall, $k_p = 65(8)\ s^{-1}$, is in agreement with the low k_p obtained from triplet decay data. Absorbance changes resulting from formation of the charge-separated intermediate **I** are proportional to the rate constant k_t, and thus k_t can be calculated independently of any other contributions to triplet-state quenching, provided the quantum yield for the formation of

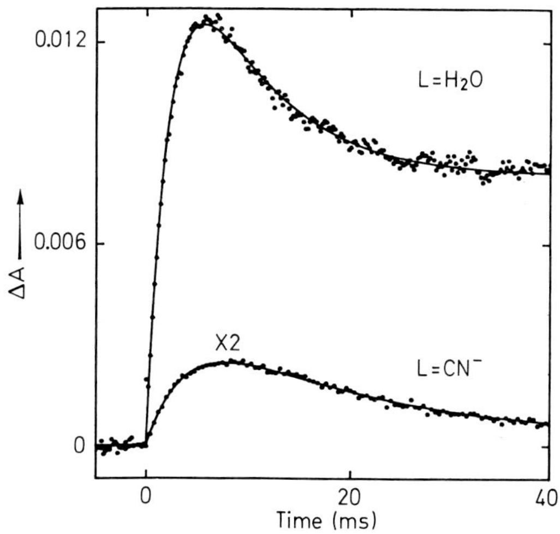

Fig. 5. Kinetic progress curves as in Fig. 3 at 435 nm for [β(ZnP), α(Fe^{3+}(L)P]. Experimental points and non-linear least squares fits for L = H_2O and L = CN^- are shown, with absorbance changes normalized to a zero-time triplet concentration $(A_o^*) = 10^{-6}$ M. For [β(ZnP), α($Fe^{3+}(H_2O)$P], $k_b = 345(45)\ s^{-1}$, $k_p = 134(15)\ s^{-1}$; for [β(ZnP), α($Fe^{3+}(CN^-)$P], $k_b = 240(30)\ s^{-1}$, $k_p = 65(8)\ s^{-1}$

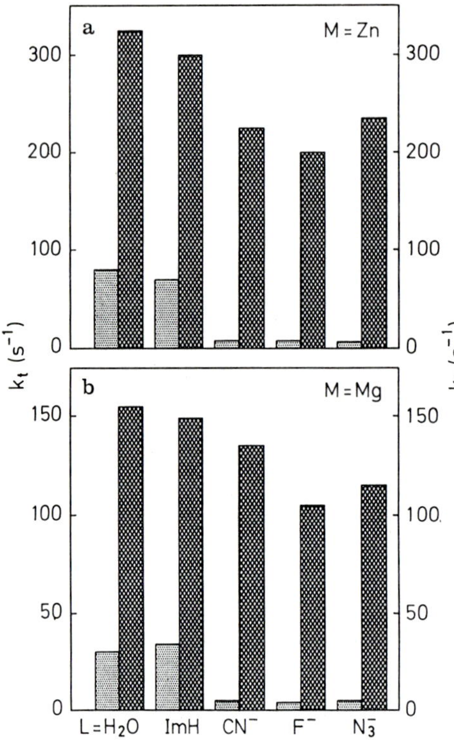

Fig. 6. Ligand dependence of rate constants for photoinitiated ET (k_t, speckled) and thermal ET (k_b, crosshatched) within the hybrids (**a**) [ZnP, Fe^{3+}(L)P] and (**b**) [MgP, Fe^{3+}(L)P]. For M = Zn, the data refer to 1-methylimidazole rather than imidazole

I can be determined. Using this procedure, analysis of the relatively large absorbance changes observed with the intermediate in the L = H_2O hybrid gives $k_t^{Zn}(H_2O) = 90(30)\ s^{-1}$, confirming the previous assignment $k_t = k_p - k_D$ for L = H_2O [12]. On the other hand, k_q for [β(ZnP), α(Fe^{3+}(CN^-)P] is 14(4) s^{-1}, and analysis of absorbance changes associated with [β$(ZnP)^+$, α(Fe^{2+}(CN^-)P)] gives an even smaller value, $k_t^{Zn}(CN^-) = 6(3)\ s^{-1}$. Thus, replacement of H_2O by CN^- in the heme coordination sphere reduces k_t by over an order of magnitude. Quantitation of I for M = Mg, Zn, and for the ligands L = H_2O, ImH, CN^-, F^-, and N_3^- gives the k_t^M(L) shown in Fig. 6. Replacement of H_2O with another neutral ligand (imidazole) does not significantly alter k_t, while replacement with an anionic ligand (CN^-, N_3^-, F^-) causes a ten-fold reduction in the ET rate.

The effect of anion binding on k_b is not nearly as great. The data in Fig. 6 show that for both metals, there is less than a 50% decrease in the thermal ET rate constant between hybrids containing neutral and anionic ligands.

2.3 *Some Mechanistic Aspects of ET within [MP, Fe^{3+} (L) P] Hb Hybrids*

The single most important mechanistic aspect of long-range ET in the hemoglobin hybrids to be confirmed is that the ET process under consideration is a single-step event and does not involve a sequence of steps with one (or more) real intermediate states, such as one with an oxidized or reduced amino acid residue. This is easily proved. Indirect hopping of an electron from ^{3}(MP) to Fe^{3+}(L)P would correspond to oxidative triplet quenching by an amino acid, with subsequent reduction of Fe^{3+}P by the amino acid anion. However, any such quenching of ^{3}(MP) would occur equally in the reduced (Fe^{2+}P) hybrids, since this process does not involve the FeP. Consequently, electron transfer by this mechanism would not give rise to an increase in triplet decay, and the ET triplet quenching in the Fe^{3+}P hybrids must be associated with a direct process.

Our data also indicate that the **I** → **A** electron transfer process is direct. If the **I** → **A** charge recombination process were indirect, it would imply the existence of an amino acid, y, which mediated electron flow from Fe^{2+}P to $(MP)^+$ via an internal electron transfer. The only thermodynamically accessible discrete intermediate would then be [(MP), y^+, (Fe^{2+}P)], which would decay by a second electron transfer process back to the ground state. If the $(MP)^+ \rightarrow y$ ET process were rapid, and $Fe^{2+}P \rightarrow y^+$ electron transfer were rate limiting, changing M would not affect the observed rate constant, and k_b^{Mg} would equal k_b^{Zn}. However, if $(MP)^+ \rightarrow y$ electron transfer were rate limiting, succeeded by rapid $Fe^{2+}P \rightarrow y^+$ ET, changing the heme ligand L would not affect the ET rate, and $k_b^{Zn}(H_2O)$ would equal $k_b^{Zn}(CN^-)$. Since $k_b^{Mg}(H_2O) \neq k_b^{Zn}(H_2O) \neq k_b^{Zn}(CN^-)$, we conclude that a two-step electron hopping mechanism does not occur, and that k_b describes direct $Fe^{2+}P \rightarrow (MP)^+$ electron transfer.

What is the fate of the heme-ligand, L, during the ET cycle of Scheme **II**? For $L = H_2O$, reduction of the aquo-bound heme yields the five-coordinate ferroheme, Fe^{2+}P, while for CN^-, it is known that ligand dissociation from $Fe^{2+}(CN^-)$P is slow, indicating the **I** → **A** process involves reoxidation of the CN^--bound species. For N_3^- and F^-, the data in Fig. 6 show that these ligands also remain bound on the ET timescale. If ET-induced ligand loss were rapid compared to the **I** → **A** electron transfer process, one would predict that for a given metal, $k_b^M(X^-) = k_b^M(H_2O)$, contrary to observation with both metals. The data also show that for a given anion, $k_b^{Mg}(X^-) < k_b^{Zn}(X^-)$. This metal dependence indicates that $k_b^M(X^-)$ cannot represent the alternative of a rate-limiting ligand dissociation from the $Fe^{2+}(X^-)$P partner of **I**, followed by fast ET (i.e., ligand "gating" of **I** → **A**). Thus, the differences in the thermal electron transfer rate, k_b^M(L), as a function of ligand and metal indicate that all anionic ligands remain bound during the entire electron transfer cycle of Scheme **I**. In contrast, the extremely slow reduction of ligated ferrimyoglobin [Mb^{3+}(L)] by exogenous $S_2O_4^{2-}$ requires anionic dissociation for most ligands, notably fluoride [17]. Taken together, these data suggest a difference in the mechanisms for reduction by exogenous dithionite and by an "internal" ^{3}(MP).

Clearly, the simple variation of ligands and metals within [MP, Fe^{3+}(L)P] allows a heretofore unparalleled view of the mechanistic aspects of long-range electron transfer between proteins.

2.4 Temperature Dependence of ET Rate Constants

The structural stability of mixed-metal hemoglobin hybrids also has allowed us to study low-temperature electron transfer in this system. We first reported the temperature dependence of triplet-state quenching within the [^{3}(ZnP), $Fe^{3+}(H_2O)P$] hybrids, which we attributed to the $^3ZnP \rightarrow Fe^{3+}P$ ET reaction [7d]. The rate constant dropped smoothly as the temperature was lowered from room temperature to ~ 200 K. Below this temperature the rate constant remained roughly constant with a tunnelling rate constant of $k_t \sim 9\ s^{-1}$ (Fig. 7).

Recently, by following absorbance changes at the ^{3}(MP)/MP isosbestic point as described above, we have been able to directly monitor the charge separated intermediate **I** at low temperatures for [MP, Fe^{3+}(L)P] (M = Mg, Zn; L = H_2O, F^-, CN^-, Imid), thereby allowing measurements of both k_b and k_t [18].

Our data show that in all hybrids, the thermal electron transfer process (Eq (2)) is remarkably insensitive to temperature, suggesting that ET proceeds by quantum mechanical tunnelling to quite high temperatures. Figure 8 shows the temperature dependence of k_b, the rate constant for the thermal $Fe^{2+}(CN^-)P \rightarrow (MP)^+$ electron transfer for M = Mg and Zn. For [$(ZnP)^+$, $Fe^{2+}(CN^-)P$], k_b decreases by less than a factor of three from 300 K to 100 K;

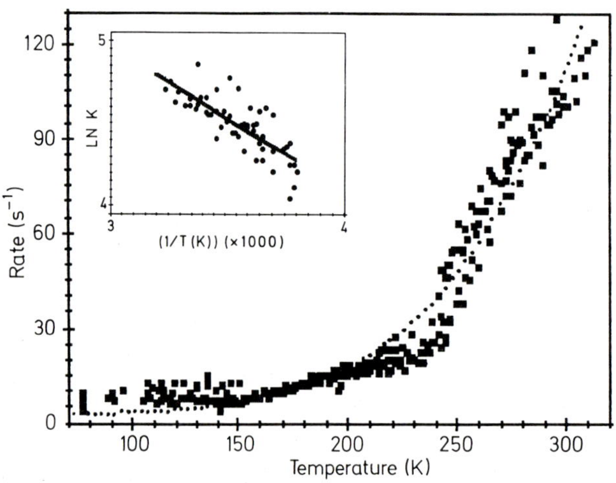

Fig. 7. Temperature dependence of the triplet-state quenching rate constant (k_q) for the [α(Zn), β($Fe^{3+}H_2O$)] hybrid. Adapted from Ref. [7d]

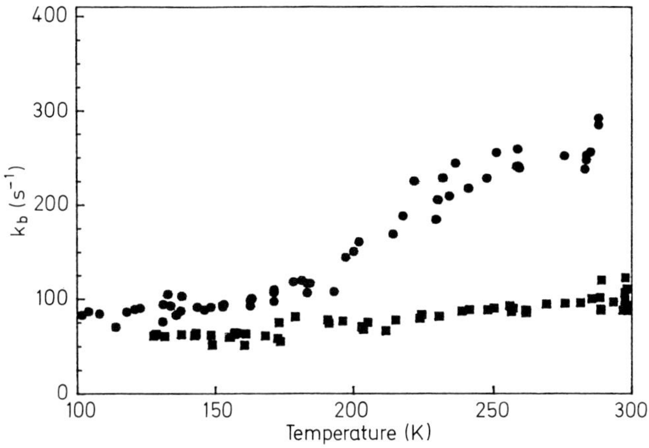

Fig. 8. Temperature dependence of the thermal $Fe^{2+}P(CN^-) \rightarrow MP^+$ electron transfer rate (k_b) for [MgP, $Fe^{3+}(CN^-)P$] (■) and [ZnP, $Fe^{3+}(CN^-)P$] (●) hybrids in 70% cryosolvent

over the same temperature regime, k_b for [$(MgP)^+$, $Fe^{2+}(CN^-)P$] varies only by a factor of two. To further understand this process, we have begun to study ET at temperatures between 4 and 77 K.

3 [Zn Cytochrome *c* Peroxidase, Cytochrome *c*] Complexes

We have used the metal replacement technique to study complexes between Zn-substituted yeast C*c*P and various C*c*, thereby permitting investigation of the electron-transfer cycle within a conformationally mobile complex as envisaged in Scheme **III**. McLendon and coworkers have made measurements with the complementary [FeC*c*P, ZnC*c*] complex [6b, c]. The coordination sphere of the metal is not under experimental control in either partner of this system; the two axial heme-ligands of FeC*c* are fixed as Met-80 and His-18 by the protein while the ZnP of ZnC*c*P is five-coordinate, with His-175 as the endogenous axial ligand. Thus, we have taken the entire C*c* protein as our experimental vehicle, by comparing results from C*c* isolated from various species, and then, on a finer scale, by using proteins modified by site-directed mutagenesis at an individual residue. To date, our work with genetically engineered cytochromes has focused on mutations at residue 82 of C*c* which is phylogenetically conserved. In the WT protein from yeast C*c*, as well as in C*c* from other species, this position is occupied by Phe.

The most stable structure of the [C*c*P, C*c*] complex deduced from computer modelling studies by Poulos and coworkers [19] looks much like that of the [α_1, β_2] ET complex of the Hb hybrids, with roughly parallel heme planes

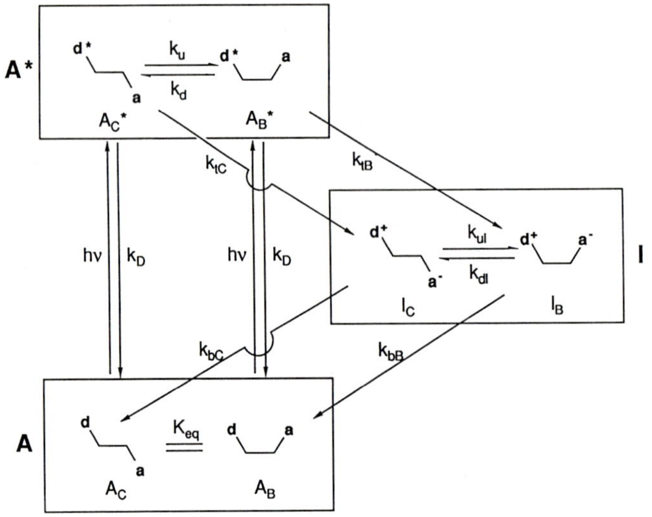

Scheme III

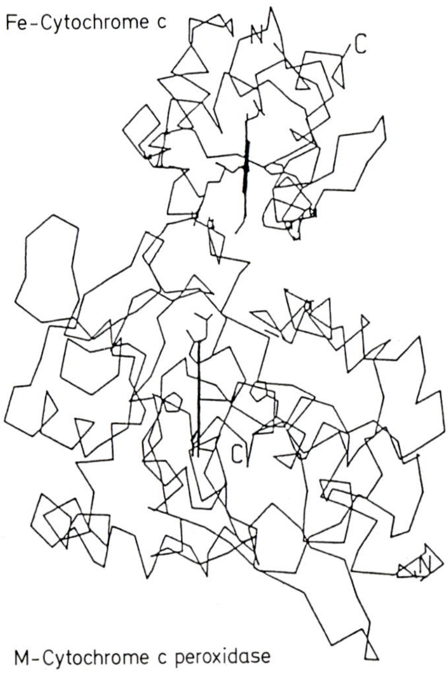

Fig. 9. α-carbon skeleton of the hypothetical model of the [Fe^{3+}C*c*P, Fe^{3+}C*c* (tuna)] complex. Adapted from Ref. [19]

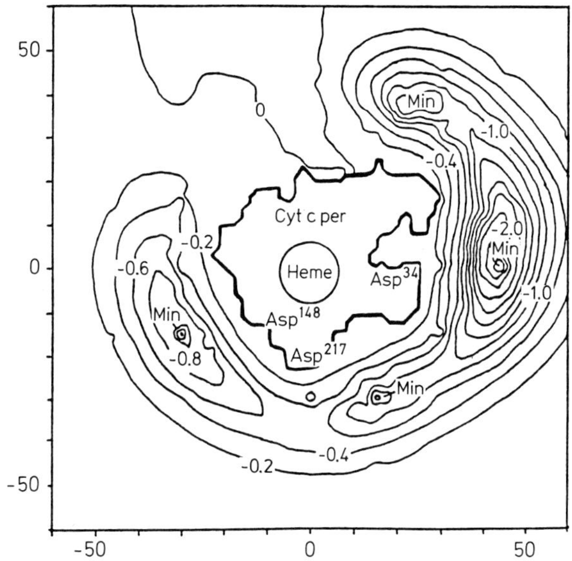

Fig. 10. Electrostatic potential energy for interaction of Fe^{3+}CcP with Fe^{3+}Cc (tuna) as a function of the center of mass position of Cc. Adapted from Ref. [20]

separated by ~ 17 Å. The intervening medium is comprised of portions of the two protein polypeptides and includes Phe-82 of Cc (Fig. 9). However, the electrostatic calculations of Northrup and coworkers [20] suggest that more than one conformation may be accessible. As seen in Fig. 10, several discrete and energetically accessible energy minima are suggested. The most stable minimum is that found by Poulos and Kraut (Fig. 9), but it appears not to be that with the smallest M-M separation. Thus, the possibility of conformational influence and/or control of ET must be considered when exploring this and other protein complexes that associate through electrostatic and hydrophobic interactions, such as [cytochrome *c*, cytochrome b_5] [21], [plastocyanin, cytochrome *c*] [22], [flavodoxin, cytochrome *c*] [23], and [hemoglobin, cytochrome b_5] [24]. We first summarize our conclusions about the possible manifestations of conformational influences in such systems, then describe our results for the [ZnCcP, Cc] complexes, where these influences have been detected.

3.1 "Gated" Electron Transfer

We recently examined a two-state case that represents the simplest model for any "gated" reaction [10a, b]. Consider a donor-acceptor pair, [d, a], as part of a system that exhibits two conformations, *B*(boat) and *C*(chair). For concreteness, we may imagine **d** and **a** as attached at the 1, 4 positions on a cyclohexane ring; the same formal situation will be realized in far more interesting ways with protein complexes. In this case, each of the three system states shown in Scheme

III, (**A**, **A***, and **I**) is composed of two conformational substates ($\mathbf{A} = \mathbf{A}_B + \mathbf{A}_C$; $\mathbf{A}^* = \mathbf{A}_B^* + \mathbf{A}_C^*$; $\mathbf{I} = \mathrm{I}_B + \mathbf{I}_C$) and introduction of the conformational interconversion expands kinetic Scheme **I** to Scheme **III**.

This scheme includes ET rate constants only for the $d^* \rightarrow d^+$ electron-transfer processes, in which the system conformation is conserved, and conformational and ET steps only occur sequentially. Intuitively, it might be expected that the kinetic scheme must include ET that is synchronous with a conformational change in the medium coordinate. However, we showed [10a] that it is not necessary to include the "diagonal" processes (e.g., $\mathbf{A}_C^* \rightarrow \mathbf{I}_B$) when considering *stable* substates.

The fact that ET and conformational reactions thus are sequential (Scheme **III**), and not concerted, is an important factor in efforts to disentangle conformational and electron-transfer influences, because standard detection methods monitor only the ET event, and not conformational changes within one electronic state. In many, if not most, instances the measured time course of a single gated ET reaction is likely to be indistinguishable from a reaction without gating.

Fortunately, the partial decoupling of the ET and conformational processes afforded by the absence of synchronous events in principle and in practice allows for the identification of an observed decay rate constant. For example, if one constructs a series of systems in which the ET energetics (or electronic coupling) is modified without change in the conformational equilibrium, thus leaving the conformational rates unchanged, then the observed rate constants will be unchanged if the reaction is controlled by a conformational rate, but will vary if this is not so.

We reported the general solutions for the concentrations **A***(t) and **I**(t) for kinetic Scheme **III** and discussed several relevant limiting cases [10b]. In the limit where the conformational substates interconvert rapidly compared with the ET rates, the functional forms for both **A***(t) and **I**(t) will reduce to those predicted by the simple cycle of Scheme **I**. However, the apparent ET rate constants (k_t^{app}, k_b^{app}) will be the weighted average of the intrinsic ET rate constants of the conformational substate. In the other extreme, the intrinsic ET rate constants are much faster than the conformational interconversions, which become rate limiting. Again, **A***(t) and **I**(t) can have a simple form that hides such underlying facts. For example, if the excited state of the system is created initially in an unreactive state ($\mathbf{A}_c^*$) and only substate $\mathbf{A}_B^*$ is reactive, then, if $k_{ET} \gg k_{conf}$, (say $k_{tB} \gg k_u, k_d$; $k_{tc} = 0$), **A***(t) will decay by a single exponential but the apparent ET rate will be the rate of conformational interconversion, k_{conf}, not the ET rate, k_{ET}. This is the "gating" limit. Between these extremes, it is possible to observe complex, multiexponential behavior.

3.2 Triplet-State Quenching in Aqueous Solutions

The triplet decay curves of ^{3}ZnC*c*P and of all [^{3}ZnC*c*P, Fe^{2+}C*c*] complexes, where electron transfer has been blocked by prior reduction of the C*c* heme, are exponential with the decay rate $k_D \approx 120\,s^{-1}$ at ambient, Upon titration of ZnC*c*P with Fe^{3+}C*c*, [8e] the decay rate constant increases until a 1:1 ratio is reached and then remains constant at a plateau value (k_p). We have compared the triplet-state quenching within [ZnC*c*P, Fe^{3+}C*c*] complexes where the C*c* employed were isolated from a variety of species or were mutants of the yeast protein [8b–e]. The quenching rate constants at 0°C, $k_q = k_p - k_D$, were found to vary by 12-fold, from an upper value of 166(4) s^{-1} for the complex with yeast C*c*(WT) to a low of 13(2) s^{-1} for the C*c* (Gly-82) mutant. Homologous complexes with fungal cytochromes, e.g., C*c*(WT), C*c* (*C. krusei*), and C*c* (*S. cerevisiae* iso-1) fall at the high end of this range, whereas heterologous C*c*, e.g., C*c*(horse) and C*c*(tuna) are at the low end. Among the position 82 mutants of C*c*(yeast iso-1), the quenching rates decrease in the order C*c*(Tyr-82), C*c*(Phe-82) > C*c*(Ser-82) > C*c*(Leu-82) > C*c*(Ile-82) > C*c*(Gly-82). We attributed these differences to variations in the docking geometry among the complexes with the various C*c* but correctly predicted that the C*c*(Gly-82) mutant must have a perturbed structure. Interestingly, recent data suggests that a percentage of the quenching is due to energy transfer, and that the percentage varies with the C*c* employed.

A difference in the temperature dependence of the quenching rate among species further suggests that the kinetic behavior of the complexes varies sharply with C*c* [25]. The complexes with homologous fungal C*c* are characterized by a relatively large and temperature-independent quenching rate, $k_q \sim 200\,s^{-1}$ whereas in those with the heterologous C*c* the quenching rate is low at ambient temperature, $k_q \sim 60\,s^{-1}$, and strongly decreases when the temperature is lowered.

As with the hemoglobin hybrids, we have also examined the kinetics for complexes with Mg-substituted C*c*P [26]. At 293 K, the triplet state with MgC*c*P and the [MgC*c*P, Fe^{2+}C*c*] complex decays with the identical intrinsic decay rate of $k_D = 25\,s^{-1}$. Surprisingly, we find that the quenching rates for the [MgC*c*P, Fe^{3+}C*c*] complexes are virtually identical to those for the analogous complexes with ZnC*c*P within each class. This observation is consistent with conformationally-gated ET according to Scheme **III**, as follows: The dominant conformation of the [M, Fe^{3+}] ground state complex ($\mathbf{A}_c$) and thus the initially formed **A*** state ($\mathbf{A}^*_C$) are in an unreactive conformation; slow conformational conversion to the reactive $\mathbf{A}^*_B$ state is rate limiting, whereas the actual $\mathbf{A}^*_B \rightarrow \mathbf{I}_B$ ET step is fast [27].

3.3 Triplet-State Quenching in Cryosolutions

We have recently shown [8a] that measurements of triplet-state quenching within the homologous (ZnC*c*P, Fe^{3+}C*c*(yeast)] complex can be used to probe the dynamics of docking rearrangements within an ET complex. Upon extending the temperature range well below ambient, we found that within a transition range of 250–220 K the complex undergoes a remarkable conformational transition from a form that shows strong triplet quenching to one that has none.

The intrinsic triplet decay traces for ZnP incorporated into C*c*P, as measured with the [ZnC*c*P, Fe^{2+}C*c*] complex in ethylene glycol (EG)/KPi buffer, are rigorously single-exponential for ≥ 5 half-lives and the decay rate constant (k_D) decreases smoothly from $k_D = 126(4)\ s^{-1}$ at 293 K to $k_D = 66(2)\ s^{-1}$ at 190 K. Quenching of ^{3}ZnP by the C*c* ferriheme within the [ZnC*c*P, Fe^{3+}C*c*] complex (rate constant, k_q) increases the triplet-decay rate constant at 293 K to $k_p = 279(16)\ s^{-1}$ in 30% EG (Fig. 11) giving $k_q = 153\ s^{-1}$. The quenching rate constant is roughly independent of temperature down to 250 K, but upon further lowering the temperature by only 30 K, the quenching vanishes ($k_q \rightarrow 0$) (Fig. 12A). The triplet decay traces for the [ZnC*c*P, Fe^{3+}C*c*] complex also exhibit single-exponential kinetics for ≥ 5 half-lives outside the transition range of $220 \leq T \leq 250$ K (Fig. 11). However, within the transition range, heterogeneous kinetics are observed (Fig. 11) and the decay traces are well described with either a two-exponential kinetic equation:

$$A = A_0[(1 - f)\exp(-k_D t) + f\exp(-k_p t)] \tag{4}$$

or a stretched exponential equation in which the quenching rate constant (k_q) is distributed [28], and all complexes have the same intrinsic rate constant (k_D):

$$A = A_0 \exp[-(k_D t) - (k_q t)^n] \tag{5}$$

Equation (4) describes a partition between two forms of the complex: one, with fraction f, that exhibits quenching and one that does not. Equation (5) corresponds to a distribution of forms with a range of quenching rates whose breadth varies inversely with n; the occurrence of distributions is well-established for proteins at low temperatures [29]. In either case, the non-exponential kinetics necessitate that conformational interconversion in the transition range is slow compared to the lifetime of the triplet state and the gating limit is applicable.

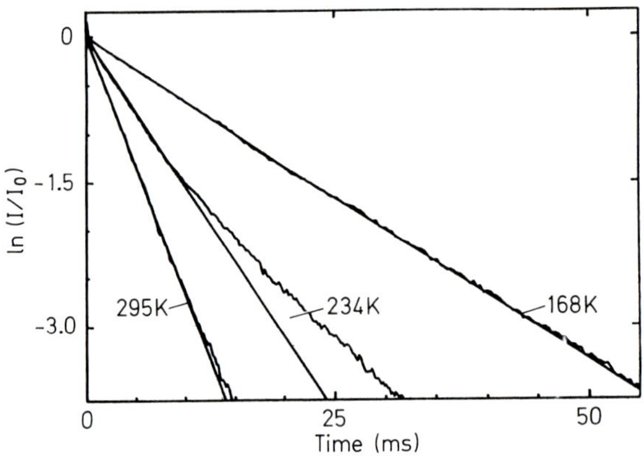

Fig. 11. Triplet-state decay traces for the [ZnC*c*P, Fe^{3+}C*c*] complex above (295 K), below (168 K), and near $T_{mid} = 234$ K. The *solid lines* are fits to a single-exponential decay. Conditions: 30% EG/10 mM KP_i buffer; pH = 6 at 4°C

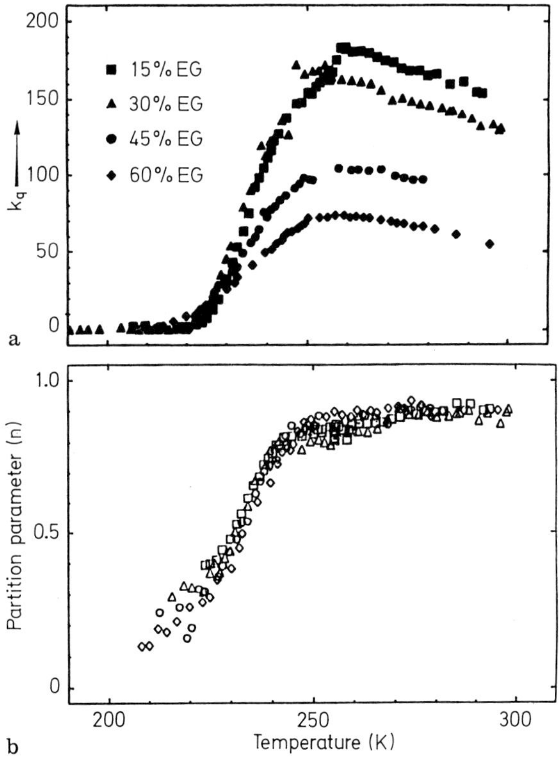

Fig. 12. Temperature dependence for the parameters that characterize the partitioning of the [ZnCcP, Fe^{3+}Cc] complex into different forms. A) Intracomplex quenching rate constant, k_q, obtained from Eq. (5): 15% (■), 30% (▲), 45% (●), and 60% (◆) EG/10 mM KP_i buffer at pH 6. B) Distribution parameter, n, from Eq. (5): 15% (□), 30% (△), 45% (○) and 60% (◇) EG/10 mM KP_i buffer

The partition parameter, f, from the two-state fit (Eq.(4)) and those from the distributed fit, Eq. (5) (Fig. 12) change abruptly with temperature. The value of f drops in a sigmoidal fashion as the temperature is lowered through the transition range; for the distributed fit, both the "1/e"-rate, k_q, and the distribution parameter, n, drop correspondingly through this transition. The decrease in n corresponds to broadening of the distribution in k_q.

The two-state model was used to test whether characteristics of the low-temperature cryosolvent cause the equilibrium constant for complex formation, K(T), to fall precipitously as the temperature is lowered through T_{mid}. In this case, the slow phase that appears below ~ 250 K would correspond to uncomplexed ZnCcP. This interpretation fails because within the transition range the fraction, f(T), is unaffected by a ten-fold reduction in the ratio, R = [Cc]/[CcP], whereas use of K(T) calculated from f(T) would predict a larger shift of f(T). Alternatively, the two-state model would apply if a low-temperature form of the complex were created by a change in ligation of either ZnP or FeP.

This is ruled out by optical and MCD spectra [30], which sharpen smoothly throughout the transition range, but do not show any shifts or anomalies that would accompany changes in the coordination state of either protein. Instead, we tentatively interpret the abrupt disappearance of quenching as signalling a transition in the interfacial docking geometry *within* the protein complex [31]. An alternate possibility is that one or both proteins freezes into a set of inactive conformational substates [29]. The former interpretation is supported by the data in aqueous solution that suggests that intracomplex ET near ambient temperatures involves a conformational conversion from inactive to active forms (conformational "gating") [19].

Remarkably, the transition temperature is independent of the solvent composition (Fig. 12), occurring in the same temperature range for solutions that glass below T_{mid} (60% EG) as for those solutions that crystallize at a temperature comparable to (45% EG) or greater than (15% and 30% EG) T_{mid} [32]. Clearly, the abrupt change in kinetics is an intrinsic molecular phenomenon that, unlike the CO-rebinding to myoglobin [29], is not "slaved" to the solvent. However, the intracomplex quenching rate constant does vary with the solvent composition in the high-temperature region (Fig. 12), which may reflect a more subtle change in the mode of docking for the [ZnC*c*P, Fe^{3+}C*c*] complex. To elucidate the structural basis for this fluctional process, we are extending these studies to include non-homologous complexes as well as complexes in which the intracomplex interface is modified by site-directed mutagenesis.

3.4 Conformational Control of $Fe^{2+}P \rightarrow ZnP^{+}$ ET

As noted above, photo-initiated ET within the [ZnC*c*P, Fe^{3+}C*c*] complex (Eq. (1)) behaves in a manner consistent with gated ET over the temperature range 0–30°C. What about the thermal, $Fe^{2+}P \rightarrow ZnP^{+}$ process?

Our early reports of the **I** → **A** reaction in [ZnC*c*P, C*c*] complexes were interpreted within the context of the simple ET cycle, Scheme **I** [8b, c]. As noted above, there are two limiting cases for this scheme. i) The **I** → **A** reaction is slow; **I**(t) appears with the triplet decay rate constant k_p and disappears with the thermal ET rate constant k_b. ii) It is rapid; here **I**(t) appears with k_b and disappears with k_p. We reported that ZnC*c*P complexes with heterologous C*c*'s [8d] and with aliphatic Phe-82 mutants of yeast C*c* [8b] fell into the former class, while the intermediate for fungal C*c* appears rapidly [8d].

From the kinetic traces for the "slow k_b" complexes, we precisely mapped out the difference spectrum for the intermediate state relative to the ground state [8b]. Comparison with the static spectrum expected for the intermediate later revealed that for these complexes, the size of the kinetic ET intermediate was only ca. 20% of the size predicted from knowledge of the static spectra. We therefore remeasured the **I** → **A** kinetics with improved instrumentation [27]. Figure 13 shows the time evolution of state **I** for the [ZnC*c*P, C*c*(horse)] complex. The intermediate [**I**] not only decays more slowly than k_p (Fig. 13

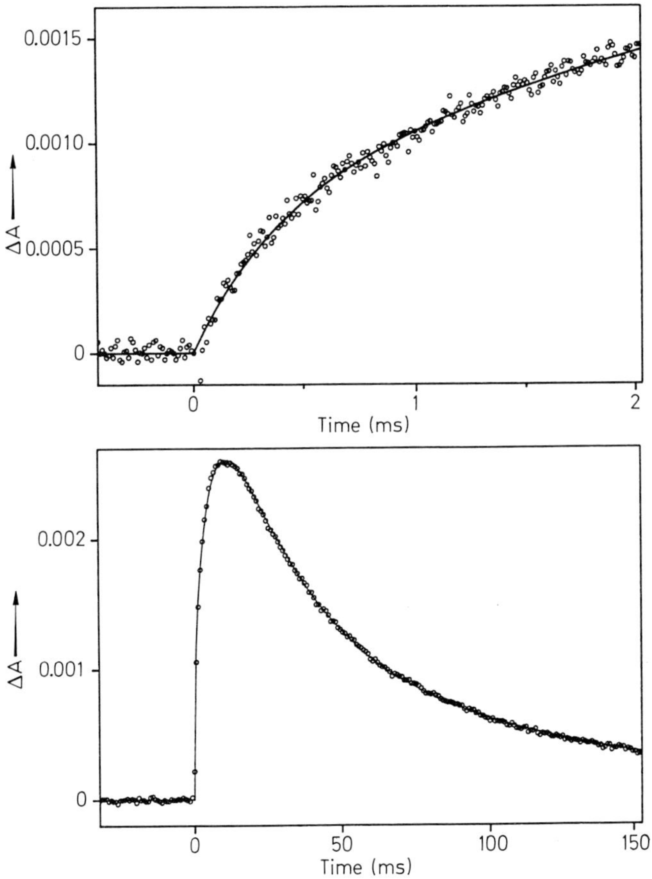

Fig. 13. Kinetic progress curves at 20°C for the electron transfer intermediate (**I**) formed upon flash photolysis of the [ZnC*c*P, C*c* (horse)] complex. The absorbance was monitored at the 549 nm **A*/A** isosbestic. The sample contained 4.7 μM ZnC*c*P and 14.2 μM C*c* in 5 mM KP_i, pH 7

Lower), as reported earlier, but also rises more rapidly than k_p (Fig. 13 Upper). This result immediately requires a more complex kinetic scheme than that of Scheme **I**. Excellent self-consistent fits to the time evolution of [**I**] (t) are obtained with an expression that is the sum of three kinetic phases, all having a common rate constant for triplet decay, k_p, but with differing values of the rate constants for the decay of **I**(3000 s^{-1}, 40 s^{-1}, 5 s^{-1}). We have further seen that complexes with different C*c* show similar behavior, but with the fractional contribution of these multiple phases varying with species.

This observation indicates that the multiple phases do not arise from static non-interconverting forms of the complex, and that they can best be interpreted by a kinetic scheme that involves dynamic conformational interconversion between conformational substates within state **I**. The analysis indicates that the initial product of the forward ET reaction (substate $\mathbf{I_B}$) undergoes the thermal $\mathbf{I_B} \rightarrow \mathbf{A_B}$ ET to the ground state with rate constant $\mathbf{k_b} \sim 3000\ s^{-1}$. Competing with this ET process is conformational interconversion to two unreactive conformational substates $\mathbf{I_C}$ and $\mathbf{I_D}$ with rate constant of ca. $500\ s^{-1}$. The slow kinetic phases associated with the decay state **I** reflect the rate-limiting activation of $\mathbf{I_C}$ and $\mathbf{I_D}$ to the reactive state $\mathbf{I_B}$, with rate constants $\sim 40\ s^{-1}$ and $5\ s^{-1}$, followed by rapid return to the ground state, **A**. Thus, the thermodynamically most stable conformers ($\mathbf{I_C}$ and $\mathbf{I_D}$) of the [C*c*P, C*c*] do not undergo facile ET. Rather, both photoinitiated and thermal ET within the complex involve activation to a less thermodynamically favored, but highly reactive structure ($\mathbf{I_B}$).

4 Discussion

4.1 Hemoglobin Hybrids

Metal-substituted hemoglobin hybrids, [MP, $Fe^{3+}(H_2O)P$] are particularly attractive for the study of long-range electron transfer within protein complexes. Both photoinitiated and thermally activated electron transfer can be studied by flash excitation of Zn- or Mg-substituted complexes. Direct spectroscopic observation of the charge-separated intermediate, [$(MP)^+$, $Fe^{2+}P$], unambiguously demonstrates photoinitiated ET, and the time course of this ET process indicates the presence of thermal ET. Replacement of the coordinated H_2O in the protein containing the ferric heme with anionic ligands (CN^-, F^-, N_3^-) dramatically lowers the photoinitiated rate constant, k_t, but has a relatively minor effect on the thermal rate, k_b.

Because metal substitution and ligand variation affect ET kinetics without perturbing the structure of the ET complex, such changes were used to probe mechanistic aspects of ET. The data show that both photoinitiated and thermal ET are direct processes. The measurements of k_t in the [ZnP, $Fe^{3+}(H_2O)P$] hybrids provided the first detection of ET tunnelling in a structurally-defined protein system.

The ease with which electron transfer can be studied in mixed-metal hemoglobin hybrids suggests this system will continue to be of value in addressing longstanding problems in this field. For example, mixed-metal mutant hemoglobins, in which amino acids between porphyrins have been changed from aliphatic to aromatic and vice versa, are being used to assess the role of electron transfer pathways and of hole superexchange in long-range electron transfer. Finally, a more complete picture of the role of energetics in long-range ET is being realized through expanded metal substitution and ligand variation. By altering the protein environment, the solvent, the temperature, and the ET sites themselves we hope to greatly increase our understanding of this important biological process.

4.2 [MCcP, Cc] Complexes

In the [MC*c*P, C*c*] system, both the photoinitiated quenching process and the thermal ET processes are dominated by conformational interconversions. Low-temperature measurements of triplet-state quenching rates have disclosed a novel intra-complex transition that likely reflects the "freezing-out" of the conversion from a stable, but inactive, conformer to one that is thermodynamically less favored but nonetheless is active. Transient absorption studies have directly revealed the conformational control of the thermal ET reaction. This process may well prove to be the paradigm for interfacial regulation of interprotein interactions. Through a combination of metal-substitution and site-directed mutagenesis, we shall strive to unravel the structural features important for maintaining the delicate balance between conformational interconversion and ET.

Acknowledgments. This work has been supported by the NIH (Grants HL13531 and HL40453) and by NSF (Grant DMB-8907559). We wish to thank Profs. A. G. Mauk, E. Margoliash and M. A. Ratner for their inspiring collaborations.

5 References

1. (a) Gray HB, Malmström BG (1989) Biochem 28:7499 (b) Marcus RA, Sutin N (1985) Biochim Biophys Acta 811:265 (c) Williams RJP (1990) In: Johnson MK, et al (eds) Adv Chem Ser 226, Amer Chem Soc, Washington, DC, 3
2. (a) McLendon G (1988) Acct Chem Res 21:160 (b) Mayo SL, Ellis WR Jr, Crutchley RJ, Gray HB (1986) Science 233:948
3. (a) Bowler BE, Meade TJ, Mayo SL, Richards JH, Gray HB (1989) J Amer Chem Soc 111:8757 (b) Meade TJ, Gray HB, Winkler JR (1989) J Amer Chem Soc 111:4353 (c) Elias H, Chou MH, Winkler JR (1988) J Amer Chem Soc 110:429 (d) Cowan JA, Upmacis RK, Beratan DN, Onuchic JN, Gray HB (1988) Ann NY Acad Sci 550:68 (e) Axup AW, Albin M, Mayo SL, Crutchley RJ, Gray HB (1988) J Amer Chem Soc 110:435
4. (a) Jackman MP, McGinnis J, Powls R, Salmon GA, Sykes AJ (1988) J Amer Chem Soc 110:5880 (b) Osvath P, Salmon GA, Sykes AG (1988) J Amer Chem Soc 110:7114
5. Yuan X, Songcherg S, Hawkridge FM (1990) J Amer Chem Soc 112:5380
6. (a) McLendon G, Pardue K, Bak P (1987) J Amer Chem Soc 109:7540. (b) Conklin KT, McLendon G (1988) J Amer Chem Soc 110:3345 (c) Cheung E, Taylor K, Kornblatt JA, English AM, McLendon G, Miller JR (1986) Proc Nat Acad Sci (USA) 83:1330 (d) Hazzard JT, Poulos TL, Tollin G (1987) Biochem 26:2836 (e) Tollin G, Meyer TE, Cusanovich MA (1986) Biochim Biophys Acta 853:29
7. (a) Natan MJ, Hoffman BM (1989) J Amer Chem Soc 111:6468 (b) Natan MJ, Kuila D, Baxter WW, King BC, Hawkridge FM, Hoffman BM, J Amer Chem Soc 112:4081 (c) McGourty JM, Peterson-Kennedy SE, Ruo WY, Hoffman BM (1987) Biochem 26:8302 (d) Peterson-Kennedy SE, McGourty JL, Kalweit JA, Hoffman BM (1986) J Amer Chem Soc 108:1739
8. (a) Nocek JM, Liang N, Wallin SA, Mauk AG, Hoffman BM (1990) J Amer Chem Soc 112:1623 (b) Liang N, Mauk AG, Pielak GJ, Johnson JA, Smith M, Hoffman BM (1988) Science 240:311 (c) Liang N, Pielak GJ, Mauk AG, Smith M, Hoffman BM (1987) Proc Natl Acad Sci USA 84:1249 (d) Liang N, Kang CH, Ho PS, Margoliash E, Hoffman BM (1986) J Amer Chem Soc 108:4665 (e) Ho PS, Sutoris C, Liang N, Margoliash E, Hoffman BM (1985) J Amer Chem Soc 107:1070

9. Marden MC, Hazard ES, Gibson QH (1986) Biochem 25:7591
10. (a) Hoffman BM, Ratner MR (1987) J Amer Chem Soc 109:6237 Erratum: ibid (1988) 110:8267 (b) Hoffman BM, Ratner MA, Wallin SA (1990) Adv Chem Ser 226, Johnson MK, et al (eds) Amer Chem Soc, Washington, DC, p 125 (c) Brunschwig BS, Sutin N (1989) J Amer Chem Soc 111:7454. (d) Sutin N, Brunschwig BS (1990) Adv Chem Ser 226, Johnson MK, et al (eds) Amer Chem Soc, Washington, DC, 65
11. Gingrich DJ, Hoffman BM, manuscript in preparation
12. Kuila D, Natan MJ, Rogers P, Gingrich DJ, Baxter WW, Arnone A, Hoffman BM (1991) J Amer Chem Soc (submitted)
13. Arnone A, Rogers P, Blough NV, McGourty JL, Hoffman BM (1986) J Mol Biol 188:693
14. Gingrich DJ, Nocek JM, Natan MJ, Hoffman BM (1987) J Amer Chem Soc 109:7533
15. Antonini E, Brunori M (1971) Hemoglobin and Myoglobin in Their Reactions with Ligands, North Holland Publishing Co., Amsterdam, Netherlands
16. Moffat JK, Deatherage JF, Seybert DW (1979) Science 206:1035
17. (a) Cox RP, Holloway MR (1977) Eur J Biochem 74:575 (b) Olivas E, deWaal DJA, Wilkins RG (1977) J Biol Chem 252:4038
18. Kuila D, Baxter WW, Natan MJ, Hoffman BM (1991) J Phys Chem (in press)
19. Poulos TL, Kraut J (1980) J Biol Chem 255:10322
20. Northrup SH, Boles JO, Reynolds JCL (1988) Science 241:67
21. (a) McLendon G, Miller JR (1985) J Amer Chem Soc 107:7811 (b) Mauk MR, Mauk AG, Weber PC, Matthew JB (1986) Biochem 25:7085 (c) Salemme FR (1976) J Mol Biol 102:563
22. Peerey LM, Kostic NM (1989) Biochem 28:1861
23. Weber PC, Tollin G (1985) J Biol Chem 260:5568
24. (a) Simolo KP, McLendon GL, Mauk MR, Mauk AG (1984) J Amer Chem Soc 106:5012 (b) Poulos TL, Mauk AG (1983) J Biol Chem 258:7369 (c) Mauk MR, Mauk AG (1982) Biochem 21:4730
25. Liang N, Thesis, Northwestern University
26. Unpublished results
27. Wallin SA, Netzel TL, Hoffman BM (1991) J Amer Chem Soc (in press)
28. (a) Marshall DB (1989) Anal Chem 61:660 (b) Siebrand W, Wildman TA (1986) Accts Chem Res 19:238 (c) Alcala JR, Gratton E, Prendergast FG (1987) Biophys J 51:587 (d) As shown by Marshall, it is not possible to differentiate between fits to Eq. (4) and Eq. (5) under normal circumstances. (e) Lindsey CP, Patterson GD (1980) J Chem Phys 73:3348
29. (a) Frauenfelder H, Young RD (1986) Comments Mol Cell Biophys 3:347 (b) Ansari A, Berendzen J, Braunstein D, Cowen BR, Frauenfelder J, Hong MK, Iben IET, Johnson JB, Ormos P, Sauke TB, Scholl R, Schulte A, Steinbach PJ, Vittitow J, Young RD (1987) Biophys Chem 26:337 (c) Austin RH, Beeson KW, Eisenstein L, Frauenfelder H, Gunsalus IC (1975) Biochem 14:5355
30. Nocek JM, Stemp EDA, Finnegan MG, Koshy TI, Johnson MK, Margoliash E, Mauk AG, Smith M, Hoffman BM, J Amer Chem Soc (submitted)
31. Hazzard JT, McLendon G, Cusanovich MA, Tollin G (1988) Biochem Biophys Res Commun 151:429.
32. (a) Published glassing points for EG cryosolvents are: 15%, $\sim$ 266 K; 30%, 256 K; 45%, $\sim$ 238 K; 60%, 204 K (Ref. [32b]). However, the addition of buffer and protein to these solvents lowers the freezing transition by 5–10° (b) Douzou P (1977) Cryobiochemistry: An Introduction, Academic Press, New York

Long-Range Electron Transfer in Metalloproteins

Michael J. Therien, Jeffrey Chang, Adrienne L. Raphael, Bruce E. Bowler, and Harry B. Gray

Arthur Amos Noyes Laboratory, California Institute of Technology, Pasadena, California 91125, USA

Studies of the dependence of the rate of electron transfer (ET) on donor–acceptor separation and reaction driving force ($-\Delta G^{\circ}$) in ruthenated derivatives of myoglobin (Mb) and cytochrome *c* (cyt *c*) have been performed. Measurements of $MP^* \rightarrow Ru^{3+}$ ET (MP^* is an electronically excited metalloporphyrin) at four fixed distances (His48, 81, 116, 12) in modified sperm whale Mb have defined limits for long-range heme:Ru electronic couplings. The $\ln k_{ET}$ values decrease exponentially with the donor to acceptor edge–edge distance; from a plot of $\ln k_{ET}$ *vs.* d, a β of ~ 0.9 Å^{-1} is obtained. The $^3ZnP^* \rightarrow Ru^{3+}$ and $Ru^{2+} \rightarrow ZnP^+$ rates in Ru(His33)Zncyt *c* derivatives have been analyzed in terms of $\lambda \sim 1.2$ eV and electronic couplings of roughly 0.1 cm^{-1}. The electronic couplings for Ru(His39)Zncyt *c* reactions (~ 0.2 cm^{-1}) are twice as large as those for Ru(His33)Zncyt *c*, but the experimentally derived electronic coupling in Ru(His62)Zncyt *c* (~ 0.01 cm^{-1}) is much smaller than the couplings in either Ru(His33)Zncyt *c* or Ru(His39)Zncyt *c*. The donor–acceptor electronic coupling order, His39 > His33 $\gg$ His62, accords well with the effective lengths of calculated ET pathways for these derivatives. The evidence suggests that shortcuts in ET pathways that involve hydrogen bonds lead to stronger couplings than those requiring through-space jumps.

1 Introduction

Electron transfer (ET) is a key reaction in biological processes such as photosynthesis and respiration [1]. Photosynthetic and respiratory chain redox proteins contain one or more redox-active prosthetic groups, which may be metal complexes or organic species. Since it is known from crystal structure analyses that the prosthetic groups often are located in the protein interior, it is likely that ET in protein–protein complexes will occur over large molecular distances (≥ 10 Å) [2–4].

Much is known about the factors that influence the rates of long-range ET reactions in organic and inorganic donor (spacer) acceptor molecules containing covalently coupled spacers (saturated or unsaturated hydrocarbons, peptides) [3]. The area of interest to us here centers around the important special case in which a protein itself serves as the spacer. Donor (protein-spacer) acceptor molecules can be constructed by attaching a redox-active species to a specific amino acid at the surface of a structurally characterized redox protein [5]. Surface modification of a protein is expected to be nonperturbative [6–8], so it can be assumed that the structure of the modified protein is the same as that of the native protein. Hence, the distance and the intervening medium involved in electron transfer between the native and synthetic protein redox sites are known. Altering the site of attachment allows both the distance and the intervening medium for ET to be varied. Orientation effects on electron transfer can be examined by varying the site of attachment of the surface redox complex relative to the protein redox center, while maintaining the same site–site separation. Changing the modification reagent also permits driving-force effects on the rate of the reaction to be studied. Modified proteins thus are extremely versatile molecules for studying biologically relevant electron transfer [1, 4, 9–12].

2 Ruthenium-Modified Proteins

The standard ruthenium modification procedure involves the reaction of aquopentaammineruthenium(II) (a_5Ru^{2+}) with the imidazole of a surface histidine of a protein [5, 13, 14]. The a_5Ru(histidine)-modified proteins are stable in both the Ru(II) and the Ru(III) oxidation states and, although a_5Ru^{2+} slowly dissociates from surface histidines [15], the a_5Ru^{3+} complex stays attached for at least two months under appropriate conditions [16].

Several methods have been used to determine the number and position of metal atoms affixed to the protein surface. The number of metal atoms is commonly determined by atomic absorption analysis [16] or by inductively coupled plasma (ICP) atomic emission analysis [15]. Under favorable circumstances, the metal ratios in modified derivatives can be determined by UV-vis spectroscopy [13]. Another method for quantifying ruthenium attached to histidines is to compare the reactions of the native and modified proteins with diethyl pyrocarbonate [15], which is a histidine specific reagent.

Two methods have been used extensively to determine the site of modification, peptide mapping [5, 14, 16–20] and NMR spectroscopy [5, 15, 17, 18, 20]. Peptide mapping typically involves cleavage of the protein with a protease, followed by HPLC purification of the fragments. A change in mobility indicates a fragment that has a metal complex attached to it. The fragment is then identified by amino acid analysis. The presence of the metal complex often is determined by UV-vis spectroscopy [5, 16, 17, 19]. The attachment of a Ru(III) complex to a histidine causes paramagnetic shifting and broadening of the C-2 and C-4 protons in the ^{1}H NMR spectrum of a protein. This results in loss of these histidine protons from the aromatic chemical shift region and their reappearance upfield of TMS [18, 19]. However, for proteins containing multiple surface histidines, such as myoglobin [18] and *Candida krusei* cytochrome *c* [19], ambiguities can remain in the assignment of which histidine is modified based solely on the NMR spectrum.

A variety of physical methods has been used to ascertain whether or not surface ruthenation alters the structure of a protein. UV-vis, CD, EPR, and resonance Raman spectroscopies have demonstrated that myoglobin [14, 18], cytochrome *c* [5, 16, 19, 21], and azurin [13] are not perturbed structurally by the attachment of a ruthenium complex to a surface histidine. The reduction potential of the metal redox center of a protein and its temperature dependence are indicators of protein structure as well. Cyclic voltammetry [5, 13], differential pulse polarography [14, 21], and spectroelectrochemistry [12, 14, 22] are commonly used for the determination of the ruthenium and protein redox center potentials in modified proteins.

High-resolution structural data for modified proteins can be obtained from NMR and X-ray crystallographic studies. Extensive NMR studies of [a_5Ru]$_3$myoglobin [23] and *Chromatium gracile* high-potential iron protein ruthenated at either His20 or His42 [24] demonstrate at most only slight chemical-shift changes. The X-ray crystal structure of a_5Ru(His48)Mb also has been determined; it is virtually identical with that of the native protein [25]. These observations confirm that surface ruthenation does not significantly alter the conformation of a protein.

3 Fe/Ru ET Kinetics

Flash photolysis and pulse radiolysis techniques have been developed to study Fe $\leftrightarrows$ Ru ET in Ru-modified proteins [21, 26, 27]. A method that allows study of electron transfer from a surface a_5Ru(III)(histidine) to a protein redox center is outlined in the Scheme [21]. The ET reaction is initiated by photogenerated

$Ru(bpy)_3^{2+*}$, which rapidly reduces the surface ruthenium. The $Ru(bpy)_3^{3+}$ is scavenged by EDTA before it can back react with $a_5Ru(II)$(histidine).

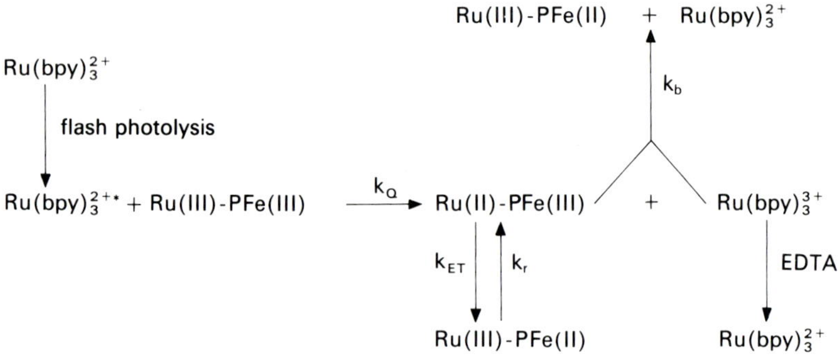

Scheme: Flash photolysis method for studying $Ru^{II} \rightarrow Fe^{III}$ ET.

Electron transfer to the protein metal center is monitored spectroscopically. In the case of a heme (FeP), a fast increase in absorbance due to direct reduction of Fe(III)P by $Ru(bpy)_3^{2+*}$ is followed by a slower increase in absorbance due to reduction of Fe(III)P by the Ru(II) on the protein surface. Control flash experiments with unmodified proteins show only the fast initial increase in absorbance due to Fe(III)P reduction by $Ru(bpy)_3^{2+*}$. Such control experiments demonstrate for horse heart cytochrome *c* [21], azurin [28], and sperm whale myoglobin [14] that slow reduction of the heme by the EDTA radical produced in the scavenging step does not occur in competition with intramolecular ET. However, for *Candida krusei* cytochrome *c*, the control experiment shows evidence for slow EDTA radical reduction of the heme after initial fast reduction by $Ru(bpy)_3^{2+*}$ [19].

A method for the study of ET from a protein metal center to a surface ruthenium has been developed by Lieber [26]. In this method, $Ru(bpy)_3^{2+*}$ acts as an oxidant, selectively removing an electron from a surface $a_5Ru(II)$(histidine). A Ni/RBr scavenger system (Ni(II)hexamethyltetraaza-cyclododecane and an alkyl bromide) oxidizes the $Ru(bpy)_3^{+}$ before it can back react with the $a_5Ru(III)$(histidine) complex. ET from the reduced protein metal center to the oxidized ruthenium can be monitored spectroscopically.

ET rates for a_5Ru(His33) cytochrome *c* and a_5Ru(His48)-myoglobin are set out in Table 1. Variable temperature kinetic studies have shown that the activation enthalpy for ET in cytochrome *c* is lower than that for myoglobin. The larger ΔH^* for myoglobin was attributed to the loss of the axially ligated water upon reduction of the heme, whereas both axial ligands are retained upon reduction of cytochrome *c* [14, 21, 22].

Table 1. ET [Ru(II) → Fe or Cu acceptor] in $a_5Ru(HisX)$-Modified Proteins

Protein	X	Acceptor	d(Å)	$-\Delta G°$(eV)	$k_{ET}(s^{-1})$
Cytochrome *c* (horse heart)[a]	33	Fe(III)	11.7	0.18	30 ± 3 (53 ± 2)[b]
Cytochrome *c* (*Saccharomyces cerevisiae*)[c]	62	Fe(III)	15.6	0.20	1.7 ± 1
Cytochrome c_{551} (*Pseudomonas stutzeri*)[d]	47	Fe(III)	7.9	0.20	13 ± 2
Myoglobin (sperm whale)[e]	48	Fe(III)	12.7	−0.02	0.019 ± 0.002
Azurin (*Pseudomonas aeruginosa*)[f]	83	Cu(II)	11.8	0.24	1.9 ± 0.4 (2.5 ± 0.8)[g]
Plastocyanin (*Anabaena variabilis*)[g]	59	Cu(II)	11.9	0.26	< 0.08
Plastocyanin (*Scenedesmus obliquus*)[g]	59	Cu(II)	10–12	0.29	< 0.26
Stellacyanin (*Rhus vernicifera*)[h]	32/100	Cu(II)	—	0.104	0.05
HIPIP (*Chromatium vinosum*)[i]	42	Fe_4S_4	7.9	0.27	18 ± 2

[a] Ref. 21; [b] Ref. 27; [c] Ref. 11; [d] Ref. 15; [e] Ref. 22; [f] Ref. 28; [g] Ref. 39; [h] Ref. 43; [i] Ref. 38.

4 Distance Dependence of ET Rates

In semiclassical ET theory, three parameters govern the reaction rates: the electronic coupling between the donor and acceptor (κ_E); the free-energy change for the reaction (ΔG°); and a parameter (λ) related to the extent of inner-shell and solvent nuclear reorganization accompanying the ET reaction [29]. Additionally, when intrinsic ET barriers are small, the dynamics of nuclear motion can limit ET rates through the frequency factor ν_N. These parameters describe the rate of electron transfer between a donor and acceptor held at a fixed distance and orientation (Eq. 1),

$$k_{ET} = \nu_N \kappa_E \exp\left[\frac{-(\Delta G^\circ + \lambda)^2}{4\lambda RT}\right] \tag{1}$$

where R is the gas constant and T is the absolute temperature.

It is commonly assumed that for long-range ET the electronic factor (κ_E) (Eq. 1) will decrease exponentially with the donor–acceptor edge–edge distance (d) [29]:

$$\kappa_E = \kappa_E^\circ \exp[-\beta(d - d_0)] \tag{2}$$

The closest contact distance, d_0, is normally taken to be 3 Å (van der Waals contact of the edges of the donor and acceptor). The value of β is a measure of the effectiveness of the intervening medium in coupling the donor and acceptor [3, 30].

The distance dependence of ET rates in proteins has been studied in ruthenated sperm whale myoglobin, where there are four surface histidines at different edge–edge distances from the metal porphyrin (Fig. 1). Electron transfer from the Fe(II) heme to $a_5Ru(III)$ bonded to any one of the three distant histidines (His81, His116, His12) is very slow, owing to the low driving force (0.02 eV) for the reaction. Much faster rates are observed in derivatives in which the iron porphyrin is replaced by zinc mesoporphyrin IX diacid (ZnP) or magnesium mesoporphyrin IX (diacid or diester); ET in these derivatives is from an electronically excited metalloporphyrin (^{3}MP*) to a a_5Ru(III)(histidine) on the protein surface ($-\Delta G^\circ \sim 0.8$ eV) [31–33]. The ET rates scale with the edge–edge distance, d, giving β values in the 0.8–0.9 Å^{-1} range for the ruthenated myoglobins (Fig. 2). It is of interest that the rates of related ET reactions in a_5Ru(His33)cytochrome *c* [34] and Zn, Fe-hybrid hemoglobin [35] fall near the lines in Fig. 2. It also has been found that the $Ru(bpy)_3^{2+*} \rightarrow Fe(III)$ and $Fe(II) \rightarrow Ru(bpy)_3^{3+}$ ET rates in $Ru(bpy)_3$(lysine) derivatives of horse heart cytochrome *c* scale roughly with edge–edge distance [36].

One notable observation is that $^3ZnP^* \rightarrow a_5$Ru(III)(His12) ET is faster than expected based on edge–edge distance. Since Trp14 lies directly between His12 and the porphyrin, it may play a role in enhancing the ET rate [31–33]. One possibility is that β is approximately 0.1 Å^{-1} less for a_5Ru(His12)Mb than for

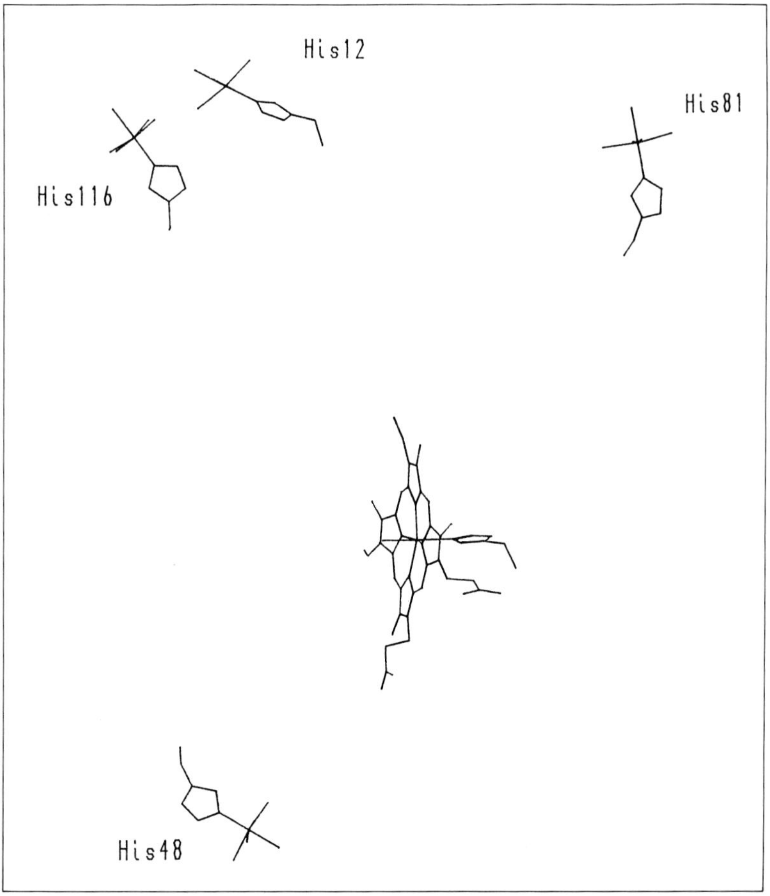

Fig. 1. Relative positions of the surface histidines (12, 48, 81, and 116) and the heme with its axial histidine in ruthenated sperm whale myoglobin. The edge–edge ET distances are 12.7 (His48), 19.1 (His81), 20.1 (His116), and 22.1 Å (His12) [12]

the other myoglobin derivatives (Fig. 2) [33]. However, the rate of ET from a_5Ru(II)(His62) to Fe(III) in a yeast *iso*-1-cytochrome *c* mutant (produced by site-directed mutagenesis) apparently is not enhanced despite having the polarizable Trp59 and Met64 side chains in the intervening medium [11].

Several unusually slow ET rates at short edge–edge separation distances have been reported for ruthenium-modified proteins [15, 37–40]. A case in point is a_5Ru(His47)cytochrome c_{551}, where the Ru(II) to Fe(III) ET rate is 13 s^{-1} [15]. The driving force for ET is the same as for horse heart cytochrome *c* modified at His33, and yet ET is slower (13 versus 30 s^{-1}) at a 3.8 Å shorter distance (Fig. 3). The most striking examples have come from experiments involving plastocyanins modified with a_5Ru^{3+} at the exposed His59 (Fig. 4) [39, 40]. The ET rates in these proteins are much lower than would be expected

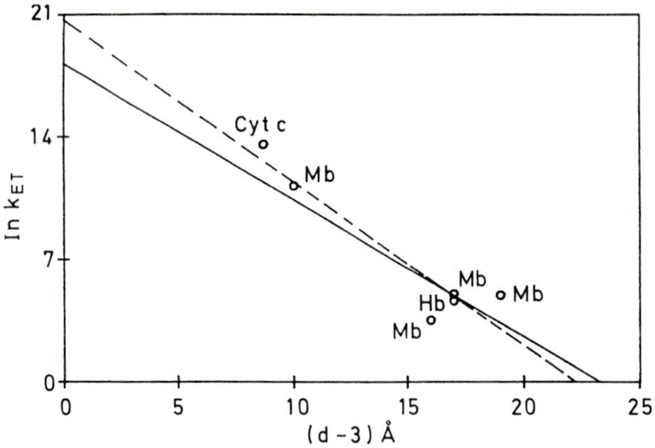

Fig. 2. Ln k_{ET} *vs.* distance plot for $^3ZnP^* \rightarrow a_5Ru(III)(HisX)$ ET reactions in ruthenated sperm whale myoglobin [31, 33]: Mb, X = 48, 81, 116, 12; also shown are points for cyt *c* [$^3ZnP^* \rightarrow a_5Ru(III)(His33)$] [34] and Zn,Fe–Hb [$^3ZnP^* \rightarrow Fe(III)P$] [35]. *Solid line*, $\beta = 0.78$ Å^{-1}; *dashed line*, $\beta = 0.93$ Å^{-1}

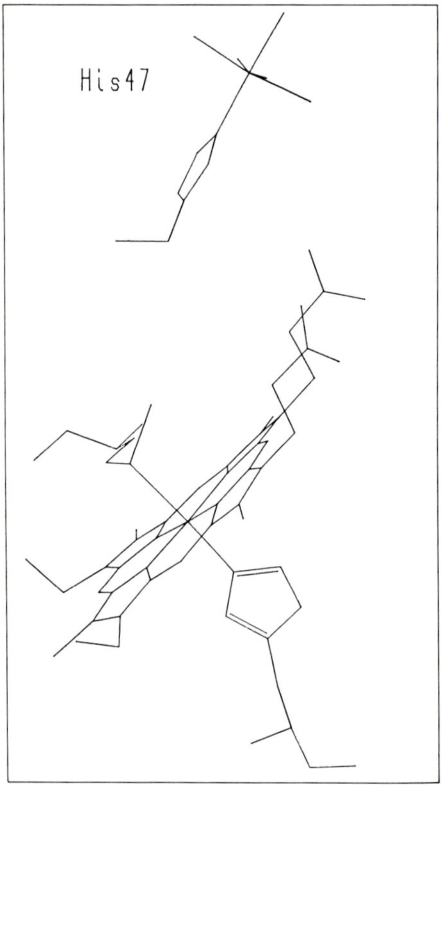

Fig. 3. View of the heme and $a_5Ru(His47)$ centers in ruthenated *Pseudomonas stutzeri* cytochrome c_{551}. The edge–edge distance is 7.9 Å [15]

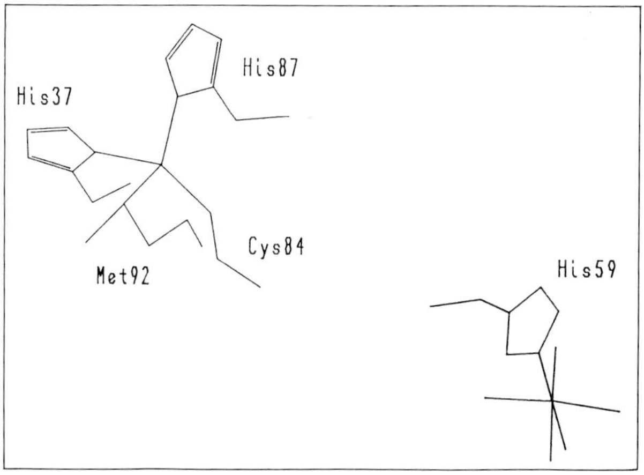

Fig. 4. View of the blue copper and a_5Ru(His59) centers in ruthenated *Anabaena variabilis* plastocyanin. The edge–edge distance is 11.9 Å [39]

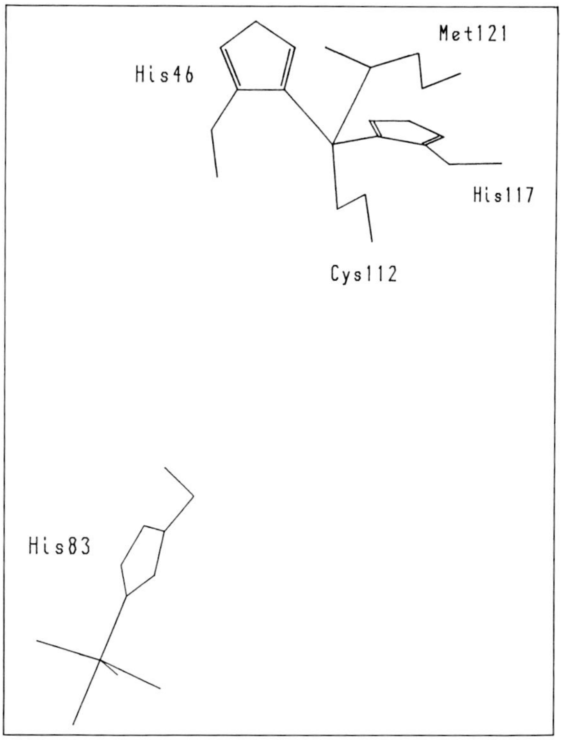

Fig. 5. View of the blue copper and a_5Ru(His83) centers in ruthenated *Pseudomonas aeruginosa* azurin [28]

for edge–edge distances in the 10–12 Å range (Table 1). The inner-sphere reorganization energy for blue copper proteins should be small, since the geometry at the copper site is intermediate between Cu(I) and Cu(II) [9, 41]. The outer-sphere reorganization energy is expected to be small as well, since the Cu site is buried (and no solvent molecules are proximal to the metal). In addition, the ruthenium-labeled histidine is thought to be similar in structure to that of the modified histidines in other proteins. Thus it is likely that the slow rates are attributable to very poor Cu–Ru electronic coupling, although it will require additional experimental and theoretical work to settle the matter.

Intramolecular Ru(II) to Cu(II) ET rates have been measured in two other blue copper proteins, stellacyanin [42, 43] and azurin [9, 13, 28]. *Pseudomonas aeruginosa* azurin has been ruthenated at His83 [13] (Fig. 5). The intramolecular Ru(II) to Cu(II) ET rate of 1.9 s^{-1} was found to be independent of temperature [28]. The Cu reorganization enthalpy was estimated to be < 7 kcal/mol [13, 28], a value confirming that blue copper is structured for efficient ET. Again, a blue copper ET rate is low in comparison with heme protein rates over similar distances (at similar driving forces) (Table 1).

The high potential iron–sulfur protein (HIPIP) from *Chromatium vinosum* has been modified with a_5Ru at His42. The $a_5Ru(III)(His42)$ protein has a Fe_4S_4 reduction potential of 350 mV. The estimated edge–edge distance from the Fe_4S_4 cluster to $a_5Ru(III)(His42)$ is only 7.9 Å, and the through-bond pathway also is relatively short [37, 38]. At a driving force of ~ 0.27 eV, the ET rate (Ru to the Fe_4S_4 cluster) was found to be 18 s^{-1}. The ET rate is ~ 2 orders of

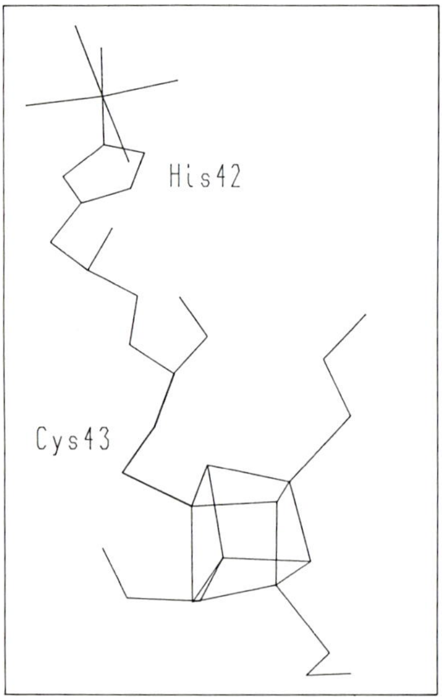

Fig. 6. View of the Fe_4S_4 and $a_5Ru(His42)$ centers in *Chromatium vinosum* HIPIP. The edge–edge distance is 7.9 Å [38]

magnitude *lower* than expected by comparison with cytochrome *c* ET rates with $\beta = 0.9$ Å^{-1}. ET in HIPIP also is disturbingly slow in view of the short through-peptide distance; His42 is adjacent to Cys43, which is bonded directly to the iron–sulfur cluster (Fig. 6).

5 Driving-Force Dependence of ET Rates

The ET driving force in modified proteins is generally varied in one of two ways. In one method, the reduction potential of ruthenium is changed by replacing an ammine ligand with a substituted pyridine [44]; the second method exploits the fact that the energy of $^3MP^*$ is a strong function of M [33]. A self-consistent method of determining $E(MP^+/^3MP^*)$ has been utilized in estimating $\Delta G°$ values for photoinduced ET in proteins [45].

Driving-force dependences of ET rates have been determined for both His33-modified horse heart cytochrome *c* [34, 44] and His48-modified sperm whale myoglobin [18, 33, 46]. For horse heart cytochrome *c*, fits of $\log(k_{ET})$ versus $-\Delta G°$ to Eq. 1 (in which $\nu_N \kappa_E^\circ = 10^{13}\ s^{-1}$ was assumed) gave either $\beta = 1.8$ Å^{-1} for $\lambda = 1.2$ eV or $\lambda = 1.85$ eV for $\beta = 1.2$ Å^{-1} [34]. Reasonable values for λ or β gave unreasonable values for the other parameter. Similar evaluation of driving-force data for myoglobin, setting $\nu_N \kappa_E^\circ = 10^{11}$–$10^{13}\ s^{-1}$ and $\beta = 0.91$ Å^{-1}, gave $\lambda = 1.9$–2.45 eV [46].

In more detailed analyses, $\nu_N \kappa_E^\circ$ has been varied freely so as to achieve the best fit of Eq. 1 to the $\log(k_{ET})$ versus $-\Delta G°$ data. With a series of metalloporphyrins substituted into myoglobin, a driving-force study of photoinduced ET ($^3MP^* \rightarrow a_5Ru(III)(His48)$) was made (Table 2) [33]. Analysis of the $\log(k_{ET})/-\Delta G°$ data gave $\lambda = 1.3 \pm 0.3$ eV and $\nu_N \kappa_E^\circ = 10^7$–$10^9\ s^{-1}$. The maximum ET rate, $\nu_N \kappa_E$, is proportional to the square of the matrix element, H_{AB}, that describes the electronic coupling between $a_5Ru(His48)$ and the metalloporphyrin. In the nonadiabatic limit ($H_{AB} \ll k_B T$), the proportionality constant is $(\pi/\hbar 2\lambda RT)^{\frac{1}{2}}$ [29], which gives $H_{AB} = 0.006\ cm^{-1}$ for the $a_5RuMb(MP)$ system ($d = 12.7$ Å); extrapolation to $d_0 = 3$ Å gave $0.3\ cm^{-1}$ [33], indicative of

Table 2. Donor to Ru(III) ET in $a_5Ru(III)(His48)Mb(^3MP^*)$[a]

Donor	$-\Delta G°$(eV)	$k_{ET}(s^{-1})$
$^3ZnP^*$	0.88	$7.0 \pm 1.0 \times 10^4$
$^3MgP^*$	0.87	$5.7 \pm 0.4 \times 10^4$
$^3CdP^*$	0.85	$6.3 \pm 0.5 \times 10^4$
$^3PtP^*$	0.73	$1.2 \pm 0.1 \times 10^4$
$^3PdP^*$	0.70	$0.91 \pm 0.1 \times 10^4$
$^3H_2P^*$	0.53	$0.076 \pm 0.006 \times 10^4$
Fe(II)	0.02	$0.041 \pm 0.003 \times 10^{-2}$

[a] Ref. 33.

Table 3. ET in Ru(His33)cytochrome c[a]

Reaction	$-\Delta G^{\circ}$(eV)	$k_{ET}(s^{-1})$	λ(eV)	$\nu_N\kappa_E(s^{-1})$	$H_{AB}(cm^{-1})$
Charge separation					
$^3ZnP^* \rightarrow a_5Ru(III)$	0.70	7.7×10^5			
$^3ZnP^* \rightarrow a_4(py)Ru(III)$	0.97	3.3×10^6	1.15	3.9×10^6	0.13
$^3ZnP^* \rightarrow a_4(isn)Ru(III)$	1.05	2.9×10^6			
Charge recombination					
$a_4(isn)Ru(II) \rightarrow ZnP^+$	0.66	2.0×10^5			
$a_4(py)Ru(II) \rightarrow ZnP^+$	0.74	3.5×10^5	1.24	2.5×10^6	0.10
$a_5Ru(II) \rightarrow ZnP^+$	1.01	1.6×10^6			
Ru/Fe ET					
$a_5Ru(II) \rightarrow Fe(III)P$	0.18	3.0×10^1	1.20	2.0×10^5	0.03

[a] Ref. 44.

a highly nonadiabatic ET reaction at close contact. For horse heart cytochrome c, an extensive driving-force study using both photoinduced ET and charge-recombination reactions has been completed (Table 3) [44]. It was found that the plots of $\log(k_{ET})$ versus $-\Delta G^{\circ}$ for photoinduced ET, charge recombination, and Ru(II) to Fe(III) ET were best fit separately to Eq. 1. The values of λ range from 1.15 to 1.25 eV, which are slightly smaller than the λ for myoglobin. H_{AB} values derived from $\nu_N\kappa_E$ are 0.03 (FeP) and 0.12–0.13 cm^{-1} (ZnP) for the ET reactions. If $\beta = 0.9$ Å^{-1} is assumed, then extrapolation to close contact gives $H_{AB}(d_0 = 3$ Å$) = 1.5$–$6.5\ cm^{-1}$. Thus, ET from His33 of horse heart cytochrome c also would be nonadiabatic at close contact; however, the electronic coupling is better for ET between His33 and ZnP/FeP in cytochrome c than from $^3MP^*$ to His48 in myoglobin.

It is interesting that λ values in modified proteins are comparable to those values (0.9–1.25 eV) extracted from studies of organic donor–acceptor complexes [3, 30]. The electronic couplings estimated for close contact, however, are dramatically smaller than typical values ($\sim$300 to 1900 cm^{-1}) obtained from experiments on donor(spacer)acceptor molecules [3]. The intrinsic (close contact) electronic coupling also varies significantly from protein to protein.

In recent work, the kinetics of long-range electron transfer have been measured in a_4LRu(His39) derivatives ($L = NH_3$, pyridine, isonicotinamide) of Zn-substituted *Candida krusei* cytochrome c [47] and a_4LRu(His62) derivatives ($L = NH_3$, pyridine) of Zn-substituted *Saccharomyces cerevisiae* cytochrome c [48]. ET rates and activation parameters are set out in Table 4. The rates of both excited–state electron transfer and thermal recombination are approximately three times greater in Ru(His39)Zn-cytochrome c than the rates of the corresponding reactions in Ru(His33)Zn-cytochrome c, but analogous ET reactions in Ru(His62)Zn-cytochrome c are roughly two orders of magnitude slower than in the His33-modified protein.

Plots of the Ru(His33)Zncyt c, Ru(His39)Zncyt c, and the Ru(His62)Zncyt c data are shown in Fig. 7. Although the reorganization parameter λ is nearly the same for the ET reactions in the three proteins ($\sim$ 1.2 eV), the H_{AB} value

Table 4. ET in Ru(His39) and Ru(His62) Cytochromes *c*.

Ru(His39)Zncyt *c*[a]	$-\Delta G°$(eV)	$k_{ET}(s^{-1})$
a_4(isn)Ru(II) → ZnP^+	0.66	6.5×10^5
$^3ZnP^*$ → a_5Ru(III)	0.70	1.5×10^6
a_4(py)Ru(II) → ZnP^+	0.74	1.5×10^6
$^3ZnP^*$ → a_4(py)Ru(III)	0.97	8.9×10^6
a_5Ru(II) → ZnP^+	1.01	5.7×10^6
$^3ZnP^*$ → a_4(isn)Ru(III)	1.05	1.0×10^7
Ru(His62)Zncyt *c*[b]	**$-\Delta G°$(eV)**	**$k_{ET}(s^{-1})$**
$^3ZnP^*$ → a_5Ru(III)	0.70	6.5×10^3
a_4(py)Ru(II) → ZnP^+	0.74	2.6×10^3
$^3ZnP^*$ → a_4(py)Ru(III)	0.97	2.7×10^4
a_5Ru(II) → ZnP^+	1.01	2.0×10^4

[a] Ref. 47; [b] Ref. 48.

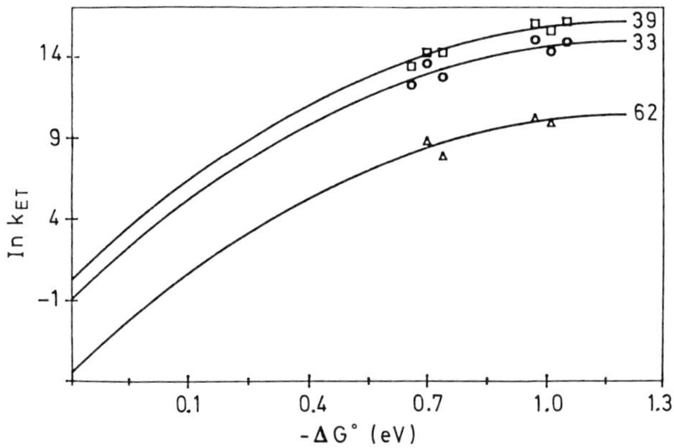

Fig. 7. Plots of $\ln k_{ET}$ *vs.* $-\Delta G°$ for the Ru(HisX)Zncytochrome *c* ET reactions. *Boxes* (His39), *circles* (His33), and *triangles* (His62) represent the experimental data (Tables 3 and 4). *Solid lines* are the best fits to Eq. 1: $\lambda = 1.21$, $H_{AB} = 0.21$ (His39); $\lambda = 1.20$; $H_{AB} = 0.12$ (His33); $\lambda = 1.20$ eV; $H_{AB} = 0.01$ cm^{-1} (His62) [48]

for Ru(His39)Zncyt *c* (0.21 cm^{-1}) is almost twice as large as that for Ru(His33)Zncyt *c* (0.12 cm^{-1}) and over twenty times larger than the H_{AB} for Ru(His62)Zncyt *c* (0.01 cm^{-1}).

Both the electronic coupling matrix element and the outer-sphere component of the nuclear reorientation parameter are thought to vary with donor–acceptor separation and orientation [29, 49]. It has been shown in studies of Os and Ru-ammines bridged by polyproline spacers that the distance dependence of λ can be greater than that of H_{AB} [50]. Dielectric continuum models of solvent reorganization predict that λ_O will increase with donor–acceptor separation. Models that describe charge transfer within low-dielectric spheres or ellipsoids embedded in dielectric continua exhibit a dependence upon ET distance as well as upon the positions of the redox sites inside the sphere or ellipsoid [51]. Modeling the Ru–Zncyt *c* systems as single spheres suggests, however, that variations in λ_O for the Ru(His33)Zncyt *c*, Ru(His39)Zncyt *c*, and Ru(His62)Zncyt *c* ET reactions will not be significant (0.57, 0.60, and 0.63 eV, respectively) [48].

Winkler has examined the maximum variation in λ_O predicted by the single-sphere continuum model [48, 52]. The cyt *c* molecule can be taken as a 34 Å diameter sphere with its metal center 5.8 Å from the origin. The Ru-ammine complex is taken as a 6 Å diameter sphere centered on the Ru atom that is assumed to be fixed 16 Å from the center of the cyt *c* sphere. The small sphere can occupy any position on the surface of the large sphere with values of the ET distance varying from 10.2 to 21.8 Å. A third sphere then encloses the two other spheres and λ_O for electron transfer between the two metals is calculated by treating the solvent as a dielectric continuum. The magnitude of λ_O varies from 0.38 to 0.63 eV almost linearly as the ET distance increases from 10.2 to 21.8 Å. The total variation of 0.25 eV is only slightly larger than the uncertainty range in the estimates of λ ($\pm$ 0.1 eV). The invariance of λ found for the ET reactions of the three modified cytochromes, is, therefore, consistent with theoretical considerations.

6 ET Pathways

In covalently coupled donor(spacer)acceptor molecules, the evidence now available suggests that ET rates depend upon the number of covalent bonds separating the donor and the acceptor, rather than upon their direct separation distance [30, 53–58]. Several investigators have begun to examine potential ET pathways in proteins [11, 33, 59–63]; interestingly, the through-peptide routes generally involve so many bonds that they cannot possibly account for the observed rates [33, 44]. Beratan and Onuchic have developed a simple model to describe the contribution of the polypeptide bridge to the electronic coupling in long-range ET in protein systems [59, 62]. The essence of the model is that H_{AB} decreases from its maximal value (at van der Waals contact of donor and acceptor) by a constant factor for each covalent bond in the ET pathway. Ionic contacts (H-bonds and salt bridges) and through-space jumps decrease H_{AB} by somewhat larger factors. It is encouraging that the Beratan-Onuchic pathway method is fully consistent with the finding that the electronic couplings in Ru(His39)Zncyt *c* and Ru(His33)Zncyt *c* are much larger than that in Ru(His62)Zncyt *c* (Fig. 7).

The shortest direct distances between the porphyrin and imidazole carbon atom of His33 (13.2 Å), His39 (13.0 Å), and His62 (15.6 Å) are much too long for any direct donor–acceptor interaction [59]. Since virtually the same donor and acceptor electronic states are found in the three proteins, the differences in H_{AB} must arise from the manner in which the intervening atoms couple the two states. If a homogeneous medium of constant tunneling-barrier height separated the donor and the acceptor in the three systems, then H_{AB} would depend primarily on the edge–edge distance and would be nearly the same for Ru(His33)Zncyt *c* and Ru(His39)Zncyt *c* and decrease only slightly for Ru(His62)Zncyt *c* relative to Ru(His39)Zncyt *c*. Clearly, this prediction is not borne out by experiment. The polypeptide medium separating the Ru-ammine and metalloporphyrin sites in these proteins is decidedly inhomogeneous, and must be responsible for the differential electronic coupling in these ruthenium-modified Zncyt *c* derivatives.

The ET pathways in Ru(His33)Zncyt *c* and Ru(His39)Zncyt *c* generated by applying the Beratan-Onuchic criteria [59, 62] are shown in Fig. 8. The best pathway from His33 to the metalloporphyrin is a 15-bond route to the Zn-atom through His18 that includes a 1.85 Å hydrogen bond between the Pro30 carboxyl oxygen and the proton on the His18 nitrogen. The shortest pathway from His39 is a 12-bond route that includes a 2.4 Å H-bond between the α-amino hydrogen atom of Gly41 and the carboxyl oxygen of a propionate side chain on the porphyrin. The key difference between these two pathways is the number of covalent bonds: The His39 pathway is built from 11 covalent bonds and 1H-bond while the His33 pathway has 15 covalent bonds and 1 H-bond.

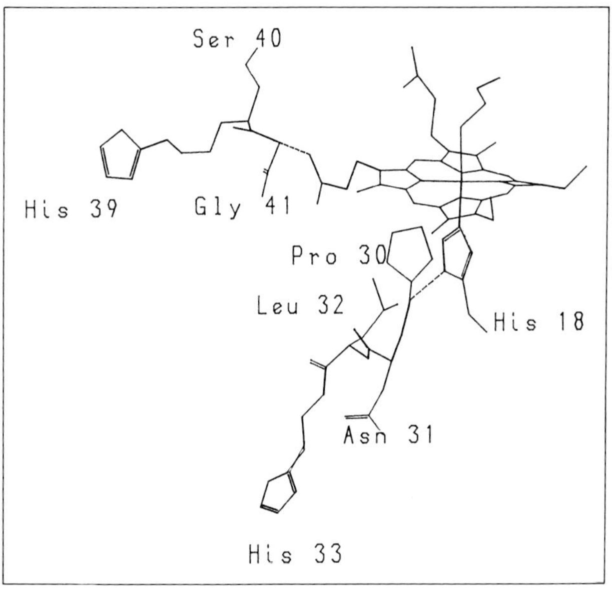

Fig. 8. ET pathways from His33 and His39 to the heme in cytochrome *c*. Edge–edge distances are as follows: His39 to the heme, 13.0; His33 to His18, 11.7; His33 to the heme 13.2 Å [47]

Hence, the experimental observation that the electronic coupling is stronger in the His39 derivative than in the His33-modified protein (even though the edge–edge distances in the two modified proteins are roughly the same) is consistent with the Beratan-Onuchic pathway analysis.

Two tunneling pathways for Ru(His62)Zncyt *c* that emerge from the analysis are shown in Fig. 9. One is a 17-bond route with 14 covalent bonds and 3 H-bonds (the third of which connects the Trp59 nitrogen atom to the carbonyl oxygen of a heme propionate side chain); the other is a 13-bond route with 12 covalent bonds and a through-space interaction between the sulfur atom of Met64 and the heme edge. The sharply lower electronic coupling in the His62 protein relative to both the His33 and His39 systems indicates that the Met64-heme through-space interaction is a poor shortcut in the His62-Met64 ET pathway. As pathway analyses are made on other structurally engineered proteins, it will be interesting to see if examples can be found in which through-space contacts are much better shortcuts. It is possible that the nature of the interacting groups and their relative orientation will greatly influence the strength of the coupling.

The very weak coupling in Ru(His62)Zncyt *c* also could be accounted for by the 17-bond pathway that includes 3 H-bonds from His62 to the heme. Since the final H-bond in this case involves Trp59, we have designed a yeast cytochrome *c* mutant with a much shorter pathway that also is completed by the Trp59-heme propionate H-bond. The mutant has His in place of Leu at position 58. In our

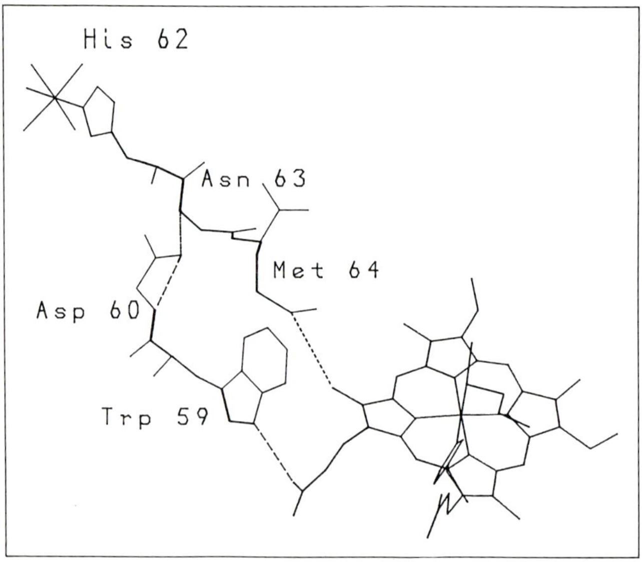

Fig. 9. ET pathways from His62 to the heme in cytochrome *c*. The His62-heme edge–edge distance is 15.6 Å [11]

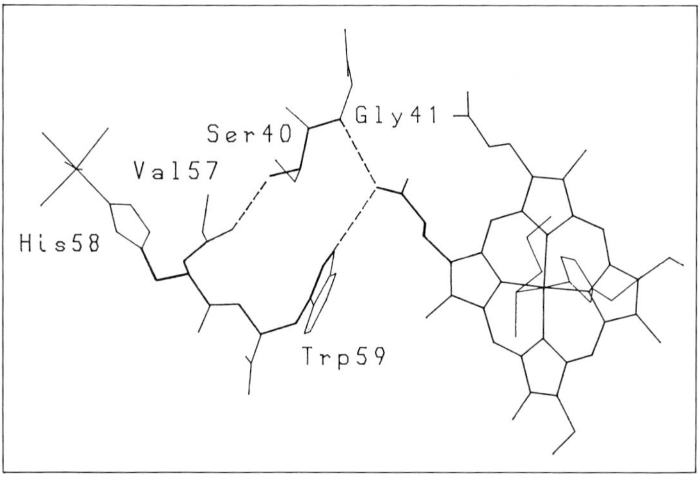

Fig. 10. ET pathways from His58 to the heme in cytochrome *c*. The His58-heme edge–edge distance is 13.4 Å [63]

laboratory, Casimiro has obtained the His58 mutant using established protocols, and a ruthenium derivative, $a_5Ru(His58)cyt$ *c*, has been characterized by standard procedures [63]. A Ru(II) → Fe(III) ET rate constant of 43 s^{-1} has been measured for $a_5Ru(His58)cyt$ *c* by flash photolysis [63].

Two ET pathways for the His58 mutant are illustrated in Fig. 10 [63]. The shorter pathway [13 covalent bonds and 1 H-bond (2.96 Å)] proceeds from His58 through Trp59 into the heme propionate group. The section of this pathway through Trp59 is the same as the final stretch of the His62 pathway (Fig. 9). The His58–Trp59 pathway is shorter than the His62–Trp59 pathway by 1 covalent bond and 2 H-bonds. The other His58 pathway shown in Fig. 10 goes through Gly41 to the heme propionate; this pathway has 12 covalent bonds and 2 H-bonds.

A calculation of the relative electronic couplings of the His58–Trp59 and His33–Pro30 pathways ($[H_{AB(\mathrm{His58-Trp59})}/H_{AB(\mathrm{His33-Pro30})}]^2 = 2.2$) predicts that the Ru(II) → Fe(III) ET rate should be greater for the His58 protein [63]. A simple edge–edge distance dependence with $\beta = 0.9$ Å^{-1} indicates that the ET rate constant will be roughly the same as that (30 s^{-1}) for $a_5Ru(His33)cyt$ *c*. Both predictions assume that the ET driving forces and reorganization energies are the same in the two reactions, an assumption that is supported by work on the His33, His39, and His62-modified proteins [48]. Both methods of estimating the variations in electronic coupling are reasonable for the His58 mutant, but the pathway approach is slightly better in the sense that it predicts correctly that the Ru(II) → Fe(III) rate in $a_5Ru(His58)cyt$ *c* will be greater than that for $a_5Ru(His33)cyt$ *c*. Perhaps more to the point, the His58–Trp59 pathway is expected to be substantially better than His62–Met64

$[(H_{AB(\text{His58}-\text{Trp59})}/H_{AB(\text{His62}-\text{Met64})}]^2 = 1.7 \times 10^2)]$ [63]. Although the rate increase falls a bit short of the prediction, the large enhancement that is observed indicates once again that a pathway that uses an H-bond to contact the heme is superior to one that is completed by a through-space interaction.

7 Protein–Protein Complexes

Another approach to studying biological ET involves the use of protein–protein complexes in which the structures of both partners are known [2, 3, 35]. Much progress has been made in this area in work on hemoglobin hybrids, $[\alpha_1(\text{Fe}), \beta_2(\text{Zn})]$ and $[\alpha_1(\text{Zn}), \beta_2(\text{Fe})]$ [35, 64, 65]. The quaternary structure adopted by the hybrids is the T form, where the metal–metal distance is 25 Å; the porphyrin planes are approximately parallel and the edge–edge distance is ~ 20 Å. In the deoxyhemoglobin structure, the subunits of closest approach are $\alpha_1\beta_2$. Since $\alpha_1\beta_1$ distances are much longer (Fe–Zn, 37 Å), the Hb is considered as two independent $\alpha_1\beta_2$ ET pairs. The rate constant for $^3\text{ZnP}^* \rightarrow \text{Fe(III)}$ ET at room temperature is about $10^2\ \text{s}^{-1}$.

The temperature dependences of the ET rates for $[\alpha_1(\text{Fe}), \beta_2(\text{Zn})]$ and $[\alpha_1(\text{Zn}), \beta_2(\text{Fe})]$Hb hybrids have been determined in the range 77–313 K [35]. In these systems, the rate constant for long-range reduction of the aquoferriheme by $^3\text{ZnP}^*$ is $k_t = k_{obsd} - k_D$, where k_D is the rate of $^3\text{ZnP}^*$ decay in reduced Hb hybrids ($\sim 55\ \text{s}^{-1}$). Decay rates (k_D) do not vary significantly over this temperature range. ET rates for the intersubunit transfer decrease rapidly from the room-temperature value ($k_t \sim 10^2\ \text{s}^{-1}$). At 140–160 K, the rates become temperature independent, and the low-temperature tunneling rate constant for $[\alpha_1(\text{Fe}), \beta_2(\text{Zn})]$ is $9\ \text{s}^{-1}$; for $[\alpha_1(\text{Zn}), \beta_2(\text{Fe})]$, it is $8\ \text{s}^{-1}$.

For $[\alpha_1(\text{Zn}), \beta_2(\text{Fe})]$, a smooth transition occurred down to the temperature-independent region, whereas for $[\alpha_1(\text{Fe}), \beta_2(\text{Zn})]$, a plateau was found between 230 and 270 K. The evidence suggests that, upon cooling, the heme of $[\alpha_1\text{Fe(III)}(\text{H}_2\text{O})\beta_2\text{Zn}]$ binds a histidine and remains low spin, whereas the $[\alpha_1\text{Zn}\beta_2\text{Fe(III)}(\text{H}_2\text{O})]$ hybrid does not change its coordination.

The reorganization energies extracted from the temperature dependence of the $^3\text{ZnP}^* \rightarrow \text{Fe(III)}$ reaction ($-\Delta G^\circ \sim 0.8$ eV) are as follows: $[\alpha_1(\text{Zn}), \beta_2(\text{Fe})]$, $\lambda = 2.06$ eV; $[\alpha_1(\text{Fe}), \beta_2(\text{Zn})]$, $\lambda = 1.79$ eV [35]. The λ values are expected to be large, since reduction of Fe(III)P results in the loss of H_2O, with change in the Fe position with respect to the porphyrin plane, as well as changes in the porphyrin geometry. It is likely, however, that the λ values obtained from analyses of the temperature dependences of ET rates are too high; substantially lower values, roughly in the 1 eV range, have been estimated from measurements of ET rates in experiments in which ΔG° was varied by changing the Fe(III) axial ligand [66].

Table 5. ET in Cyt c/Cyt b_5 Complexes[a]

Donor–Acceptor	$-\Delta G^\circ$(eV)	k_{ET} (s^{-1})
Fe(II)cyt b_5–Fe(III)cyt c	0.3	$1.6 \pm 0.7 \times 10^3$
^{3}(porph)-cyt c^*–Fe(III)cyt b_5	0.35	$5 \pm 0.5 \times 10^4$
^{3}Zncyt c^*–Fe(III)cyt b_5	0.75	$5 \pm 1 \times 10^5$
(H_2porphyrin)-cyt c^*–Fe(III)cyt b_5	1.1	$8 \pm 1 \times 10^3$

[a] Ref. 67.

Cytochrome c and cytochrome b_5 form a complex in which the two heme centers are believed to lie in parallel planes separated by a closest edge–edge distance of 8.5 Å, with a center–center separation of 16 Å [2]. Using pulse radiolysis and flash photolysis, rates were measured in various derivatives under conditions where >90% of the cytochrome b_5 was bound to cytochrome c [67, 68]. Rates were found to be independent of concentration in this regime. Fe(III)cyt b_5 decay at 428 nm and Fe(II)cyt c growth at 416 nm were coincident. ET driving forces and rates are given in Table 5. The reorganization energy was estimated to be 0.8 eV [67].

Cytochrome c and cytochrome c peroxidase (ccp) are physiological partners in the ccp reaction cycle; structural, thermodynamic, and kinetic data are available for the protein–protein interaction [69–72]. A model indicates that the cyt c/ccp complex is stabilized by specific salt bridges with the hemes in parallel planes; the Fe–Fe distance is ~ 24 Å, and the edge–edge distance is 16 Å [70].

Long-range ET rates have been measured in c/ccp complexes [73, 74]; the reactions between cyt c and Feccp [ES is the oxidation product of Fe(II)ccp and peroxide; it has two oxidizing equivalents, namely, Fe(IV)O and a protein-based organic radical cation] are given in Eq. (3):

$$\text{Mcyt}\ c + \text{ccp(ES)} \underset{k_r}{\overset{k_f}{\rightleftarrows}} [\text{Mcyt}\ c/\text{ES}] \xrightarrow{k_{obsd}} \text{Mcyt}\ c^+/\text{ccp(III)} \tag{3}$$

Electron transfer from Fe(II)cyt c to ccp(ES) proceeds with a rate of 800 s^{-1} at $-\Delta G^\circ = 0.90$ eV [73]. From measurements of ET rates at other driving forces, the cyt c/ccp reorganization energy was estimated to be ~ 1.5 eV; the relatively large λ value may be a result of redox-dependent fluctuations of the protein–protein orientation, since the primary binding mode is electrostatic (salt bridges) [73].

In all protein–protein complexes studied to date in which cytochrome c has been a partner, it has been shown that the ET rates depend strongly on the reaction driving force. It follows that variations in the reorganization energy could control ET rates in these cases [12]. In redox enzymes with two or more active centers, ET between two centers could be turned on by lowering λ at roughly constant $-\Delta G^\circ$ [1]. Indeed, a proposal has been advanced that this type of mechanism would be an efficient way to gate the electron flow in a redox-linked proton pump such as cytochrome oxidase [75].

Another attractive way to control ET rates in protein–protein complexes is by variations in electronic coupling. Conformational changes conceivably could form or break up pathways involving H-bonds or through-space interactions. Examination of potential ET pathways in selected protein–protein complexes is underway in collaboration with Beratan and Onuchic [76]. The foundation we have built in our investigations on Ru-modified proteins should be helpful in this work.

Acknowledgement. We thank David Beratan and Jay Winkler for helpful discussions. Our research on electron transfer in proteins is supported by grants from the National Science Foundation and the National Institutes of Health. NRSA/NIH postdoctoral fellowships were held by M. J. T., J. C., and A. L. R.; and a Medical Research Council (Canada) postdoctoral fellowship was held by B. E. B. This is contribution no. 8115 from the Arthur Amos Noyes Laboratory.

8 References

1. Gray HB, Malmström BG (1989) Biochemistry 28: 7499
2. McLendon G (1988) Acc Chem Res 21: 160
3. Bowler BE, Raphael AL, Gray HB (in press) Prog Inorg Chem
4. Mayo SL, Ellis WR, Jr, Crutchley RJ, Gray HB (1986) Science 233: 948
5. Yocom KM, Shelton JB, Shelton JR, Schroeder WA, Worosila G, Isied SS, Bordignon E, Gray HB (1982) Proc Natl Acad Sci USA 79: 7052
6. Alber T, Dao-pin S, Nye JA, Muchmore DC, Matthews BW (1987) Biochemistry 26: 3754
7. Baldwin RL, Eisenberg D (1987) In: Oxender DL, Fox CF (eds) Protein Engineering, Alan R Liss, New York, p 127
8. Go M, Miyazawa S (1980) Int J Peptide Protein Res 19: 211
9. Gray HB (1986) Chem Soc Rev 15: 17
10. Scott RA, Mauk AG, Gray HB (1985) J Chem Ed 62: 932
11. Bowler BE, Meade TJ, Mayo SL, Richards JH, Gray HB (1989) J Am Chem Soc 111: 8757
12. Lieber CM, Karas JL, Mayo SL, Albin M, Gray HB (1987) Proceedings of the Robert A. Welch Conference on Chemical Research. Design of Enzymes and Enzyme Models, Nov. 2–4, 1987, p 9, Robert A. Welch Foundation, Houston, TX
13. Margalit R, Kostic NM, Che C-M, Blair DF, Chiang H-J, Pecht I, Shelton JB, Shelton JR, Schroeder WA, Gray HB (1984) Proc Natl Acad Sci USA 81: 6554
14. Crutchley RJ, Ellis WR, Jr, Gray HB (1986) In: Xavier AV (ed) Frontiers in Bioinorganic Chemistry, VCH, Weinheim, FRG, p 679
15. Osvath P, Salmon GA, Sykes AG (1988) J Am Chem Soc 110: 7114
16. Yocom KM, Winkler JR, Nocera DG, Bordignon E, Gray HB (1983) Chemica Scripta 21: 29
17. Bechtold R, Gardineer MB, Kazami A, van Hemelryck B, Isied SS (1986) J Phys Chem 90: 3800
18. Karas JL (1989) Ph.D. Dissertation, California Institute of Technology
19. Selman M (1989) Ph.D. Dissertation, California Institute of Technology
20. Margalit R, Pecht I, Gray HB (1983) J Am Chem Soc 105: 301
21. Nocera DG, Winkler JR, Yocom KM, Bordignon E, Gray HB (1984) J Am Chem Soc 106: 5145
22. Crutchley RJ, Ellis WR, Jr, Gray HB (1985) J Am Chem Soc 107: 5002
23. Toi H, La Mar GN, Margalit R, Che C-M, Gray HB (1984) J Am Chem Soc 106: 6213
24. Sola M, Cowan JA, Gray HB (1989) Biochemistry 28: 5261
25. Mottenen J, Ringe D, Petsko G, unpublished results
26. Lieber CM, Karas JK, Gray HB (1987) J Am Chem Soc 109: 3778
27. Isied SS, Kuehn C, Worosila G (1984) J Am Chem Soc 106: 1722
28. Kostic NM, Margalit R, Che C-M, Gray HB (1983) J Am Chem Soc 105: 7765
29. Marcus RA, Sutin N (1985) Biochim Biophys Acta 811: 265

30. Closs GL, Miller JR (1988) Science 240: 440
31. Axup AW, Albin M, Mayo SL, Crutchley RJ, Gray HB (1988) J Am Chem Soc 110: 435
32. Cowan JA, Gray HB (1988) Chemica Scripta 28A: 21
33. Cowan JA, Upmacis RK, Beratan DN, Onuchic JN, Gray HB (1988) Ann New York Acad Sci 550: 68
34. Elias H, Chou MH, Winkler JR (1988) J Am Chem Soc 110: 429
35. Peterson-Kennedy SE, McGourty JL, Kalweit JA, Hoffman BM (1986) J Am Chem Soc 108: 1739
36. Durham B, Pan LP, Long JE, Millett F (1989) Biochemistry 28: 8659
37. Jackman MP, Lim M-C, Salmon GA, Sykes AG (1988) J Chem Soc Chem Commun, 179
38. Jackman MP, Lim M-C, Salmon GA, Sykes AG (1988) J Chem Soc Dalton Trans, 2843
39. Jackman MP, McGinnis J, Powls R, Salmon GA, Sykes AG (1988) J Am Chem Soc 110: 5880
40. Jackman MP, Sykes AG, Salmon GA (1987) J Chem Soc Chem Commun, 65
41. Gray HB, Malmström BG (1983) Comments Inorg Chem 2: 203
42. Farver O, Pecht I (1989) FEBS Lett 244: 376
43. Farver O, Pecht I (1989) FEBS Lett 244: 379
44. Meade TJ, Gray HB, Winkler JR (1989) J Am Chem Soc 111: 4353
45. Cowan JA, Gray HB (1989) Inorg Chem 28: 2074
46. Karas JL, Lieber CM, Gray HB (1988) J Am Chem Soc 110: 599
47. Therien MJ, Selman MA, Gray HB, Chang I-J, Winkler JR (1990) J Am Chem Soc 112: 2420
48. Therien MJ, Bowler BE, Selman MA, Gray HB, Chang I-J, Winkler JR, ACS Symp Ser (in press)
49. Brunschwig BS, Ehrenson S, Sutin N (1984) J Am Chem Soc 106: 6858
50. Isied SS, Vassilian A, Wishart JW, Creutz C, Schwarz H, Sutin N (1988) J Am Chem Soc 111: 635
51. Brunschwig BS, Ehrenson S, Sutin N (1986) J Phys Chem 90: 3657
52. Winkler JR, unpublished results.
53. Closs GL, Calcaterra LT, Green NJ, Penfield KW, Miller JR (1986) J Phys Chem 90: 3673
54. Schmidt JA, McIntosh AR, Weedon AC, Bolton JR, Connolly JS, Hurley JK, Wasielewski MR (1988) J Am Chem Soc 110: 1733
55. Heitele H, Michel-Beyerle ME, Finckh P (1987) Chem Phys Lett 134: 273
56. Balaji V, Ng L, Jordan KD, Paddon-Row MN, Patney HK (1987) J Am Chem Soc 109: 6957
57. Oliver AM, Craig DC, Paddon-Row MN, Kroon J, Verhoeven JW (1988) Chem Phys Lett 150: 366
58. Oevering H, Paddon-Row MN, Heppener M, Oliver AM, Cotsaris E, Verhoeven JW, Hush NS (1987) J Am Chem Soc 109: 3258
59. Beratan DN, Onuchic JN (1989) Photosynthesis Research, 22: 173
60. Kuki A, Wolynes PG (1987) Science 236: 1647
61. Larsson S (1983) J Am Chem Soc 103: 4034, (1981); J Chem Soc Faraday Trans 79: 1375
62. Beratan DN, Onuchic JN, Betts JN, Bowler BE, Gray HB, J Am Chem Soc, submitted for publication
63. Casimiro D, Chang J, unpublished results
64. McGourty JL, Peterson-Kennedy SE, Ruo WY, Hoffman BM (1987) Biochemistry 26: 8302
65. Gingrich DJ, Nocek JM, Natan MJ, Hoffman BM (1987) J Am Chem Soc 109: 7533
66. Natan MJ, Kuila D, Baxter WW, King B, Hawkridge FM, Hoffman BM (1990) J Am Chem Soc 112: 4081
67. McLendon G, Miller JR (1985) J Am Chem Soc 107: 7811
68. McLendon GL, Winkler JR, Nocera DG, Mauk MR, Mauk AG, Gray HB (1985) J Am Chem Soc 107: 739
69. Northrup SH, Boles JO, Reynolds JCL (1988) Science 241: 67
70. Poulos T, Finzel B (1984) In: Hearn MTW (ed) Peptide and Protein Reviews, Marcel Dekker, Inc, New York, p 115
71. Taniguchi V, Ellis W, Cammarata V, Webb J, Anson F, Gray HB (1982) In: Kadish K (ed) Electrochemical Studies of Biological Redox Components, American Chemical Society, Washington, DC, ACS Adv Chem Ser 201, p 51
72. Yonetani T (1976) In: Boyer PD (ed) The Enzymes, vol 13, Academic Press, New York, p 345
73. Conklin KT, McLendon G (1988) J Am Chem Soc 110: 3345
74. Nocek JM, Liang N, Wallin SA, Mauk AG, Hoffman BM (1990) J Am Chem Soc 112: 1623
75. Brzezinski P, Malmström BG (1987) Biochim Biophys Acta 894: 29
76. Beratan DN, Onuchic JN, Gray HB, in preparation

Electron Transfer in Genetically Engineered Proteins. The Cytochrome *c* Paradigm

A. Grant Mauk

Department of Biochemistry, University of British Columbia, Vancouver, British Columbia V6T 1W5, Canada

The investigation of metalloprotein electron transfer mechanisms through structural and functional characterization of modified proteins produced by site-directed mutagenesis provides a systematic means of gaining new, often quantitative, insight into the structural characteristics that regulate the electrochemical, kinetic and protein-binding properties of such proteins. This review discusses aspects of the experimental strategies that have been employed in the application of site-directed mutagenesis as a means of addressing these mechanistic issues, some of the known limitations of this approach, and strategies for expression of recombinant electron transfer proteins. In addition, a detailed summary of the current literature concerning functional studies of engineered mutant forms of cytochrome *c* is presented to illustrate the manner in which the technique has been employed in studies of this protein as a representative member of this large category of proteins.

1 Introduction

With the application of molecular genetic techniques to the study of electron transfer proteins, it has become possible to develop experimental systems and design experiments that are not accessible through traditional, non-genetic techniques or through the use of random mutagenic techniques. At present, the most widespread application of this approach to the study of electron transfer proteins has been the introduction of highly specific structural modifications into such proteins through use of site-directed mutagenesis. The experimental considerations involved in the use of site-directed mutagenesis have been discussed extensively elsewhere (e.g. Refs. [1–6]) and will not be included here. The first part of this review addresses the various strategies involved in use of molecular genetic techniques in the study of electron transfer proteins and the limitations and implications of such strategies while the second part reviews the manner in which this approach has been applied to cytochrome *c*. The use of classical mutagenic techniques or of the proteins derived from them will not be considered here. Reviews of some of the literature on this topic as related to cytochrome *c* have been presented elsewhere [7, 8].

2 Objectives in the Application of Site-Directed Mutagenesis to the Study of Electron Transfer Metalloproteins

Since the introduction of oligodeoxyribonucleotide directed site-specific mutagenesis in 1978 [9, 10], this technique has been used to study the relationship between the structure and function of a large number of proteins. One useful means of considering the various mutations that have been introduced into proteins is in terms of the functional characteristics of the protein that are the target of alteration by genetic manipulation. For example, the structural elements that are responsible for the stability and folding characteristics of a given protein may be addressed through replacement or introduction of prolyl residues, introduction of disulfide linkages at critical positions in the protein sequence [11], modification of residues in critical regions between structural domains, or introduction or deletion of peptide segments or loops (perhaps selected on the basis of alignment of several species of protein sequences). Another general class of mutations would include mutations that alter the electrostatic properties of a protein, often through perturbation of the pK_a of one or more critical residues.

In the case of metalloproteins, at least two additional categories of mutation may be considered: mutations that affect the coordination environment of the metal center and mutations that affect the ligand binding and/or electron transfer properties of the protein without modification of the coordination environment. The reasons for substitution of the amino acid residues that provide ligands to the central metal are relatively apparent. Such mutants are of interest because they (1) offer an opportunity to investigate the structural basis for a variety of spectroscopic properties, (2) perturb the ligand binding or catalytic properties of the active site in a controlled manner, (3) modify the kinetics and thermodynamics of the electron transfer properties of the protein. The electron transfer or ligand binding properties of metalloproteins can also be modified through more subtle types of mutations. For example, the electrostatic properties of an active site can be perturbed through modification of residues involved in critical hydrogen bonding interactions with metal center ligands. Alternatively, charged amino acid residues in the active site of a metalloprotein that do not interact directly with the metal ligands may be eliminated or introduced; substitution of hydrophobic residues present in the active site of a metalloprotein may be of value insofar as they influence the dielectric environment of the metal center or the stereochemistry of the active site and thereby the ligand binding properties of the central metal, the reduction potential of the metal center, or the stability of the active site. Examples of each of these types of study as they have been applied to cytochrome *c* are discussed further below.

Aside from these relatively direct applications of site-directed mutagenesis, combination of recombinant DNA techniques with other experimental strategies will no doubt prove to be of increasing importance. If the gene of interest can be expressed with sufficient efficiency in auxotrophs, then proteins in which selected amino acids are isotopically enriched [12] may be produced to increase the sensitivity and selectivity of magnetic resonance techniques. Alternatively, amino acid analogues that are recognized as substrates by aminoacyl tRNA synthetases may be incorporated randomly in place of the true substrate amino

Table 1. Representative metalloproteins that bind redox-active metals and that have been studied by site-directed mutagenesis

Protein	Source	Ref.
Cytochrome b_5	Rat	[16–19]
	Bovine	[20]
Cytochrome *d*	*Escherichia coli*	[21]
Flavocytochrome b_2	*Saccharomyces cerevisiae*	[22–24]
Cytochrome P-450	*Pseudomonas putida*	[25–28]
	Rat	[29–32]
	Rabbit	[29, 30]
	Mouse	[33]
Cytochrome *c* peroxidase	*Saccharomyces cerevisiae*	[12, 34–44]
Myoglobin	Sperm whale	[45–49]
	Human	[50–53]
	Bovine	[54]
Hemoglobin	Human	[49, 55–60]
Azurin	*Pseudomonas aeruginosa*	[61–63]
Ferredoxin I	*Azotobacter vinelandii*	[64]
Tyrosinase	*Neurospora crassa*	[65]
Ribonucleotide reductase	*Escherichia coli*	[66]
Superoxide dismutase	Human	[67–72]

acid into recombinant proteins for spectroscopic purposes [13]. Site-directed mutagenesis can also be used to introduce mutations suitably placed to facilitate site-specific chemical crosslinking [14], chemical modification [15], chemical or proteolytic cleavage, or semisynthetic substitution of non-naturally-occurring amino acid residues. From these examples, it is clear that the availability of recombinant DNA techniques offers a wide range of experimental possibilities and that when these techniques are combined with the more traditional methods of protein chemistry the possible experimental strategies are limited primarily by the ingenuity of the investigator and the structural tolerance of the protein being manipulated. The list of metalloproteins that bind redox-active metal centers and that have been studied through use of site-directed mutagenesis is currently growing at a considerable rate. This phenomenon is illustrated by the representative list of such proteins other than cytochromes *c* that is provided in Table 1.

3 Limitations of Site-Directed Mutagenesis and Caveats

As with any technique, site-directed mutagenesis possesses clear limitations in the types of experiment that can be performed. The most obvious limitation is that the range of modification available is limited to replacement of a given residue with one of the remaining 19 naturally-occurring amino acids. This restriction may be relaxed by employing a combination of site-directed mutagenesis and semi-synthetic strategies as discussed above or eliminated entirely through combined use of site-directed mutagenesis and chemically acylated aminoacyl-tRNAs in a strategy [73–75] simulating that employed normally by a strain of *E. coli* to incorporate selenocysteine into a specific position of hydrogenase [76, 77]. Also, the structural consequences of substituting a single amino acid residue are impossible to predict with certainty. Go et al. have argued that the greater evolutionary variability of surface amino acid residues relative to internal residues suggests that substitutions of surface residues are less likely to introduce structural (or functional) changes [78]. While this generality may have a statistical validity, there are undoubtedly instances in which this assumption will prove to be hazardous particularly with reference to relatively subtle changes in hydrogen bonding interactions or solvation. For this reason, it is important that structural characterization of mutant proteins be incorporated as a routine component of such studies so that the structural basis for any mechanistic alterations that are produced is not misinterpreted. In addition to detailed crystallographic and NMR spectroscopic characterization of mutant proteins, both of which are labor intensive, one method of quickly screening such proteins that has not been widely employed is analysis of their pH titration curves [79]. In principle, such measurements have a resolution of 0.1 proton and can, therefore, detect small, mutation-induced shifts in pK_as of titratable groups.

It is also essential that any functional properties of the mutant protein that can be assessed be assessed. Although the substitution of one particular residue for another may be made in an attempt to determine the effect of the mutation on a specific property of a protein, it is quite possible that other properties that are not of immediate concern may be modified unintentionally and that these modifications may have important, otherwise occult, implications for the functional studies that are of immediate interest (vide infra). In the case of electron transfer proteins it may be useful, for example, to produce a family of mutants the members of which differ from each other only in their reduction potentials. This result may prove to be difficult to achieve because many mutations that perturb the reduction potential of a protein may also change its electrostatic properties or its reorganizational barrier to electron transfer. Depending on the experiments to be conducted with the mutants, these other properties may prove to be more important considerations than the reduction potentials of the mutants. In summary, new mutant proteins are ideally studied as if they were altogether new proteins of the same general class as the wild-type protein, and assumptions regarding the properties of such mutants should be kept to a minimum.

4 Expression of Recombinant Redox-Active Metalloproteins

The two fundamental requirements for the application of recombinant DNA techniques to the study of a protein are the availability of a gene coding for the protein of interest and an expression system that permits production of the protein in amounts that are adequate for the experiments to be performed. Most recombinant proteins studied at present have been derived from genes cloned from the relevant organism or tissue using a variety of techniques (e.g. refs. 80–82). Alternatively, proteins for which a cloned gene is not available but for which the amino acid sequence is known may be made tractable for recombinant work through synthesis of a gene coding for the appropriate sequence. This approach has been pioneered by Sligar and co-workers in their work with rat liver cytochrome b_5 [16] and sperm whale myoglobin [45] in which they have expressed each of these proteins efficiently in *Escherichia coli* from totally synthetic genes. An obvious yet crucial requirement of this latter approach is that the protein sequence be known with absolute certainty, an assumption that is not necessarily valid for many protein sequences determined by older sequencing techniques. The importance of this assumption is illustrated in the work of Funk et al. [20] on recombinant bovine liver cytochrome b_5 produced from a synthetic gene. In this case, the reported bovine liver cytochrome b_5 amino acid sequence incorporated three errors of amidation, which led this group to produce a recombinant protein possessing functional and spectroscopic properties different from those of authentic wild-type protein.

The successful expression of cloned or synthetic genes remains largely empirical at present. Ideally, expression should be performed in a host that is or has been made genetically deficient in the protein of interest. Furthermore, proteins that possess prosthetic groups or that require post-synthetic modification may present special challenges to expression. In particular, *c*-type cytochromes, which possess covalently bound prosthetic groups, are not readily expressed in *E. coli* [83]. In this case, the problem may be that *E. coli* does not possess sufficient enzymatic activity for post-translational covalent incorporation of heme into apo-cytochrome *c*. Self et al., however, have recently reported that *E. coli* transformed with the gene coding for *Rhodospirillum rubrum* cytochrome c_2 and grown on L-broth plates develop a reddish-brown color that on the basis of its electronic absorption spectrum is identical to that of authentic cytochrome c_2 [84]. Unfortunately, these bacteria could not be grown in liquid culture [84], an observation that these authors attributed to toxicity of the gene product in *E. coli.* At present, cytochromes *c* from various species have been expressed in yeast [85], transfected kidney cells [86], and *Rhodobacter capsulatus* [87]. It is reasonable to expect that other prokaryotes known to possess significant levels of *c*-type cytochromes (e.g. *Thermus thermophilus* and *Pseudomonas aeruginosa*) might also prove to be useful alternatives. The general applicability of these putative hosts would, however, depend on the sequence-specificity of the heme lyase activities present in these organisms. With the availability of the genes coding for *Saccharomyces cerevisiae* [88] and *Neurospora crassa* [89] heme lyases, it is conceivable that expression of *c*-type cytochromes in *E. coli* or other prokaryotes might be possible if a suitable heme lyase gene were expressed in the host concomitantly with the gene for a compatible cytochrome. Clearly, similar considerations may be important in the heterologous expression of other electron transfer proteins possessing cofactors or prosthetic groups.

Even with expression of heterologous proteins from cloned genes in hosts possessing the required post-synthetic processing activities, it is likely that in the absence of other considerations the level of expression will not be as great as that found for the homologous protein. For example, various laboratories have found that expression of rat and human heart cytochrome *c* in yeast produces smaller quantities of cytochrome *c* [90–93] than reported [86] for expression of the yeast cytochrome and its variants. This result may arise from any number of factors, including species variation in codon bias and efficiency of yeast cytochrome *c* heme lyase in processing heterologous apocytochromes *c*. In addition to heme incorporation, yeast cytochrome *c* is modified postsynthetically by trimethyllation of Lys-72, and most other eukaryotic cytochromes *c* are acetylated at the amino terminus. Interestingly, the rat cytochrome *c* produced in *S. cerevisiae* undergoes cleavage of its N-terminal Met with subsequent acetylation of the penultimate Gly residue that is thereby exposed [91]. In addition, Lys-72 of rat cytochrome *c* expressed in yeast is trimethylated as is yeast cytochrome *c* [91, 92]. The observed extent of rat cytochrome *c* amino terminal acetylation and Lys-72 trimethylation has varied in two available reports, for reasons that have not yet been defined [91, 92].

Heme proteins possessing non-covalently bound heme prosthetic groups can be expressed in *E. coli* with less difficulty. For example, Kraut and co-workers have expressed wild-type and mutant forms of *Saccharomyces* cytochrome *c* peroxidase in *E. coli* [35] in reasonable efficiency. After initially expressing mutant cytochrome *c* peroxidase in low yields in a strain of *Saccharomyces cerevisiae* deficient in the enzyme [33], Goodin et al. expressed the enzyme in greater levels in *E coli* [94] through identification of an initiation-efficient amino terminal sequence by random mutagenesis. In this case, these workers have achieved expression levels as high as 0.125 g (3.5 μmol) of protein/L of culture on optimized media and in the presence of inducer [91]. Unlike cytochrome b_5 [16, 20], myoglobin [45, 95, 98], cytochrome P-450$_{cam}$ [97] or flavocytochrome b_2 [23] expressed in *E. coli*, however, cytochrome *c* peroxidase is produced as the apoprotein [35, 94] rather than the heme-containing protein under identical growth conditions [98] even though the mechanism of heme incorporation is believed to be non-enzymatic. The basis for the inability of *E. coli* to incorporate heme into cytochrome *c* peroxidase is currently unknown, though it is tempting to speculate that heme simply does not bind as readily to the apo-peroxidase as it does to the other proteins and, possibly, that this property might be related to the smaller degree of heme exposure to solvent in native cytochrome *c* peroxidase [99] relative to most of these other proteins. This naive explanation, however, may obscure other more relevant considerations.

For example, rat cytochrome P-450 IIA1 expressed in *Spodoptera frugiperda* cells that had been transfected with a recombinant baculovirus containing the cDNA coding for IIA1 produced primarily the apoprotein [35]. Inclusion of protoheme IX or modified hemins in the culture medium resulted in incorporation of the prosthetic group into the recombinant enzyme [100]. The success of several groups in producing cytochrome P-450 of various types as the complete, heme-containing enzyme in a variety of expression systems (reviewed in Ref. 101) argues against the role of heme exposure to solvent in the mature protein as a critical factor in the ability of hosts to produce mature heme proteins. From crystallographic studies of cytochrome P-450$_{cam}$, the heme binding site of this protein completely protects the heme prosthetic group from exposure to solvent [102].

The success of Chapman and co-workers in expression of flavocytochrome b_2 in *E. coli* [23] is encouraging in its implications for future expression of flavoproteins in this host because, in their experience both the flavin and heme groups are incorporated into the recombinant protein. Moreover, the bacterial expression system produces the protein 500-1000 fold more efficiently than the yeast from which it was cloned. The enzyme produced in *E. coli*, however, lacks the first five amino acid residues at its amino terminus, a result which presumably reflects subtle differences in protein synthesis between the two organisms.

Nagai and Thørgersen found that an alternative approach was required to achieve efficient expression of the β-chain subunits of human hemoglobin in *E. coli* [55]. In this case, β-globin was produced as a fusion protein that possessed the sequence Ile-Glu-Gly-Arg between the 31 amino-terminal residues of λ cII protein and the amino terminal residue of the human β-chain. This construct was designed to achieve efficient initiation of protein synthesis through the presence of the λ cII sequence while providing a small coupling sequence that permits specific proteolytic release of the target protein by coagulation factor X_a. More recently, this same approach has been employed to express porcine myoglobin in *E. coli* [96]. Mutant hemoglobins produced in this manner have been used primarily to evaluate the effects of mutations on ligand binding equilibria and kinetics [49, 56, 58], though it seems inevitable that these and other mutants derived in this manner will provide important insight into the electrochemical and electron transfer properties of hemoglobin.

Bacterial hosts are inappropriate choices for expression of proteins such as the blue copper proteins stellacyanin, laccase, and ceruloplasmin which are extensively glycosylated. In these cases, it may be necessary to employ tissue cultures of appropriate origin to obtain the native protein. In this regard, the amino-terminal half of human serum transferrin, which lacks carbohydrate, has been expressed in high yield in baby hamster kidney cells by Funk et al. [13], while the glycosylated carboxyl-terminus has proved to be more problematic [103].

5 Studies of Cytochrome *c* Variants Produced by Site-Directed Mutagenesis

The groups of Hall and Smith were the first to clone and sequence structural genes for metalloproteins, the iso-cytochromes *c* from bakers yeast, *Saccharomyces cerevisiae* [104, 105]. This achievement combined with the development of oligonucleotide-directed site specific mutagenesis [reviewed in Refs. 1 and 2] made it possible for Pielak et al. to prepare and express three mutants of yeast *iso*-1-cytochrome *c* in which Tyr, Ser, and Gly were substituted for Phe-82 [85]. This residue was selected for substitution for three reasons: (1) a hypothetical model for the mechanism of electron transfer from cytochrome *c* to cytochrome *c* peroxidase developed by Poulos and Kraut proposed that Phe-82 facilitates electron transfer from cytochrome *c* to the peroxidase [106]; (2) phenylalanyl residues are generally refractory to chemical modification, so mutagenesis provided the one specific and unambiguous means of studying its function; and (3) Phe-82 is phyllogenetically conserved among mitochondrial cytochromes *c* [107], so comparative studies do not provide a resonable alternative means of evaluating the functional contributions of this residue. Subsequently the Ala, Leu and Ile-82 variants were produced [108], and the remaining 13 possible substitutions have now also been made [109]. Verification of the authenticity of several of these mutations has been provided by the combined application of HPLC tryptic peptide mapping and amino acid analysis [110], an approach that may prove useful at times for analytical assessment of selected mutants.

Functional characterization of several members of this family of mutants is described in detail below.

Functional studies of yeast *iso*-1-cytochrome *c* and any point mutations based on this protein are complicated by the presence of a reactive Cys residue at position-102. This residue promotes accelerated autoreduction of ferricytochrome *c* (presumably through an intermolecular electron transfer reaction from Cys-102 on one protein to the heme iron of another), and it allows formation of cytochrome *c* dimers through intermolecular disulfide bond formation [111]. Furthermore, Cys-102 appears to cause the protein to bind to electrode surfaces and prevent reversible electrochemical behavior of the protein [85, 107]. To eliminate these problems, work in our laboratories has focused on use of a mutated gene coding for Thr-102 as a background into which other mutations are introduced [107]. Thr was selected for this purpose because it is the concensus residue at this position in the approx. 100 cytochrome *c* sequences that have been determined. Others have selected Ser-102 [15] to stabilize the wild-type protein because this residue is the best stereochemical analogue of Cys available. Based on the three-dimensional structure of wild-type *iso*-1-cytochrome *c* possessing Cys-102 [112], it is clear that this residue is inaccessible to solvent and that chemical modification of Cys-102 by sulfhydryl-specific reagents or dimerization of the protein through intermolecular disulfide bond formation must necessarily involve a conformational change in the C-terminal helix of the protein.

5.1 *Mutations of Cytochrome c that Affect Electron Transfer Properties*

5.1.1 *Phenylalanine-82*

As discussed above, substitutions at Phe-82 were originally designed as a means of evaluating the proposed role of this residue on regulation of the electron transfer properties of cytochrome *c*. This functional role for Phe-82 results from its close proximity to the heme prosthetic group on the surface of the protein (Fig. 1). In this position and orientation, Phe-82 is clearly suited to make a significant contribution to the dielectric of the heme binding site and thereby to modulate the functional properties of the heme. In the course of studying several variants of yeast *iso*-1-cytochrome *c* in which Phe-82 was replaced with other residues, it was ultimately determined that replacement of Phe-82 destabilizes the structure of the cytochrome *c* through perturbation of the stability of the interaction between Met-80 and the heme iron atom [113]. This effect of mutations at Phe-82 is observed as a reduction in the pK_a for the alkaline transition of cytochrome *c* and as a change in the mechanism of this conformational change, a consequence of mutation that was not originally intended or predicted [113]. Substitution of Phe with either Gly or Ser reduces the pK_a for the alkaline transition from the value of 8.5 observed for the wild-type protein to

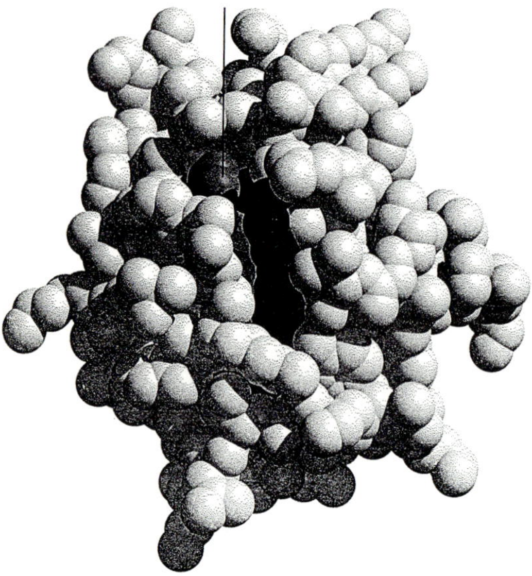

Fig. 1. Space filling model of yeast *iso*-1-cytochrome *c*. The edge of the heme prosthetic group is visible as a black linear structure in the center of the protein. Phe-82 is shaded a dark gray at the left upper side of the heme group

7.7, while replacement with Leu or Ile reduces this pK_a to 7.2. Investigation of these mutants in terms of the mechanism proposed from pH-jump experiments as described by Davis et al. [114] indicates that the modified behavior of the Leu and Ile mutants results from the reduction of the pK_a of the (unknown) residue that titrates during this transition by 1.4 pK_a units. Similar analysis of the Ser and Gly mutants indicates that this same change occurs in these proteins but that a compensatory shift in the K_{eq} for the ligand exchange equilibrium increases the overall apparent alkaline pK_a for these variants [113].

In addition to the effect of mutations at Phe-82 on the stability of the cytochrome *c* active site, the intense, negative Soret Cotton effect in the circular dichroism spectrum of ferricytochrome *c* is profoundly affected by the presence of non-aromatic amino acid residues at this position [115]. Recent examination of six position-82 *iso*-1-ferricytochrome *c* mutants establishes that while Tyr-82 exhibits a Soret CD spectrum closely similar to that of the wild-type protein, the intensity of the negative Soret Cotton affect varies with the identity of the residue at this position in the order Phe > Tyr > Gly > Ser = Ala > Leu > Ile, though the Ser, Ala, Ile, and Leu variants have effectively no negative Soret Cotton effect [108].

The midpoint reduction potential of cytochrome *c* and the kinetics of its reduction by $Fe(EDTA)^{2-}$ are also significantly influenced by substitutions at Phe-82. As measured by direct electrochemistry at pH 6 to eliminate any contribution of the alkaline form to the potential measurements, the following midpoint reduction potentials were observed (25°C, $\mu = 0.1$ mol L^{-1}, mV vs. SHE): wild-type, 290(2); Tyr-82, 280(2); Leu-82, 286(2); Ile-82, 273(2); Ala-82, 260(2); Ser-82, 255(2); Gly-82, 247(2). Similarly, the second order rate constant for reduction of these variants by Fe(EDTA)$^{2-}$ varies from 6.2×10^4 mol^{-1} s^{-1} for Tyr-82 to 14.8×10^4 mol^{-1} s^{-1} for Ser-82 [108]. Analysis of these rate constants in terms of relative Marcus theory, primarily as a means of correcting them for the effect of mutations on the thermodynamic driving force of the reaction, suggests that the mutations exert a greater influence on the reduction kinetics. Specifically, the apparent self-exchange rates (25°C, $\mu = 0.1$ mol L^{-1}, pH 6.0) for the position-82 mutants range from a low of 10.9 mol^{-1} L^{-1} s^{-1} for the wild-type protein to a high of 165 mol^{-1} L^{-1} s^{-1} for Ser-82 and 190 mol^{-1} L^{-1} s^{-1} for the Gly-82 mutant.

Replacing Phe-82 with other aromatic and non-aromatic residues does not prevent the steady-state turnover of cytochrome *c* by its peroxidase, though the efficiency of the mutants as peroxidase substrates is compromised to some extent [85]. The initial observations of Peilak et al. indicated that the turnover numbers for the Ser, Tyr, and Gly-82 variants were 70%, 30% and 20% of the value observed for the wild-type protein (pH 7.0, 0.2 M Tris-HCl buffer) [85]. On the other hand, initial studies of intramolecular electron transfer within the electrostatically-stabilized complex formed by ferricytochrome *c* and Zn-substituted cytochrome *c* peroxidase at low ionic strength, indicated that mutations at position-82 do have dramatic inhibitory effects [116, 117]. Specifically, the thermal oxidation of wild-type protein and the Tyr-82 mutant by Zn-CCP.$^+$ occurs with a first order rate constant of ca. 1×10^4 s^{-1} while the Ser, Gly, Leu, and Ile mutants are oxidized at a rate of 2–4 s^{-1}. Similar discriminatory rates are not exhibited by these mutants in the photoexcited electron transfer from ^{3}Zn-CCP to ferricytochrome *c*, however. More recent studies employing an enhanced flash photolysis apparatus have suggested that the effects of these mutations arise from introduction of a rate limiting conformational change that involves either cytochrome *c* alone, or a rearrangement of the cytochrome *c*-cytochrome *c* peroxidase complex [118, 119]. Further discussion of the mechanistic implications of these findings is presented in another chapter of this volume.

Steady state kinetics and protein-protein binding measurements have also been reported for the interaction of these mutant cytochromes with bovine heart cytochrome *c* oxidase [120]. The binding of cytochrome *c* variants to the oxidase occurred with increasing values of K_d in the order Ile (3×10^3 Mol L^{-1}) < Leu = Gly < wild-type < Tyr < Ser (3×10^5 mol L^{-1}). Steady-state kinetic analysis indicated that the rate of electron transfer with cytochrome *c* oxidase increased in the order Ser < Ile < Gly < Leu < Tyr < wild-type, an order notably different from that observed for a related analysis of the oxidation of these mutants by cytochrome *c* peroxidase [85]. This difference in order of mutant turnover by the oxidase and peroxidase may arise from differences in the mode of interaction of the cytochrome with these two enzymes.

The non-congruence of the K_d values for interaction of the mutants with cytochrome *c* oxidase with the K_{ml} values calculated from the steady-state kinetic analysis included in this study suggests that the rate of cytochrome *c* oxidation by the oxidase is not limited by the rate of product dissociation.

2D ^{1}H-NMR studies of the reduced and oxidized Thr-102 yeast *iso*-1-cytochrome by Pielak et al. indicate that this protein is quite similar in structure to the extensively studied tuna and horse heart cytochromes [121] despite a sequence identity between the yeast and vertebrate proteins of only about 60%. At the same time, these investigators also compared the 2D NMR properties of the Tyr-82 mutant of *iso*-1-cytochrome *c* (also possessing Thr-102) by difference and overlay techniques as a means of assessing the structural consequences of the position-82 mutation [122]. Based on this analysis, Pielak et al. conclude [122] that the dynamics of the mutant protein in solution are similar to the dynamics of the parent Thr-102 cytochrome and other species of cytochrome *c* because the flip rates of the aromatic residues are unchanged in the mutant protein. Furthermore, the flip rate of Tyr-82 appears to be rapid, as observed for Phe-82. The chemical shifts of six amino acid residues and some of the heme resonances were found to change by 0.04–0.13 ppm. Only the resonances from one of the thioether linkages to the heme were shifted more than 0.13 ppm. Although this work indicated that changes in chemical shift as small as 0.01 Å could be detected by this spectroscopic method, such small changes are not readily translated into structural information. However, based on interpretation of the NOESY spectra for the two proteins and on molecular graphics model building, these authors suggest that Tyr-82 occupies the same position as Phe-82 in the wild-type cytochrome but that Leu-85 rotates away from Tyr-82 in the mutant to accommodate the hydroxyl group of this residue [122].

Detailed structural assessment of selected position-82 mutant proteins has been provided by X-ray crystallography. With the successful crystallographic determination of the structure of wild-type *iso*-1-ferrocytochrome *c* from yeast [112], similar analysis of a number of mutants and of the oxidized wild-type protein has proceeded relatively quickly. The structure of the reduced wild-type protein bears a strong resemblance to that of other cytochromes *c* for which three-dimensional atomic structures are available [112]. Interestingly, this analysis found Cys-102 to be completely inaccessible to solvent, which means that for reaction with exogenous sulfhydryl reagents, protein-protein dimer formation, or autoreduction, the C-terminal helix must undergo a significant conformational reorganization to provide adequate access to this residue.

In the 2.8 Å structure of the reduced Ser-82 mutant, it is clear that the space normally occupied by the phenyl group of Phe-82 (Figs. 1 and 2A) is converted into a solvent channel adjacent to the heme prosthetic group (Fig. 2B) [123]. The residues lining this channel are hydrophobic in nature, and the floor of the channel is derived from the sulfur ligand to the heme iron that is provided by Met-80. In addition, this structure reveals a change in the conformation of the indole ring of Trp-59 in which the χ_1 torsion angle has changed by 15°. Aside from the solvent channel and the movement of Trp-59, the remainder of

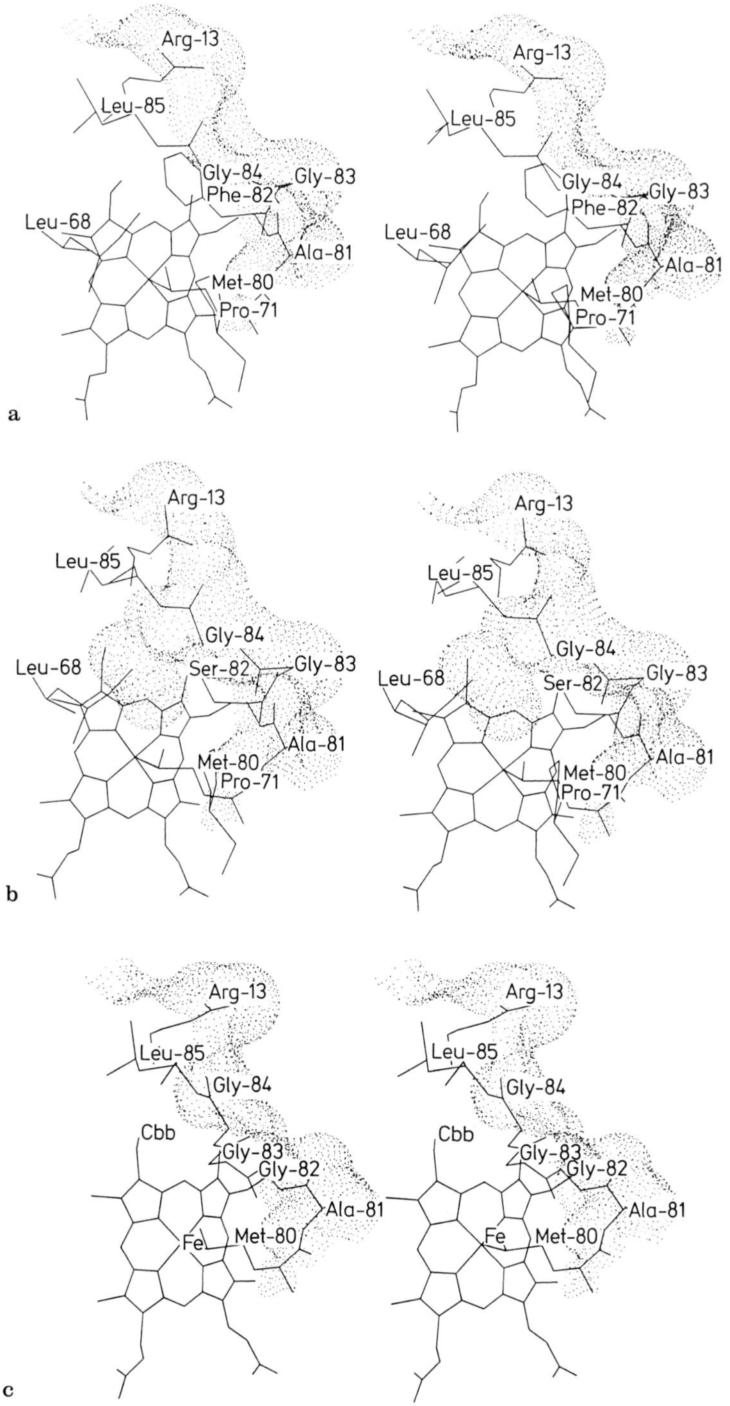

Fig. 2a–c. Stereodiagram of the yeast *iso*-1-cytochrome c surface. (**a**) Surface of the wild-type protein; (**b**) surface of the Ser-82 mutant; (**c**) surface of the Gly-82 mutant. (Modified from Refs. [123, 124])

the Ser-82 structure is very similar to that of the wild-type protein. Notably, the residues positioned between Ser-82 and Trp-59 do not exhibit any change in conformation that might explain the mechanism by which the substitution at position-82 transmits its influence to Trp-59.

In the 2.6 Å structure of the reduced Gly-82 mutant, the situation is quite different. In this case, the region of the main chain that contains residue-82 collapses toward the heme group to prevent the formation of the solvent channel seen in the Ser-82 structure [124] (Fig. 2C). Evidently, the flexibility introduced by a Gly residue at position-82 is sufficiently great that the wild-type structure is effectively destabilized and can no longer be maintained. One result of this conformational change is that the solvent accessibility of the heme group (44.4 Å^2) is closely similar to that of the wild-type protein (48.4 Å^2) and much lower than that of the Ser-82 mutant (70.0 Å^2) [124]. The increase in heme solvent accessibility in the Ser-82 mutant correlates with the increased apparent self-exchange rate for the mutant in its reaction with $Fe(EDTA)^{2-}$ [108]. On the other hand, the increase in the apparent self-exchange rate calculated for the Gly-82 variant based on the $Fe(EDTA)^{2-}$ reduction kinetics [108] may result from improved interaction of this hydrophilic reductant with the mutant that results from the greater hydrophilic nature of its heme binding pocket. The reduction potentials of the Gly-82 and Ser-82 mutants are both low as the result of increasing polarity in the heme binding pocket. In the Gly-82 structure, this change results from a movement of the main chain in the region surrounding position-82 toward the heme group that increases the number of hydrophilic functionalities that are in contact with the heme. In the Ser-82 mutant, on the other hand, the dielectric of the heme environment is increased by the creation of the solvent channel adjacent to the prosthetic group.

5.1.2 Tyrosine-67

In the model for redox-linked conformational change of cytochrome *c* developed by Takano and Dickerson on the basis of their crystallographic analyses of reduced and oxidized tuna cytochrome *c*, the principal structural rearrangement centers on an internally bound water molecule [125]. This water molecule is bound to Tyr-67, Thr-78, and Asn-52 and moves about 1 Å toward the heme iron when the protein is oxidized by pivoting on Thr-78. Although this model involves no movement of Thr-78, the side chain of Tyr-67 does move in a manner that produces a slight lengthening of its apparent hydrogen-bond with the Met-80 sulfur atom (Fig. 3). This involvement in the proposed redox-linked conformational change for cytochrome *c* combined with the hydrogen bond that it forms with Met-80, makes Tyr-67 a critically important target for modification through mutagenesis.

Luntz et al. have addressed this issue by preparing the Phe-67 mutant of rat cytochrome *c* and have found that although it demonstrates several remarkable properties, this variant functions quite reasonably as a cytochrome *c* [91].

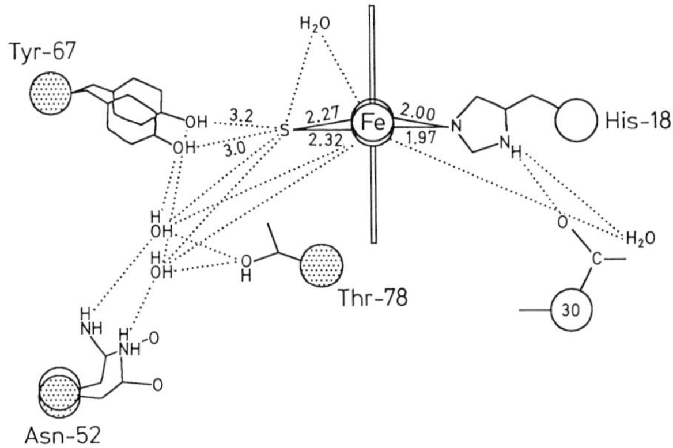

Fig. 3. The amino acid side chains that form hydrogen bonds with the internally-bound water molecule that moves in the redox-linked conformational change of cytochrome c originally described by Takano and Dickerson (modified from Ref. [125])

Notably, this mutant exhibits an increase in the pK_a for conversion to the alkaline form of 1.2 pK_a units and a decrease of 0.7 units in the pK_a for conversion to the acidic form. These observations reflect stabilization of the Fe-S bond of 1.7 and 1.0 kcal/mol respectively. The reduction potential of the protein was found to be 35 mV lower than that of the wild-type protein, while polarographic assays of the reaction of this mutant with cytochrome *c* oxidase found the K_m for the high-affinity phase to be the same as for the wild-type protein, with V_{max} for the mutant 25% greater. Preliminary one-dimensional NOE and 2D-NOESY experiments suggested a movement of Phe-67 downward with a concomitant movement of Leu-68 into the space normally occupied by the internally-bound water molecule discussed above [91]. On the basis of these observations and related thermodynamic arguments, these authors conclude that conversion of Tyr-67 to Phe results in loss of the internally-bound water molecule that is normally hydrogen bonded to this residue followed by a conformational change involving Leu-68 that results in an increased hydrophobicity of the heme binding site. While this is a plausible explanation of the available results, definitive evaluation of changes in solvation related to this mutation must be deferred until crystallographic analysis of this mutant has been performed.

5.1.3 *Histidine-62*

One example of the combined use of site-directed mutagenesis with chemical modification has been provided by the work of Bowler et al. with yeast *iso*-1-cytochrome *c* [15]. These workers introduced a His residue at position 62 to provide a site for attachment of a pentammineruthenium complex. Introduction of a second redox-active metal center to the protein at this position permitted assessment of the effect of an aromatic residue (Trp-59) and a sulfur (from Met-62) on the rate of electron transfer between the ruthenium and iron centers when these residues lie in the apparent path of electron transfer between the metal centers. This mutant exhibited a reduction potential similar to that of the wild-type protein (268 mV vs. SHE, pH 7.0, 50 mol L^{-1} sodium phosphate, 0.4 mol L^{-1} NaCl), and the heme center of the ruthenated derivative had a potential of 275 mV under identical conditions. Interestingly, the rate of electron transfer from the ruthenium to the iron in this modified mutant was fully consistent with the distance dependence for electron transfer between two such sites based on results obtained for pathways not involving aromatic groups or sulfur atoms. This finding suggests that the presence of aromatic or σ-donor atoms in the path of electron transfer between two metal centers is not a generally critical determinant of the rate of electron transfer.

5.2 Mutations of Cytochrome c that Primarily Affect Protein Stability and Folding

5.2.1 Proline-71

Cytochrome *c* possesses three phyllogenetically-conserved proline residues that are presumably involved in the correct folding of the protein to form the native structure. The effects of substitutions at one of these sites, Pro-71 (Fig. 4), on the equilibrium and kinetics of yeast *iso*-2-cytochrome *c* unfolding have been studied by Nall and co-workers through comparison of the properties of a Thr

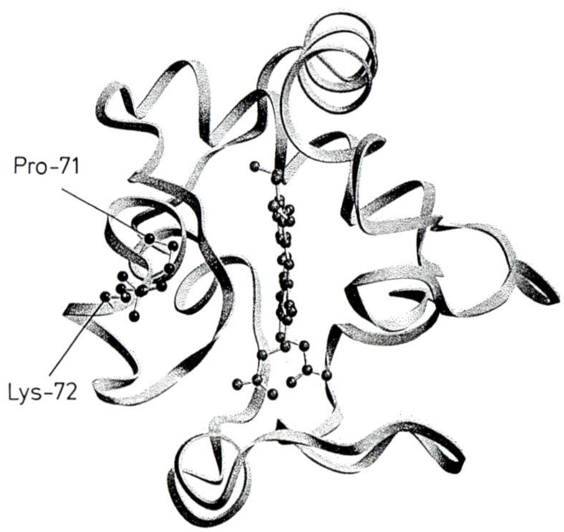

Fig. 4. Yeast *iso*-1-cytochrome *c* (reduced) with side chains of Pro-71 and Lys-72 illustrated. Lys-72 is trimethylated in the yeast cytochrome. The orientations of the protein structure in Figs 4 and 5 are identical

mutant at this position with those of the wild-type protein [126]. The electronic spectrum of this mutant (pH 7.2) exhibited a shift in the Soret maximum from 408 nm to 404 nm and a reduced intensity of the 695 nm band that is generally assigned as an S → Fe charge transfer band. These observations combined with an increased intensity of the 695 nm band at lower pH suggested that a significant fraction of this mutant exists as the alkaline form of the protein at near-neutral pH. Second derivative UV spectroscopy of the mutant and wild-type proteins was employed to establish that at this lower pH, the degree of tyrosyl exposure to solvent in the mutant and wild-type proteins is similar.

Titration of the wild-type and Thr-71 proteins with guanidine HCl indicated that the stability of the mutant to unfolding by this denaturant was within error of the wild-type protein [126]. Differences in the kinetics of cytochrome *c* refolding as studied by rapid dilution of the protein from 3 mol L^{-1} guanidine HCl to 0.3 mol L^{-1} denaturant in a stopped-flow spectrophotometer could be attributed to a reduction in the pK_a for formation of the alkaline conformation of cytochrome *c* in the mutant protein. Recently, the replacement of Pro-71 with Gly residue has been shown to reduce the alkaline pK of the protein from 8.45 observed for the wild-type protein to 6.71. Analysis of this mutant by pH-jump kinetics as described by Davis et al. [114] indicated that this change arises primarily from a substantial increase in the rate of formation of the alkaline conformation rather than through a change in pK_a of the group that presumably undergoes a conformationally-linked deprotonation or through a change in the rate of alkaline conformation conversion to the deprotonated, native protein conformation [127]. This effect of Gly substitution for Pro-71 presumably explains the altered folding/unfolding kinetics of this mutant that have been reported in some detail [128].

Initial attempts at producing the Thr mutant of Pro-35 were unsuccessful in producing sufficient quantities of the mutant protein for functional analysis [129], suggesting that cytochrome *c* is significantly more sensitive to modifications at this prolyl residue than at Pro-71. Either the Thr-35 mutant is not properly processed to mature cytochrome *c*, it is less thermally stable than wild-type *iso*-2-cytochrome *c*, or its functional properties are sufficiently perturbed that it cannot function adequately under physiological conditions to support yeast growth.

5.2.2 Proline-30

Mutation of Pro-30 is of interest because the backbone carbonyl group of this residue forms a hydrogen bond with the ND1 nitrogen of the axial His ligand to the heme iron that may be involved in determining the reduction potential of the protein as well as influencing the conformation of the protein. To evaluate this possibility, Gooley and MacKenzie have replaced this residue with alanine in *Rhodobacter capsulatus* cytochrome c_2 and have studied the electrochemical and NMR properties of the resulting mutant (Note: This residue occurs at position

35 in the numbering scheme for the *R. capsulatus* sequence) [87]. The effects of this mutation on the reduction potential of the protein are relatively small, with only a 10 mV decrease observed. The dynamics of the protein, however, appear to be modified to a detectable degree as determined by two criteria. First, the flip rates of Tyr-48 and Phe-46 (tuna numbering scheme) are clearly increased in the mutant protein relative to the wild-type even though the proximity of these two residues to each other and the hydrogen bond formed by Tyr-48 with heme propionate-6 remain intact in the mutant. This finding suggests that the time-average structure of the protein in this region is unaffected by the replacement of Pro-30 while the dynamic attributes of the protein are increased. This view is further supported by increases in His-18 ND1 (38-fold) and Gly-34 NH (> 100-fold) proton exchange rates produced by the mutation combined with chemical shift data that suggest that the positions of these residues have not been significantly affected by the mutation [87].

5.2.3 Omega Loop Deletions

At present, few applications of site-directed mutagenesis to the study of electron transfer metalloproteins have considered the functional roles of larger structural elements or domains. Comparison of the three-dimensional structures of cytochromes *c* from a number of evolutionarily-diverse sources, however, suggests that evolutionary changes in this protein often involve significant alterations in loop structures [130]. Use of this principle as a guide to the use of site-directed mutagenesis in the study cytochrome *c* has been developed by Fetrow et al. in two ways [131]. Both lines of study were coordinated with parallel molecular graphics modelling work in an attempt to increase the probability of constructing viable mutant proteins. Furthermore, mutations involving the His ligand to the iron and involving the cysteinyl residues that are involved in formation of thioether linkages to the heme were avoided.

In the first line of study, four Ω-loops were deleted one at a time to evaluate their functional significance: 22–28, 74–78, 37–40, and 43–50. Second, one of these loop regions, 22–29, was replaced with five loops of varying size, both longer and shorter. Attempts to express these mutants revealed that deletion of either 22–28 or 74–78 resulted in failure of the transformed yeast to produce cytochrome *c* for reasons that are difficult to identify at present. On the other hand, deletion of 37–40 simply reduced the level of cytochrome *c* expression to 40% of wild-type while deletion of 43–50 reduced the level of expression to 60% (Fig. 5). The ability of yeast to produce these latter two mutants and to grow on lactate, establishes that while deletion of these loop structures may reduce the efficacy of the cytochrome in supporting cell growth, they are not absolutely required. Replacement of the 22–28 loop with peptides corresponding to those present in this region in three other species of cytochrome *c* (i.e. tuna, *R rubrum*, *P. denitrificans*) results in a protein that appears to function adequately in yeast.

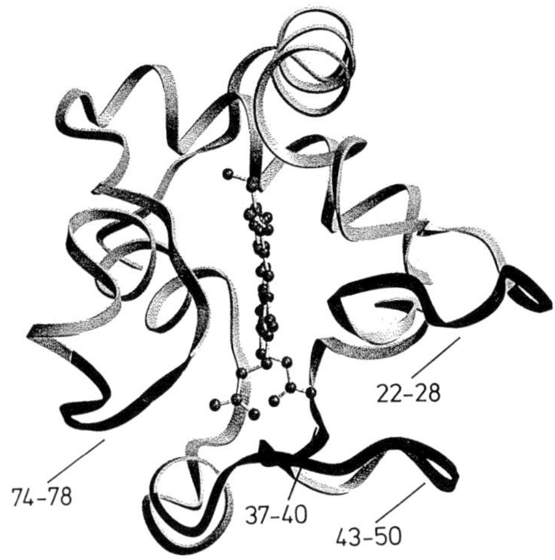

Fig. 5. Ribbon diagram of yeast *iso*-1-cytochrome *c* (reduced). The peptide loops studied by Fetrow et al. [131] are rendered in black. Deletion of the 43–50 and 37–40 loops led to production of mutant cytochromes, while attempts to express genes coding for proteins containing deletion of residues 22–28 or 74–78 failed to produce protein

In fact, the tuna-analogue was indistinguishable from wild-type iso-1-cytochrome *c* under the conditions studied. On the other hand, replacement of the 22–28 loop with one corresponding to the sequence found in *P. aeruginosa* cytochrome c_{551} resulted in production of cytochrome *c* to a level 30% of that of the yeast protein as determined spectroscopically, though the transformed yeast could not grow on glycerol or lactate. Replacement of this same loop with an unrelated sequence (residues 216–224 from porcine pancreatic esterase, which was determined to be structurally compatible with replacement at this position in cytochrome *c*) resulted in production of about 80% the normal level of cytochrome *c* and with limited ability of the resulting mutant to support yeast growth.

While the results of this work are encouraging, it is clear that the structural definition of mutant proteins of this type is critical to development of rational interpretation of the results if for no other reason than that the structural perturbation introduced is presumably greater than for simple point mutations. Moreover, it would be particularly interesting to compare the functional properties of mutants compared in this manner in assays involving protein-protein reactions relevant to the species of cytochrome *c* on which the mutagenesis is based. For example, comparison of the activities of wild-type yeast cytochrome *c* with that of a loop-insertion mutant modelled on a photosynthetic cytochrome *c* in the reaction with the photosynthetic reaction center could help define the structural elements involved in the cytochrome *c* binding domain for the reaction center.

5.3 Mutations of Cytochrome c that Primarily Affect Electrostatic Properties of the Protein

5.3.1 Arginine-38

Arginine-38 has been of considerable interest to a number of investigators, because the guanidino side chain of this residue occupies an internal location that results in formation of a hydrogen bond with heme propionate-6 [132, 133]. This interaction (Fig. 6) has been suggested to be involved in regulating the ionization behavior of this propionate group and in variation of the cytochrome *c* reduction potential with pH. Mutations at this position seemed likely to produce an increase in the pK_a of heme propionate-6 so that this group might titrate within the range of pH that the protein is stable. Such a change in electrostatic properties should be detectable either directly by NMR or indirectly through its effect of the mutation on the variation in reduction potential with pH in the mutant proteins.

To evaluate the functional role(s) of this residue, six mutations, Lys, His, Glu, Asn, Leu, and Ala, were introduced at this site, and the electrochemical and NMR properties of the resulting proteins examined [134]. Contrary to expectation, removal of Arg-38 did not result in a change in the dependence of the cytochrome *c* reduction potential on pH. Instead, as the electron-withdrawing ability of the residue substituted at position-38 decreased, the reduction potential also decreased, with the greatest decrease (50 mV) observed for the Ala mutant. The variation of reduction potential with pH, however, remained essentially the same as that previously observed for the wild-type protein.

Some of these mutants (Asn-38, Glu-38 and Ala-38) were examined by NMR spectroscopy to assess the influence of substitutions at position-38 on the pK_a of the adjacent His-33 and His-39 residues. The results of these experiments indicated that the behavior of these two residues was minimally perturbed by these substitutions. Interestingly, theoretical electrostatic calculations, based on

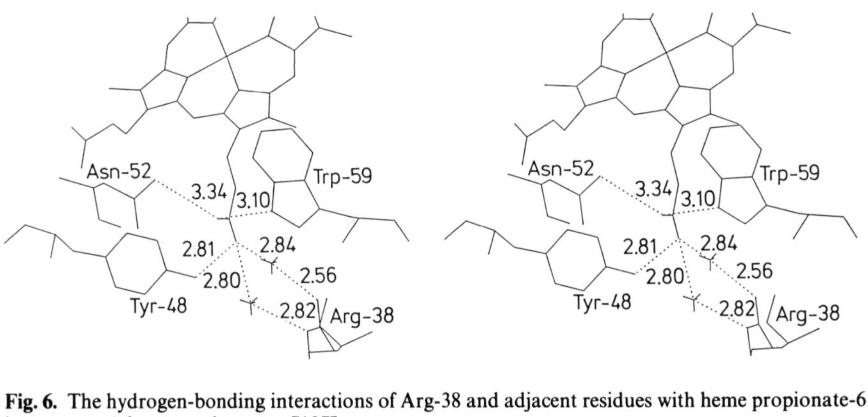

Fig. 6. The hydrogen-bonding interactions of Arg-38 and adjacent residues with heme propionate-6 in yeast *iso*-1-ferrocytochrome *c* [107]

the method, of the change in reduction potential resulting from substitution of Leu for Arg at position-38 were consistent with the experimentally observed potential of this mutant. With this series of mutants, however, there is no significant change in the alkaline transition, which indicates that the observed changes in reduction potential are not a secondary consequence of a decrease in the alkaline pK_a. We conclude that these mutations alone are insufficient to increase the pK_a of heme propionate-6 or that the pK_a of this group is abnormally high in the wild-type protein.

5.3.2 Lysine-72

Lysine-72 is a surface residue in cytochrome *c* (Fig. 4) that is of particular interest for two related reasons. First, in yeast *iso*-1-cytochrome *c*, this residue undergoes post-synthetic modification to trimethyllysine (see references provided in Refs. [86 and 87]), a modification that has both steric and electrostatic implications for the surface properties of the protein. Second, this residue has been implicated as having a critical role in the electrostatic recognition of cytochrome *c* by redox-active heme proteins with which it interacts physiologically (e.g., Ref. [101]). Mutagenic investigation of the functional role of this residue in yeast *iso*-1-cytochrome *c* was initiated by the work of Holzschu et al. in which the Arg-72 mutant of this protein was prepared and characterized in terms of its stability to guanidine HCl denaturation and its steady-state activity with flavocytochrome b_2, cytochrome *c* peroxidase and cytochrome *c* oxidase [135]. While titration of the mutant protein with guanidine was consistent with slight destabilization of the variant relative to the wild-type protein, the steady-state kinetics parameters for ferrocytochrome *c* oxidation by cytochrome *c* peroxidase [pH 6.0, 100 mmol L^{-1} potassium phosphate buffer, 25°C] and by yeast cytochrome *c* oxidase [spectrophotometric assay; pH 7, 20 mmol L^{-1} phosphate buffer, 25°C] were identical for the two proteins. On the other hand, the kinetics for steady-state reduction of cytochrome *c* by flavocytochrome b_2 exhibited a near four-fold increase in K_m and two-fold increase in V_{max} (pH 7, 20 mmol L^{-1} potassium phosphate buffer, 10 mmol $^{-1}$ lactate, 25°C). Overall, however, the effects of replacing trimethyllysine-72 with an arginyl residue as revealed in this study are strikingly minimal. As the authors note, this finding is surprising insofar as the postsynthetic modification of Lys-72 by yeast constitutes a substantial metabolic investment, the value of which is currently not apparent.

As part of a subsequent study concerning primarily second-site revertant yeast *iso*-1-cytochrome *c* variants, Hazzard et al. evaluated the effect of converting Lys-72 to an aspartyl residue by site-directed mutagenesis on the electron transfer kinetics of the cytochrome *c*-cytochrome *c* peroxidase complex [136]. Lys-72 was of interest for this purpose, because it is involved in the hypothetical model for the complex formed by these two proteins that was proposed by Poulos and Kraut on the basis of molecular graphics docking [106]. In these experiments it was found that the rate of cytochrome *c* reduction by lumiflavin semiquinone is not affected significantly by conversion of Lys-72 to Asp and that the rates of reduction of both proteins by this reductant are decreased to a similar degree by binding of the cytochromes to cytochrome *c* peroxidase (pH 7.0, 3 mmol L^{-1} phosphate buffer, 0.5 mmol L^{-1} EDTA, μ, = 8 mM). These results are interpreted in terms of a similar degree of heme exposure to solvent (and, by implication, to flavin semiquinone) in the mutant and wild-type protein. In the same study, however, it was demonstrated that yeast *iso*-1 and *iso*-2-cytochromes differ significantly in their rate of reduction by flavin semiquinone even though the amino acid sequences of these two proteins in the vicinity of the heme group exhibit a high degree of identity. As a result, the interpretation of differences in reaction rates of this type solely in terms of heme accessibility to solvent is not altogether unambiguous.

In this same study [136], the rate of electron transfer from the ferrocytochrome to cytochrome *c* peroxidase compound I within the electrostatically stabilized complex formed by the two proteins was shown to increase from about 245 s^{-1} to 440 s^{-1} (pH 7, μ = 8 mmol L^{-1}) on replacement of Lys-72 by Asp. This observation was interpreted in terms of a mutation-induced destabilization of the cytochrome *c*-cytochrome *c* peroxidase complex that results in a more flexible complex that exhibits more facile electron transfer. This conclusion is consistent with previous arguments by these authors based on the finding that the limiting rates of electron transfer observed for this same reaction between the wild-type proteins at high ionic strength is greater than that observed at low ionic strength [137]. It seems equally possible, however, that the relatively small changes in rate that occur with increasing ionic strength (2–3 fold) could arise from one or more perturbations to this system that are dependent on ionic strength. Such effects could include relatively minor adjustments in the geometry of the docked complex to optimize donor-acceptor orbital overlap or variation in the reduction potentials of the two heme centers within the protein-protein complex that in turn could be linked to small changes in the docking geometry.

5.4 *Mutations of Cytochrome c that Affect Axial Ligation*

Relatively little attention has been directed at perturbation of the heme iron ligands in cytochrome *c* through site-directed mutagenesis. In part, the reason for this apparent oversight is that the expression systems currently employed to produce this protein require that any mutant produced provide some minimal level of physiological competence for the host yeast to grow in liquid culture. As most modifications of the cytochrome *c* heme iron ligands can be expected to result in substantial changes in the reduction potential of the protein, relatively few such mutations are likely to support adequate yeast growth.

Sorrell and co-workers have used site-directed mutagenesis of yeast *iso*-2-cytochrome *c* to identify what may be the first functional axial ligand mutant of cytochrome *c* [138]. In this work, the His-18 ligand has been replaced with an Arg residue to produce a mutant that supports yeast growth on glycerol. This protein exhibits the 695 nm band associated with coordination of the Met-80 sulfur, and does not bind carbon monoxide or dioxygen. Interestingly, cyclic voltammetry of this protein at a tin-doped indium oxide electrode demonstrated that this variant has the same midpoint potential as the wild-type protein while reacting much less efficiently with the electrode. This result may suggest that the reorganization energy for the mutant is significantly greater than that of the wild-type protein and its rate of electron transfer correspondingly low. Currently, however, the authors acknowledge that evidence verifying that Arg-18 is in fact coordinated to the heme iron is limited, presumably as the result of difficulties in obtaining sufficient quantities of the mutant for thorough physical analysis.

6 Concluding Observations

Prior to the advent of site-directed mutagenesis as a viable technique for the production of specifically modified proteins, the last major event to exert a major influence on the study of protein structure and function was the development of X-ray diffraction analysis for the detailed structural analysis of macromolecules. In the intervening thirty years, the availability of protein structures obtained in this manner combined with a wide range of physical and chemical studies of these proteins allowed development of substantial insight into the relationship between the structure of a protein and its functional attributes. There was some reason to expect, therefore, that functional characterization of specifically mutated proteins based on understanding developed with more classical techniques should permit efficient confirmation of existing hypotheses, particularly for proteins for which the available literature is as extensive as that for cytochrome *c*.

The studies described above, and particularly those concerning cytochrome *c*, demonstrate that this expectation was somewhat naive. In the case of cytochrome *c*, several mechanistic lessons have been learned from the extensive studies that have already been undertaken, though these lessons were often of a different nature than originally anticipated. It is now apparent that the challenge is to design studies using mutants based on the available information and to be prepared for the unanticipated. Rather than negating the value of site-directed mutagenesis as an implement for surgical intervention in protein rennovation, this situation provides a route to observations that are no doubt more stimulating than any that our preconceived notions could contrive.

Acknowledgements. Work in the author's laboratory concerning studies of recombinant heme proteins has involved a continuing collaboration over the past several years with Professors Gary D. Brayer, Ross MacGillivray and Michael Smith in the Department of Biochemistry at the

University of British Columbia, Professor Brian M. Hoffman (Northwestern University), Dr. Geoffrey R. Moore (University of East Anglia) and Professor Cyril M. Kay (University of Alberta) without whose enthusiastic contributions this work would not have been possible. The experimental skills and insights of Gary Pielak, David Goodin, Walter Funk, Robert Cutler, Gordon Louie, Terence Lo, Linda Pearce, Bhavini Sishta, Paul Barker, Steven Rafferty and Marcia Mauk have been critical to the success of this work. I thank Terence Lo for assistance in preparation of the figures. Financial support for work in the author's laboratory has been provided by NIH Grant GM-33804 and MRC Grant MT-7182.

7 References

1. Smith M (1985) Ann Rev Genet 19: 423
2. Smith M (1986) Phil Trans R Soc Lond A317: 295
3. Carter P (1986) Biochem J 237: 1
4. Zoller MJ, Smith M (1987) Methods Enzymol 154: 329
5. Higuchi R, Krummel B, Saiki R (1988) Nucleic Acids Res 16: 7351
6. Nelson RM, Long GL (1989) Anal Biochem 180: 147
7. Ernst JF, Hampsey DM, Stewart JW, Rackovsky S, Goldstein D, Sherman F (1985) J Biol Chem 260: 13225
8. Hampsey DM, Das G, Sherman F (1988) FEBS Lett 231: 275
9. Hutchison CA, III, Phillips S, Edgell MH, Gillam S, Jahnke P, Smith M (1978) J Biol Chem 253: 6551
10. Razin A, Hirose T, Itakura K, Riggs AD, (1978) Proc Natl Acad Sci USA 75: 4268
11. Sowdhamini R, Srinivasan N, Shoichet B, Santi DV, Ramakrishnan C, Balaram P (1989) Prot Eng 3: 95
12. Sivaraja M, Goodin DB, Smith M, Hoffman BM (1989) Science 245: 738
13. Funk WD, MacGillivray RTA, Mason AB, Brown SA, Woodworth RC (1990) Biochemistry 29: 1654
14. Milligan DL and Koshland DE, Jr (1988) J Biol Chem 263: 6268
15. Bowler BE, Meade TJ, Mayo SL, Richards JH, Gray HB (1989) J Am Chem Soc 111: 8757
16. von Bodman SB, Schuler MA, Jollie DR, Sligar SG (1986) Proc Natl Acad Sci USA 83: 9443
17. Sligar SG, Egeberg KD, Sage JT, Morikis D, Champion PM (1987) J Am Chem Soc 109: 7896
18. Stayton PS, Fisher MT, Sligar SG (1988) J Biol Chem 263: 13544
19. Rodgers KK, Pochapsky TC, Sligar SG (1988) Science 240: 1657
20. Funk WD, Lo TL, Mauk MR, Brayer GD, MacGillivray RTA, Mauk AG (1990) Biochemistry 29: 5500
21. Fang H, Lin R-J, Gennis RB (1989) J Biol Chem 264: 8026
22. Reid GA, White S, Black MT, Lederer F, Mathews FS, Chapman SK (1988) Eur J Biochem 178: 329
23. Black MT, White SA, Reid GS, Chapman SK (1989) Biochem J 258: 255
24. White Sa, Black MT, Reid GA, Chapman SK (1989) Biochem J 263: 849
25. Atkins WM, Sligar SG, (1988) J Biol Chem 263: 18842
26. Atkins WM, Sligar SG (1989) J Am Chem Soc 111: 2715
27. Imai M, Shimada H, Watanabe Y, Matsushima-Hibiya Y, Makino R, Koga H, Horiuchi T, Ishimura Y (1989) Proc Natl Acad Sci USA 86: 7832
28. Atkins WM, Sligar SG (1990) Biochemistry 29: 1271
29. Sadeque AJM, Shimizu T, Hirano K, Hatano M, Fujii-Kuriyama Y (1988) Inorg Chim Acta, 153: 161
30. Furuya H, Shimizu T, Hatano M, Fujii-Kuriyama Y (1989) Biochem Biophys Res Commun 160: 669
31. Shimizu T, Hirano K, Takehashi M, Hatano M, Jujii-Kuriyama Y (1988) Biochemistry 27: 4138
32. Furuya H, Shimizu T, Hirano K, Hatano M, Fujii-Kuriyama Y, Raag R, Poulos TL (1989) Biochemistry 28: 6848
33. Lindberg RLP, Negishi M (1989) Nature 339: 632

34. Goodin DB, Mauk AG, Smith M (1986) Proc Natl Acad Sci USA 83: 1295
35. Fishel LA, Villafranca JE, Mauro JM, Kraut J (1987) Biochemistry 26: 351
36. Goodin DB, Mauk AG, Smith M (1987) J Biol Chem 262: 7719
37. Mauro JM, Fishel LA, Hazzard JT, Meyer TE, Tollin G, Cusanovich MA, Kraut J (1988) Biochemistry 27: 6243
38. Miller MA, Hazzard JT, Mauro JM, Edwards SL, Simons PC, Tollin G, Kraut J (1988) Biochemistry 27: 9081
39. Smulevich G, Mauro JM, Fishel LA, English AM, Kraut J, Spiro TG (1988) Biochemistry 27: 5477
40. Smulevich G, Mauro JM, Fishel LA, English AM, Kraut J, Spiro TG (1988) Biochemistry 27: 5486
41. Edwards SL, Mauro JM, Fishel LA, Wang JM, Miller MA, Xuong NH, Kraut J (1988) Prog Clin Biol Res 274: 463
42. Scholes CP, Liu Y, Fishel LA, Farnum MF, Mauro JM, Kraut J (1989) Israel J Chem 29: 85
43. Erman JE, Vitello LB, Mauro JM, Kraut J (1989) Biochemistry 28: 7992
44. Miller MA, Coletta M, Mauro JM, Putnam LD, Farnum MF, Kraut J, Traylor TG (1990) Biochemistry 29: 1777
45. Springer BA, Sligar SG (1987) Proc Natl Acad Sci USA 84: 8961
46. Springer BA, Egeberg KD, Sligar SG, Rohlfs RJ, Mathews AJ, Olson JS (1989) J Biol Chem 264: 3057
47. Braustein D, Ansari A, Berendzen J, Cowen BR, Egeberg KD, Frauenfelder H, Hong MK, Ormos P, Sauke TB, Scholl R, Schulte A, Sligar SG, Springer BA, Steinbach PJ, Young RD (1988) Proc Natl Acad Sci USA 85: 8497
48. Morikis D, Champion PM, Springer BA, Sligar SG (1989) Biochemistry 28: 4791
49. Olson Js, Mathews AJ, Rohlfs RJ, Springer BA, Egeberg KD, Sligar SG, Tame J, Renaud J-P, Nagai K (1988) Nature 336: 265
50. Lambright DG, Balasubramian S, Boxer SG (1989) J Mol Biol 207: 289
51. Varadarajan R, Lambright DG, Boxer SG (1989) Biochemistry 28: 3771
52. Varadarajan R, Zewert TE, Gray HB, Boxer SG (1989) Science 243: 69
53. Hughson FM, Baldwin RL (1989) Biochemistry 28: 4415
54. Shimada H, Dong A, Matsushima-Hibiya Y, Ishimura Y, Caughey WS (1989) Biochem Biophys Res Commun 158: 110
55. Nagai K, Thorgersen HC (1984) Nature 309: 810
56. Nagai K, Perutz MF, Poyart C (1985) Proc Natl Acad Sci USA 82: 7252
57. Luisi BF, Nagai K, Perutz MF (1987) Acta Haematol 78: 85
58. Olson JS, Mathews AJ, Rohlfs RJ, Springer BA, Egeberg KD, Sligar SG, Tame J, Renaud JP, Gagai K (1988) Nature 336: 265
59. Boissel J-P, Kasper TJ, Bunn HF (1988) J Biol Chem 263: 8443
60. Ishimori K, Morishima I, Imai K, Fushitani K, Miyazaki G, Shih D, Tahe J, Pagnier J, Nagai K (1989) J Biol Chem 264: 14624
61. Karlsson BG, Aasa R, Malmström BG, Lundberg LG (1989) FEBS Lett 253: 99
62. Pascher T, Bergström J, Malmström BG, Vänngard T, Lundberg LG (1989) FEBS Lett 258: 266
63. van de Kamp M, Floria R, Hali FC, Canters GW (1990) J Am Chem Soc 112: 907
64. Martin A, Burgess BK, Stout CD, Cash V, Dean DR, Jensen GM, Stephens PJ (1990) Proc Natl Acad Sci USA 87: 598
65. Huber M, Lerch K (1988) Biochemistry 27: 5610
66. Larsson A, Karlsson M, Sahlin M, Sjöberg BM (1988) J Biol Chem 263: 17780
67. Beyer, WF Jr, Fridovich I, Mullenbach Gt, Hallewell R (1987) J Biol Chem 262: 11182
68. Banci L Bertini I, Luchinat C, Hallewell RC (1988) J Am Chem Soc 110: 3629
69. Bertini I, Banci L, Luchinat C, Hallewell RC (1988) Ann N Y Acad Sci 542: 37
70. Hallewell RA, Laria I, Tabrizi A, Carlin G, Getzoff ED, Tainer JA, Cousens LS, Mullenbach GT (1989) J Biol Chem 264: 5260
71. Bertini I, Banci L, Luchinat C, Bielski BHJ, Cabelli DE, Mullenbach GT, Hallewell RA (1989) J Am Chem Soc 111: 714
72. Horton PJ, Borders CL Jr, Beyer WE Jr (1989) Arch Biochem Biophys 269: 114
73. Noren CJ, Anthony-Cahill SJ, Griffith MC, Schultz PG (1989) Science 244: 182
74. Bain JD, Glabe CG, Dix TA, Chamberlin AR (1989) J Am Chem Soc 111: 8013
75. Anthony-Cahill SJ, Griffith MC, Noren CJ, Suich DJ, Schultz PG (1989) Trends Biochem Sci 14: 400

76. Leinfelder W, Zehelein E, Mandrand-Berthelot M-A, Bock A (1988) Nature 331: 723
77. Söll D (1988) Nature 331: 661
78. Go M, Miyazawa S (1980) Int J Pept Protein Res 15: 211
79. Barker PD, Mauk MR, Mauk AG (in preparation)
80. Sambrook J, Fritsch EF, Maniatis T (1989) Molecular cloning A laboratory manual, 2nd edn Cold spring Harbor Laboratory Press
81. Scharf SJ, Horn GT, Erlich HA (1986) Science 233: 1076
82. Erlich HA (ed) PCR technology Principles and applications for DNA amplification, Stockton Press, New York
83. Sligar SG, personal communication
84. Self SJ, Hunter CN, Leatherbarrow RJ (1990) Biochem J 265: 599
85. Pielak GJ, Mauk AG, Smith M (1985) Nature 313: 152
86. Evans MJ, Scarpulla RC (1988) Mol Cell Biol 8: 35
87. Gooley PR, MacKenzie NE (1990) FEBS Lett 260: 225
88. Dumont ME, Ernst JF, Hampsey DM, Sherman F (1987) EMBO J 6: 235
89. Drygas ME, Lambowitz AM, Nargang FE (1989) J Biol Chem 264: 17897
90. Scarpulla RC and Nye SH (1986) Proc Natl Acad Sci USA 83: 635
91. Luntz TL, Schejter A, Garber EAE, Margoliash E (1989) Proc Natl Acad Sci U.S,A 86: 3524
92. Clements JM, OConnell LI, Tsunasawa S, Sherman F (1989) Gene 83: 1
93. Tanaka Y, Ashikari T, Shibano Y, Amachi T, Yoshizumi H, Matsubara H (1988) J Biochem 103: 954
94. Goodin DB, Mauk AG, Smith M (unpublished)
95. Varadarajan R, Szabo A, Boxer SG (1985) Proc Natl Acad Sci 82: 5681
96. Dodson G, Hubbar RE, Oldfield TJ, Smerdon SJ, Wilkinson AJ (1988) Prot Eng 2: 233
97. Unger BP, Gunsalus IC, Sligar SG (1986) J Biol Chem 261: 1158
98. Barker PD, Mauk AG (unpublished results)
99. Finzel BC, Poulos TL and Kraut J (1984) J Biol Chem 259: 13027
100. Asseffa A, Smith SJ, Nagata K, Gillette J, Gelboin HV, Gonzalez FJ (1989) Arch Biochem Biophys 274: 481
101. Gonzalez F (1989) Pharmacol Revs 40: 243
102. Poulos TL, Finzel BC, Howard AJ (1987) J Mol Biol 195: 687
103. MacGillivray RTA (personal communication)
104. Smith M, Leung Dw, Gillam S, Astell Cr, Montgomery DL, Hall BD (1979) Cell, 16: 753
105. Montgomery DL, Leung DW, Smith M, Shalit P, Fay G, Hall BD (1980) Proc Natl Acad Sci USA 77: 541
106. Poulos TL and Kraut J (1990) J Biol Chem 255: 10322
107. Cutler RL, Pielak GJ, Mauk AG, Smith M (1987) Prot Eng 1: 95
108. Rafferty SP, Pearce LL, Barker PD, Guillemette G, Kay CM, Smith M, Mauk AG, Biochemistry (in press)
109. Inglis SV, Smith M (unpublished)
110. Motonaga K, Miśaka E, Nakajima E, Ueda S, Nakanishi K (1965) J Biochem (Tokyo) 57: 22
111. Mauk MR and Mauk AG (1988) J Chromatog 439: 408.
112. Louie GV, Hutcheon WLB, Brayer GD (1988) J Mol Biol 199: 295
113. Pearce LL, Gärtner AL, Smith M, Mauk AG (1989) Biochemistry 28: 3152
114. Davis LA, Schejter A, Hess GP (1974) J Biol Chem 249: 2624
115. Pielak GJ, Oikawa K, Mauk AG, Smith M, Kay CM (1986) J Am Chem Soc 108: 2724
116. Liang N, Pielak GJ, Mauk AG, Smith M, Hoffman BM (1987) Proc Natl Acad Sci USA 84: 1249
117. Liang N, Mauk AG, Pielak GJ, Johnson J, Smith M, Hoffman BM (1988) Science 240: 311.
118. Nocek JM, Liang N, Wallin SA, Mauk AG, Hoffman BM (1990) J Am Chem Soc 112: 1623
119. Stemp EDA, Nocek JM, Mauk AG, Margoliash E, Hoffman BM (1989) J Inorg Biochem 36: 212 (1989)
120. Michel B, Mauk AG, Bosshard HR (1989) FEBS Lett 243: 149
121. Pielak GJ, Boyd J, Moore GR, Williams RJP (1988) Eur J Biochem 177: 167
122. Pielak GJ, Atkinson RA, Boyd J, Williams RJP (1988) Eur J Biochem 177: 179
123. Louie GV, Pielak GJ, Smith M, Brayer GD (1988) Biochemistry 27: 7870
124. Louie GV, Brayer GD (1989) J Mol Biol 209: 313
125. Takano T, Dickerson RE (1980) Proc Natl Acad Sci USA 77: 6371
126. White TB, Berget PB, Nall BT (1987) Biochemistry 26: 4358

127. Nall BT, Zuniga EH, White TB, Wood LC, Ramdas L (1989) Biochemistry 28: 9834
128. Wood LC, White TB, Ramdas L, Nall BT (1988) Biochemistry 27: 8562
129. Wood LC, Muthukrishnan K, White TB, Ramdas L, Nall BT (1988) Biochemistry 27: 8554
130. Dickerson RE (1981) In: Srinivasan R (ed) Biomolecular structure, conformation, function, evolution vol 1 Pergamon, New York
131. Fetrow JS, Cardillo TS, Sherman F (1989) Proteins 6: 372
132. Moore GR, Harris DE, Leitch Fa, Pettigrew GA (1984) Biochim Biophys Acata 764: 331
133. Churg AK, Warshel A (1986) Biochemistry 25: 1675
134. Cutler RL, Davies AM, Creighton S, Warshel A, Moore GR, Smith M, Mauk AG (1989) Biochemistry 28: 3188
135. Holzschu D, Principio L, Conklin KT, Hickey Dr, Short J, Rao R, McLendon G, Sherman F (1987) J Biol Chem 262: 7125
136. Hazzard JT, McLendon GL, Cusanovich MA, Das G, Sherman F, Tollin G (1988) Biochemistry 27: 4445
137. Hazzard JG, McLendon G, Cusanovich MA, Tollin G (1988) Biochem Biophys Res Commun 151: 429
138. Sorrell TN, Martin PK, Bowden EF (1989) J Am Chem Soc 111: 766

Control of Biological Electron Transport via Molecular Recognition and Binding: The "Velcro" Model

G. McLendon

Department of Chemistry, University of Rochester, Rochester, NY 14627, USA

Some of the factors which control molecular recognition and binding between electron transfer proteins have been elucidated from studies of the interaction between cytochrome *c* and its redox partners, particularly cytochrome *c* peroxidase (ccp). Data from a number of labs point towards a new motif for (macro)molecular recognition: not the classical lock and key of enzymology, but a variety of overlapping binding sites which create broadly complementary patches of charge and hydrophobicity on the partners, which consequently stick on contact, like "velcro". The initially formed complex may then undergo dynamic reorientation to obtain an optimal orientation for electron transfer. If slow, this reorientation may be rate limiting (gating), while if fast, may contribute to the observed reorganization energy. A biological rationale for this motif is briefly discussed.

Structure and Bonding 75

1 Introduction

Two basic problems face the cell in controlling energy flow via coupled electron transfer reactions [1, 2]. The first involves controlling the rates of these processes, so that a level of electron flow appropriate to the metabolism of the cell can be maintained.

The second problem in biological redox chemistry focuses not on rate, per se, but on specificity, (ensuring the electron is delivered between the *correct* proteins) as controlled by molecular recognition and binding. At a physico-chemical level, the control of the electron transfer rate depends on the values of key physical parameters like the distance between redox sites, the reaction free energy (ΔG^*), and the amount of motion (the reorganization energy) which accompanies the transfer of charge from the electron donor (protein) to the acceptor.

The importance of each of these parameters at this fundamental level have been reviewed previously, and are also dealt within in some detail elsewhere in this volume.

We therefore will only remark briefly on them here, to the extent that they impact on the second problem of molecular recognition, which serves as the focus of this chapter.

2 Rate Determinants

Distance: The affects of electron donor–acceptor distance on reaction rate arises because electron transfer, like any reaction, requires the wavefunctions of the reactants to mix (i.e. "orbital overlap" must occur). Unlike atom transfer, the relatively weak overlap which can occur at long distances (≥ 10 Å) may still be sufficient to allow reaction at significant rates. On the basis of work with both proteins and models, it is now generally accepted that donor–acceptor electronic coupling, and thus electron transfer rates, decrease exponentially with distance: $k_{et} \approx \nu_{el} \exp^{-(\beta R)}$. FCF where ν_{el} is the frequency of the mode which promotes reaction (previously estimated between 10^{11}–10^{14} s^{-1}) FCF is a Franck Condon Factor explained below, and β is empirically estimated to range from 0.8–1.2 $Å^{-1}$, with a value of $\beta \approx 0.9$ $Å^{-1}$ most common for proteins.

In protein–protein reactions, the donor–acceptor distance is determined by the structure of the reacting proteins, and the way(s) in which they bind and interact. For example, it is generally believed that cytochrome *c* binds to its reaction partners at or near the "exposed" heme edge, in order to minimize the reactant distance and thereby maximize the rate. The redox active centers of most proteins are sufficiently buried that the large protein imposed distances provide low intrinsic reactivity for the proteins with respect to exogenous

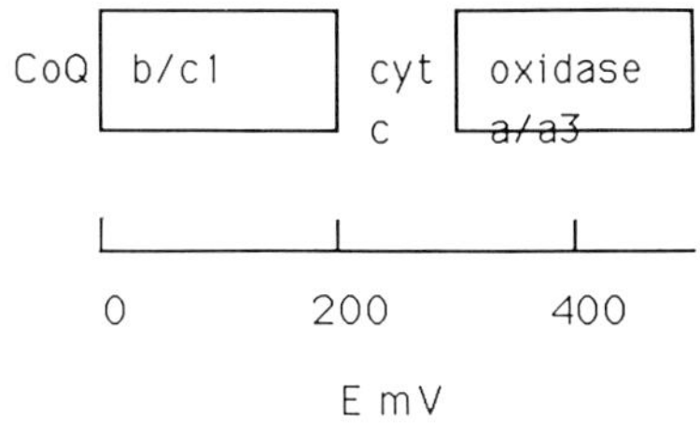

Fig. 1. Simplified diagram of the section of the electron transport system coupled to cyt *c*

mediators. This is surely no accident. Cells contain a number of redox-active components (e.g. ascorbic acid, ubiquinone) which might adventitiously provide "short circuits" between Complex I and Complex IV in the electron transport system (Fig. 1). Indeed, high concentrations of redox mediators like paraquat are often strong cell poisons. Fortunately, the bimolecular rate constants for redox proteins interacting in the non-physiological reactants are sufficiently low (10^2–10^6 M^{-1} s^{-1}) that they cannot compete with the effective first-order rates for specific reaction with a bound physiological redox partner. Thus binding can serve the *kinetic* function of maintaining sufficient interaction time to allow the otherwise improbably long distance electron transfer to occur with specificity, with the bound partner.

As discussed later, if binding can occur at more than one site, then more than one rate may occur. Alternatively, if the sites interconvert, this interconversion may modulate the electron transfer rate by modulating the distance. This would represent one limiting example of "gated" electron transfer as discussed by Hoffman and Ratner [4], and demonstrated by Pardue [5].

3 Free Energy (and Reorganization Energy)

In the previous section, we alluded to the Franck Condon factors (FCF) in controlling electron transfer rates. For this topic, detailed reviews of theory and experiment are provided elsewhere. In sum, it is now well known that the reaction free energy required to transfer charge can be reduced by the reaction free energy, ΔG°, as summarized in the famous "Marcus equation":[2] $\Delta G^{\neq} = (\Delta G^\circ - \lambda)^2/4\lambda$ where λ, the reorganization energy, is related to the degree of molecular motion both at the active sites in the surrounding protein and solvent medium which accompanies charge transfer. It was once believed that λ was likely to be small for biological redox proteins ($\lambda \leq 0.2$ V) [1, 2]. However, the most detailed recent experiments suggest for proteins are similar to values encountered for reactions in normal solvents; $1 < \lambda < 2$ V [1, 6, 7].

Continuing our focus on biological recognition and control, several points might be made here. When two reactant proteins bind to form a complex, the interactions which bind and stabilize the complex may contribute to λ, and thus

$\Delta G^{\neq}$. It is also true that complexation might affect ΔG. However, the best available data suggests that these effects are rather small [8]. For example, binding cyt c to cyt b_5 does not shift the Em cyt c^+/cyt c *at all*, and only shifts Em cyt b_5^+/cyt b_5 by 10 mV.

Finally, we offer the following caveat. In this chapter, as throughout this volume, we implicitly invoke what I call the "first law of biophysics" 'if we can observe something, it *must* be important.' There is little doubt that the fundamental physical parameters like reactant distance and reaction free energy which control rates in vitro are important in vivo as well. However, we know rather little about the range over which then parameters can be modulated in vivo in response to varying physiological conditions and demands. One recent set of results may provide an interesting and counter-intuitive example. As summarized in Fig. 1, electron transport is thought to proceed on a continuous downhill potential gradient with each step comprising a relatively small potential drop (so as not to waste biological energy). Several assumptions underly this text book diagram: the redox potential measured in vitro are valid in vivo, and these potentials, by Marcus theory, should control the rates of electron transfer.

With the advent of site directed mutagenesis, it has become possible to vary these amino acids which control the redox potential, and thereby in principle to vary the electron transfer rates.

A key result of such investigations [9–12] is that the redox potential of cytochrome c can be lowered from the normal 270 mV to below 200 mV, which is thermodynamically *uphill* from cytochrome c_1.

One might therefore predict that electron transport in a cell which contains this cytochrome c mutant would be blocked, or greatly diminished. However, it has been found that these low potential mutant cytochromes expressed in yeast can in fact support aerobic respiration ie: yeast grow on such non-fermentable substrates as lactate. This growth occurs despite the apparently unfavorable potential of transfer from cytochrome c_1 to cytochrome c, which is necessary to complete the respiration pathway. These observations minimally suggest that the kinetics of respiration in vivo can remain effective over a wide range of free energy. Thus, the control of electron transport kinetics in vivo are not so sensitive to physicochemical parameters as might have been thought.

4 Molecular Recognition and Binding

In the foregoing sections, we have emphasized that in addition to controlling rates, biological systems undertake the more difficult (and perhaps more compelling) task of controlling reaction specificity. As already mentioned specificity is largely controlled by molecular recognition of the donor and acceptor which gives rise to efficient binding. In order to understand some of the

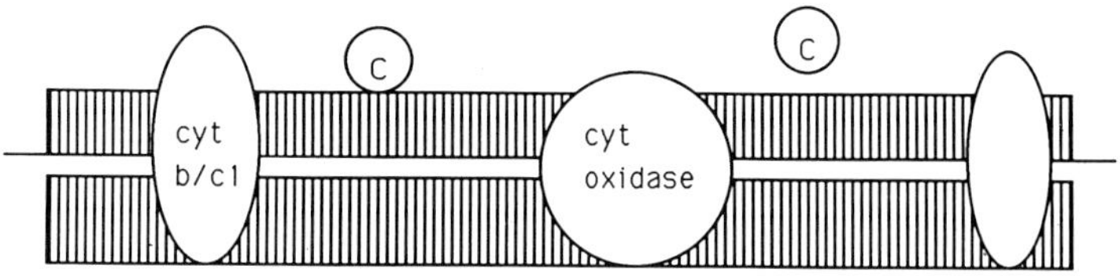

Fig. 2. Fluid mosaic model of an energy transducing membrane

requirements for binding, it is first necessary to briefly consider the structural organization of the electron transport system. In a cell, the energy of electron transport is transduced into a useable chemical potential by coupling electron transport to proton translocation across a membrane (the famous "Chemiosmotic" mechanism). This proton translocation apparently occurs via proteins like the bc_1 complex, and cytochrome oxidase, which span the mitochondrial membrane from the inner to the outer surface.

The most recent model of this membrane, the fluid mosaic model [13] is pictured in cartoon fashion in Fig. 2. In this model, the transduction proteins (complexes I–IV) are randomly dispersed in the membrane and redox equivalents are delivered from one complex to another via the mobile electron carriers cytochrome *c* and ubiquinone. It is necessary that cytochrome *c* be able to move relatively facilely from one complex to another. Thus the binding constants cannot be too high without making the associated "off" rates too slow. Conversely, to prevent unproductive short circuits via cytochrome *c* from complex I directly to IV, there must exist molecular recognition which favors selective binding of cytochrome *c* to bc_1 and cytochrome oxidase (and perhaps *disfavors* binding to complex I or II).

The details of such recognition and binding mechanisms are unclear, in part because the structures of the key enzymes in these processes are unknown at high resolution. The structural information which is available is summarized below. A large body of information from many labs has documented both kinetic and thermodynamic aspects of the binding of cytochrome *c* to cytochrome oxidase [15] (and to a lesser extent, cytochrome c_1)[16]. Both systems show binding which is sensitive to ionic strength (K decreases as μ increases) suggesting that interactions between charged groups play a role in recognition. However, since binding persists even at high ionic strength, other factors (like hydrophobic interactions) must also contribute to binding.

Elegant chemical modification studies [15, 17] by Margoliash, Millett, and others have pinpointed a few specific charged groups (Lys8, 13, 27, 77, 89) on the surface of cytochrome *c* as being involved in electrostatic recognition (Fig. 3). These few lysines are either evolutionarily invariant or highly conserved, consistent with an important biological role in recognition.

More recently, the more subtle modification techniques of site-directed mutagenesis have been applied to the problem of biological recognition of redox proteins. While the results are consistent with the previous electrostatic models,

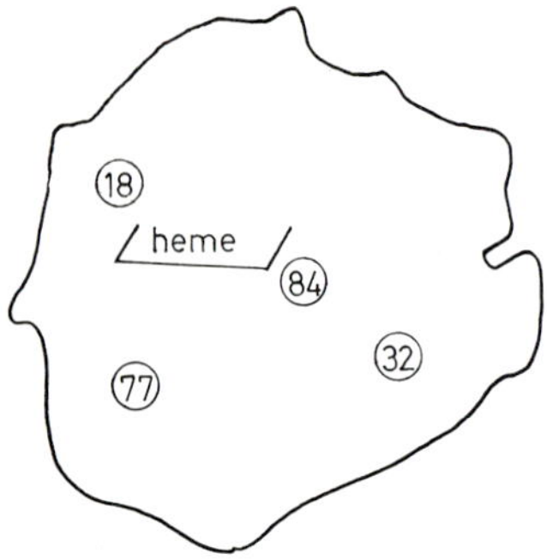

Fig. 3. Schematic of cytochrome *c*, indicating the positions of invariant charged lysines with respect to the heme reactive center

quantitative studies have shown that *individual* charged residues may play a much smaller quantitative role, in binding than had been assumed. Conversely, mutagenesis of e.g. Phe82 suggests [9] that specific van der Waals (hydrophobic) interactions can also play an important role in docking of two redox proteins.

5 Cyt *c*:Ccp a Molecular Paradigm

In order to relate structure and function at a more direct level, it is necessary to focus on systems which have better characterized structure than the complex membrane bound proteins like cyt *c* oxidase. One particularly useful paradigm in this context is the cytochrome *c*–cytochrome *c* peroxidase couple [18]. Ccp is not involved in electron transport, per se: its apparent function [19] is detoxification of hydrogen peroxide via the sequence H_2O_2 + ccp Fe(III) → H_2O + ccp Fe(IV) O (protein)$^{\cdot +}$ compound "ES"

$$\text{ES} + \text{cyt } c(\text{II}) \rightarrow \text{ccp Fe(IV)}$$

$$\text{ccp Fe(IV)} + \text{cyt } c(\text{II}) \rightarrow \text{ccp Fe(III)}$$

Although it is not an electron transport protein, its primary function involves specific electron transfer from cyt *c* to a heme center, in which binding plays a key kinetic role. It has thus become a model for understanding the structurally more complex interactions between cytochrome *c* and its partners in electron transport. This similar cyt *c*:ccp system has several advantages.

First the structures of cytochrome *c*[20] & cytochrome *c* peroxidase [21] are both known at high resolution. Although the precise three dimensional structure of the protein–protein complex is unknown (and, we shall argue, unknowable), molecular modeling has produced detailed stereochemical models for the *c*:ccp complex which are subject to experimental testing and subsequent improvement, as detailed below.

Second, these proteins are particularly stable and because of the particularly rich spectroscopy of their heme porphyrin active sites, a wide variety of physical methods are available to monitor the thermodynamics and kinetics of binding, and subsequent electron transfer reactions, ranging from 2D NMR[22] to laser flash techniques.

Third, the genes for both proteins have been successfully cloned [24]. This cloning provides an approach to test the role of specific residue interactions in binding, by synthesizing proteins with single amino acid replacements. We consider each of these points in the following sections.

Structural data for the cyt *c*:ccp, and the cyt *c*:ccp complex.

The high resolution structure of yeast cytochrome *c*, as solved by Brayer [20], is shown in Fig. 4. Several key aspects of this structure include: 1) The heme group is slightly exposed to solvent around pyrrole ring III and the thioether bridge. Because of the strong distance dependence of rate, electron transfer probably takes place near this edge. 2) This heme edge is surrounded by the group of important positively charged residues (lysines 17, 18, 21, 77, 88) already referred to. This positive charge patch, (often whimsically dubbed the "ring of fire"), provides a logical focus for electrostatic orientation and binding of cytochrome *c*. In yeast cyt *c*, the natural partner of ccp, there are two changes among these charged residues: Lys18 → Arg, and Lys77 is trimethylated (which prevents direct "salt bridge" H-bond formation).

It is interesting that there is a small negative charge cluster on the opposing face of cytochrome *c* (dubbed the proctol surface by Mauk). This charge separation leads to a huge apparent dipole moment for cytochrome *c* estimated (in vacuo) at 200 Debye![15] In addition to charged groups at the apparent

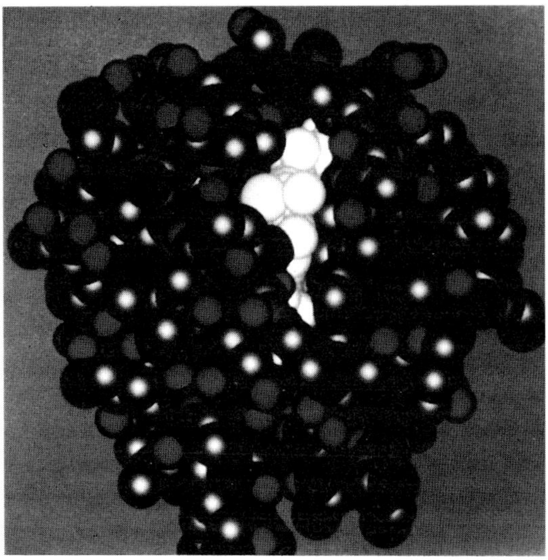

Fig. 4. Structure of tuna cytochrome *c*, based on the coordinates of Takano

binding interface, it is important to realize that a number of hydrophobic amino acids, e.g. Phe82, are exposed at the surface around the reactive heme edge. These amino acids provide a way for hydrophobic interactions to contribute to the free energy of binding of cytochrome *c*.

The axial ligands of yeast cytochrome *c*, Histidine 21 and Met 85 respectively, not only contribute to the high positive potential of cytochrome *c* but also control the electronic distribution in cytochrome *c* in such a way that charge density is concentrated at the reactive heme edge.

From the standpoint of molecular recognition, the structure of ccp is highly complementary to that of cyt *c* (Fig. 5). The reactive heme site is more buried in ccp than in cyt *c*. This "burial" may serve to shield the reactive oxidizing radical intermediates, which are formed during the peroxidase catalytic cycle, from solvents, which might attach the radical sites and thereby inactivate the enzyme. This shielding introduces a rather long distance for the closest approach which can be obtained between the heme edges. "Docking" studies in which the crystal structures of the two proteins are manipulated on an Evans and Sutherland system to simulate binding, suggest a closest heme–heme (edge to edge) distance of about 16 Å, or 24 Å between the Fe centers. This long distance suggests that intramolecular electron transfer in the cytochrome *c*:ccp complex will be slow. This is indeed true: the rate at optimal ΔG is approx. $10^3\,s^{-1}$ compared with $>10^4\,s^{-1}$ for the Hb/b_5 system, where the reaction distance is approx. 10 Å edge–edge.

The charge distribution on ccp is compellingly complementary to that on cyt *c*: negatively charged residues cluster in roughly three patches on the surface,

Fig. 5. Structure of cytochrome *c* peroxidase, based on coordinates of Poulos

Fig. 6. Proposed structure of the cytochrome *c*:ccp complex based on the model of Poulos and Kraut

any of which *could* represent a binding site for cytochrome *c*. This charge complementarily led Poulos to propose his seminal *model* [25] for the cyt *c*–ccp complex (Fig. 6). This model emphasizes stereo specific hydrogen bonding between specific pairs of oppositely charged residues on cyt *c* and ccp.

Lys18(c) – Asp217(ccp)

Lys32(c) – Asp79(ccp)

Subsequent molecular graphics studies by Lum (personal communication) produced a slightly different version of this model.

Most recently, Northrup and coworkers used molecular dynamics simulations [26] (guided by Coulomb interactions) to probe possible binding interactions between cyt *c* and ccp. Their results suggested that several different docking configurations were possible for the cytochrome *c*–ccp complex involving different charge triplets: These various conformers could rapidly interchange by an effective two dimensional diffusion to access an optimal combination of distance, angle, and medium to facilitate electron transfer.

6 The Cyt *c*/Ccp Complex: Structural Data

A seminal contribution to understanding the binding of cytochrome *c* to ccp was made by Tom Poulos, who together with Joseph Kraut, proposed a structural model for the cyt *c*:ccp complex [25]. This model was guided by several considerations. First it was known from binding studies, and the elegant chemical modification studies of Margoliash and others [15], that binding involved positively charged amino acids in cyt *c* (lys) interacting with negatively charged amino acids in ccp (asp and glu). The first principle, then, was to align the evolutionarily conserved lysines on cyt *c* with stereochemically compatible anionic groups on ccp, to produce a stereospecific complex where salt bridges with appropriate stereochemistry were formed. A second key criterion was added to distinguish the possible available alignments; the heme–heme distance was kept minimal. These primary criteria were used to guide *visual* docking of cyt *c* and ccp using a molecular graphics system, and resulted in the structure shown in Fig. 6.

Several predictions of this model have been experimentally supported, including the estimated heme–heme distance, and the protection of various (charged) groups on cytochrome *c* and ccp. However, as already noted, other theoretical models have suggested that multiple binding sites are possible within the cyt *c*:ccp complex. The available spectroscopic and crystallographic data, reviewed in this section, tend to support a multisite model, rather than a highly specific "lock and key" structural model.

A very important "negative" result comes from crystallographic studies [27]. In an attempt to unambiguously define the structure of the complex, Sheriff et al. crystallized the complex, which is impressive by itself. Depending on one's viewpoint the results of crystallographic study were disappointing, or illuminating. In essence, although ccp could be readily located in the unit cell, cyt *c* could not, even though it was clearly shown to be present in the crystal in a 1:1 ratio to ccp. The reason cyt *c* could not be visualized is that it was disordered, either due to static disorder, where cyt *c* occupies multiple sites on ccp, or dynamic disorder, in which a single site undergoes large thermal fluctuations (rms/atom > 1 Å). This "negative" result could be construed to support the multisite model, but might be an artifact of crystallization. Data from the author's lab, however, also support the multisite model. These data involve several types of experiments. The first experiments, NMR exchange experiments, (carried out in collaboration with Drs. Englander and Roder), are based on a simple idea. Amide hydrogens have a high pK, and thus exchange with (solvent) deuterium very slowly. This rate is dramatically enhanced by base catalysis. To be effective, such catalysis requires ready steric access of the base to the (exchanging) proton. Thus, internal amide protons generally exchange far more slowly than do surface amides. Thus, if a surface amide of cyt *c* is sterically protected by binding to another protein (like ccp), then those protons which lie within the binding interface should have their exchange rates significantly reduced, while protons outside this interface should be minimally affected. Exchange experiments carried out on cyt *c*, and the cyt *c*:ccp complex agree with this general assumption.

The results of exchange after a specific period of time are visualized by 2D NMR. It must be remarked that this experiment requires that *every* resonance in cyt *c* be assigned. This tour de force has been accomplished by Englander, Roder and Wand. An example is shown in Fig. 7 which compares the protons exchanged after 27 hrs. for cyt *c* alone and cyt *c* "protected" by binding to ccp.

The pattern of protection is shown in Fig. 8. The striped lines show those regions in which amide exchange is retarded by binding. The hatched area is the area predicted from the Poulos model to be solvent accessible. The area predicted by Poulos is indeed protected. However, the protected area also

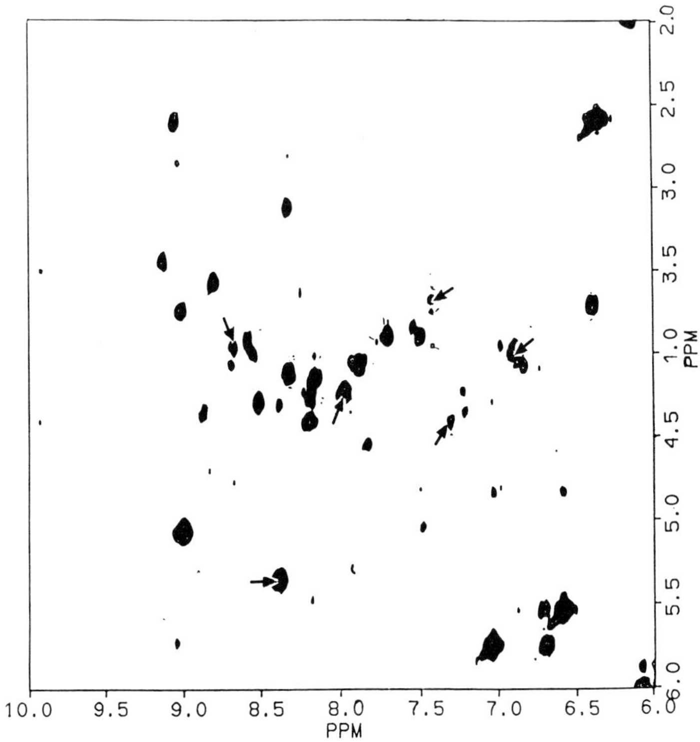

Fig. 7. Differential Hydrogen Exchange in cyt *c* in the cytochrome *c*:ccp complex. *Arrows* indicate resonances protected from exchange by ccp binding

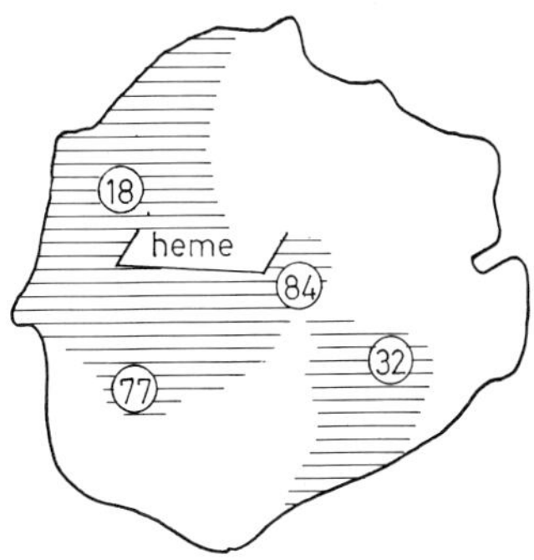

Fig. 8. Results of hydrogen exchange data, as per Fig. 3. *Lined area* shows the regions of the problem which are protected from exchange

includes regions not predicted by the Poulos model, but in general accord with the multisite models. These data thus generally support a multisite model for cyt *c*:ccp binding and recognition.

If there are multiple sites, each should have a different characteristic distance, and thus a characteristic electron transfer rate. Recent studies [28] by Hazzard et al. support this assumption. At low ionic strength, where complex formation is strongest and presumably most specific, the rate of intra complex electron transfer from cytochrome *c* to Fe(IV) ccp is relatively low ($200\,s^{-1}$) compared with the rate ($2000\,s^{-1}$) at higher ionic strength where a range of complex conformations can be accessed [28]. Fluorescence energy transfer has been used to obtain information on the distance distribution. When the heme iron in ccp is substituted by Mg^{2+} one obtains a highly fluorescent porphyrin at the active site of ccp. This fluorescence decays in a simple (roughly single exponential) decay ($\tau \approx 8$ ns). When cyt *c* binds it acts as an energy acceptor and quenches Mgccp fluorescence. At room temperature this quenching is described by a single lifetime ($\tau \approx 4.7$ ns). At low temperature however the system is more complex. At 77 K, Mg ccp still shows a single exponential emission ($\tau \approx 9$ ns) but the Mgccp/Fecyt *c* adduct shows more complex decay, involving at least three components. More likely, energy transfer shows a *distribution* of rates which correspond to a distribution of structures of the cyt *c*:ccp complex. The difference between low temperature and high temperature then just reflects the different *dynamics* at these temperatures. At low temperature, all motion is frozen, and the decay reflects the percent distribution of states. At high temperature, however, these structures can reequilibrate via two dimensional diffusion along the protein–protein surface. Furthermore, this equilibration must occur within the few nanoseconds of the fluorescent lifetime.

7 Binding of Cyt *c* and Ccp: Thermodynamic Aspects

Thus structural background suggests that the bound cyt *c*:ccp adduct may actually consist of a distribution of structures. In this section, we consider the thermodynamics of binding cyt *c* and ccp, both for the native proteins, from different species, and proteins incorporating single site replacements, as prepared by site directed mutagenesis.

A number of different approaches have been employed in different laboratories to characterize cyt *c* ccp binding. The earliest estimates of binding constants come from steady state kinetic studies by Yonetani and coworkers [19] (subsequently refined by Erman) [29]. At 50 mM phosphate, pH6, (conditions which favor maximum turnover), an apparent Km value of 3 μM is obtained using yeast isol cyt *c* as the reaction partner of ccp. Km is intrinsically a kinetic parameter, which in the complex ccp mechanism may incorporate a number of elementary rate constants unrelated to binding.

Thus, specific thermodynamic studies have been undertaken. One of the first was reported by Vitello and Erman [30], who found that mixing Fe(III) cyt *c* and Fe(III) ccp generated a small but reproducible difference spectrum around 410 nm which could be ascribed to binding. Using horse heart cyt *c*, the binding constant obtained from these spectroscopic studies proved to be quite dependent in ionic strength, ranging from $K_b = 10^8\ M^{-1}$ at $\mu = 0$ to $K_b = 10^3\ M^{-1}$ at $\mu = 0.1$ M.

At about the same time, Leonard and Yonetani[31] studied the binding of yeast isol cyt to fluorescent ccp derivatives ((porphyrin) ccp, and (ANS) ccp: ANS = aniline naphthalene sulfonic acid). In this case binding is followed by quenching the fluorescence of ccp. At $\mu = 0.3$ M, they found a binding constant of $K_B = 10^5\ M^{-1}$, which is much higher than the value suggested by Erman's studies. Although Erman suggested that this discrepancy reflects an "inner filter" artifact in Leonard's data, time resolved energy transfer (which cannot have an inner filter artifact) gives a similar result. Most recently, Billstone et al. [32] compared the binding of Mgccp to horse cyt *c* and/or yeast cyt *c* (Fig. 9). These results resolve the apparent discrepancy. It is now clear that, despite the strong homology between the horse and yeast proteins, their recognition and binding with ccp are qualitatively different. Specifically, binding of yeast cyt *c* is far less sensitive to ionic strength than is binding of the horse proteins. The qualitative difference in recognition is underscored by the fact that even at low ionic strength the rate of energy transfer differs significantly between the yeast cyt *c*:ccp complex ($k_{et} = s^{-1}$) and the horse cyt *c*:ccp complex ($k_{et} = s^{-1}$). This difference suggests that the complex structure (particularly the heme to heme distance) differs between these two complexes.

NMR has also been used to characterize binding [22, 23]. When cyt *c* binds to ccp the paramagnetically shifted heme methyl groups of cyt *c* shift further downfield by 1–2 ppm. Normally, binding occurs in the fast exchange regime, so that the fraction bound can be assessed from the frequency shift (Fig. 10). Two

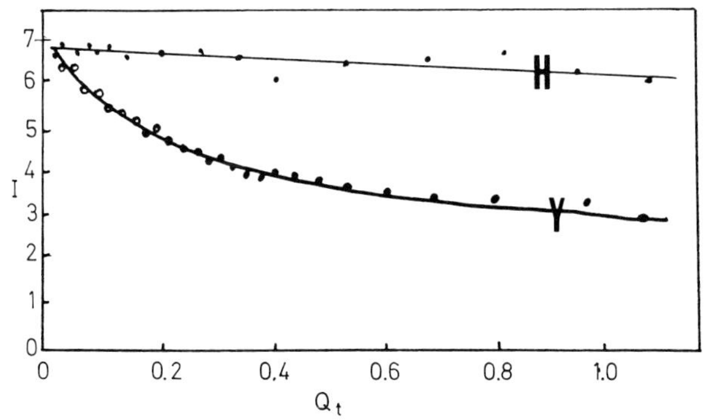

Fig. 9. Fluorescence quenching study of isol cyt *c* (Y) and horse cyt *c* (H) binding to H_2 porphyrin ccp 50 mM Pi, Ph7.
I = normalized intensity Q_t = conc. cyt × 10^6 M

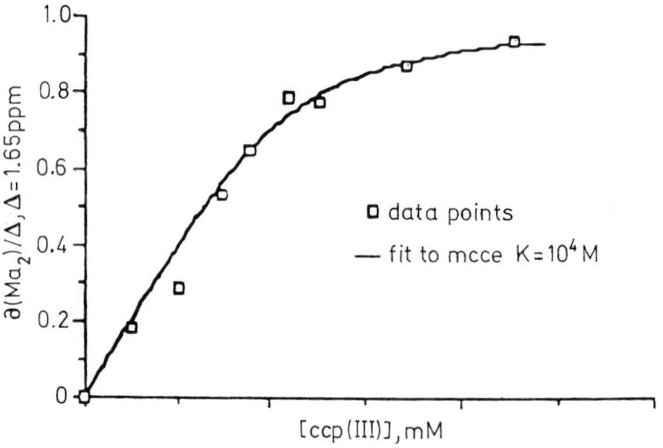

Fig. 10. NMR study of iso 1 cyt *c* binding to ccp

points emerge from the NMR studies. First, the magnitude of the shifts are sensitive to the species of cyt *c* employed: the shift for horse cyt *c* (0.9 ppm) is smaller than that for yeast cyt *c* (2 ppm) again suggesting that different modes of binding are employed by the two proteins. Second, the absolute magnitude of the binding constant obtained by NMR can differ from that obtained by e.g. fluorescence. This difference might be interpreted in terms of the existence of multiple states which are sensed differentially by different techniques. Thus only apparent binding constants are determined by any given technique which are related to the actual species distribution by technique specific weighting.

8 Site-Directed Mutagenesis Studies

Finally, we consider the effects on recognition and binding of single site replacements [34] of the individual charged residues which have been emphasized in molecular modeling studies. These residues include Lys (18, 32, 77) in cytochrome *c* and Asp (37, 79, 217) in ccp.

The following replacements have been prepared by conventional or site directed mutagenesis:

cyt *c*: Lys18 $\rightarrow$ (Ile, Leu), Lys32 $\rightarrow$ (Gln, Leu), Lys77 $\rightarrow$ (Arg, Asp)

ccp: Asp37 $\rightarrow$ Lys, Asp79 $\rightarrow$ Lys, Asp17 $\rightarrow$ Lys.

Each of these replacements has some effect on binding, but the magnitude of each effect is surprising, and even counterintuitive. Each of the single charge changes produce only small net changes in binding free energy ($<$ twofold). The Lys18 $\rightarrow$ Ile replacement even seems to produce a small net *increase* in binding of cyt *c* to ccp, as measured by NMR. Among the double charge mutations, the effects remain relatively small, and what effects there are are not equivalent: ccp (Asp217 $\rightarrow$ Lys) seems to have very little effect on binding strength or reaction kinetics, while ccp (Asp37 → Lys) has a noticeable effect on binding and a significant effect on reaction kinetics, decreasing the apparent intramolecular electron transfer rate constant by almost 40 fold.

Taken together, the results of these mutant studies suggest at least that for the yeast cyt *c*:ccp system, while charged amino acids may play a role in orienting the proteins for proper reactive binding, individual stereospecific, charge–charge interactions make only a minor contribution to the thermodynamics of complex formation. The results of a number of thermodynamic binding studies are compiled in Table 1.

9 Summary: The "velcro" Model

Molecular recognition and binding is the first step in protein electron–electron transfer. The foregoing data suggests that for the prototypical cyt *c*:ccp system, this binding does not involve specific "lock and key" recognition, but rather involves *complementary* "sticky patches" of protein surfaces (rather like two pieces of velcro adhesive).

Thus, as earlier suggested by Margoliash et al. [15] charge patches may serve to orient the proteins as they approach collision. Once a collision complex occurs, the proteins likely undergo a rapid two dimensional diffusion [26] until the most favorable conformations for electron transfer occur, at which point the reaction proceeds, followed by further diffusion and dissociation. Of course, binding must be sufficiently strong to hold the complex together long enough to react, but *not* much longer (or product inhibition would occur). Although at first glance, this "velcro" binding mechanism may seem inefficient, or at least inelegant relative to a stereo-specific "lock and key", it is actually quite a logical design, since it assures that most collisions between redox partners will ultimately produce a reaction. If high stereospecificity were required for (initial) binding then very few collisions could lead to reaction. If, for example, about 10% of the surface of each protein were involved in binding, then only one out of every one hundred collisions would produce a possible reaction. The distribution of charge patches suggested here *preorients* the proteins to favor useful "face to face" encounters, and permits a variety of such encounters to produce mutual binding followed by diffusion to the reactive conformation. A final biological note might invoke Darwins "principle of multiple utility" [35]. Cytochrome *c* must interact with multiple (redox) partners within the cell. If binding were stereochemically optimized for the protein structure of one partner, it would necessarily be poor for another partner which has a different structure. It might be postulated that nature has obtained a global optimization with respect to binding and recognition, which might well be nonoptimal for any local interaction. Thus, replacement of an invariant lysine on cyt *c* may actually increase binding to ccp, but not to other partners.

10 References

1. a. Hatefi A (1985) Ann Revs Biochem 54: 1015
   b. Tzagaloff A (1982) Mitochondria, Plenum, N. Y.
2. a. Marcus R, Sutin N (1985) Biochem Biophys Acta 811: 265
   b. McLendon G (1988) Acts Chem Res 21: 160
3. Chance B (ed) (1979) Tunneling in biological systems, Academic, NY
4. Hoffman B, Ratner M (1988) J Am Chem Soc 110
5. Pardue K, McLendon G, Bak P (1987) J Am Chem Soc 109: 7540
6. Lieber C, Karas J, Gray H (1987) J Am Chem Soc 109: 2778
7. Peterson Kennedy S, McGourty J, Ho P, Sutoris C, Liang N, Zemel H, Blough FN, Margoliash E, Hoffman B (1985) Coord Chem Rev 64: 125
8. Magner E (1988) Ph.D. Thesis, University of Rochester
9. Pielak G, Mauk AG, Smith M (1985) Nature 313: 152
10. Fetrow J, Cardillo T, Sherman F, Proteins (in press)
11. McLendon G, Hickey D, Sherman F, Nature (submitted)
12. Holzschu D, Principio L, Conklin KT, Hickey DR, Short J, Rao R, McLendon G, Sherman F (1987) J Biol Chem 262: 7125
13. Singer S, Nicholson G (1972) Science 175: 720
14. Margoliash E, in Prog Chim Biol Res 254: 208
15. An excellent review is given by Bosshard and Margoliash Trends Biochem Sci (1983) 8: 316
16. Kim C (1986) Meth Enzym 126: 238
17. Stone Lever J, O'Brien P, Geren L, Heb FM, Yu L (1985) J Biol Chem 260: 5392
18. Poulos T, Finzel B (1984) Pept Prot Revs 4: 1 115
19. Yonetani T, Ray G (1966) J Biol Chem 241: 700
20. Louis G, Hutcheon W, Brayer G (1988) J Mol Biol 199: 295
21. Finzel B, Poulos T, Kraut J (1984) J Biol Chem 259: 13027
22. Satterlee J, Moench S, Erman J (1987) Biochem Biophys Acta 912: 87
23. a. Cusanovich M, Hazzard J, Meyer T, Rollin G (1988) Prog Chim Biol Res 254: 139
    b. Liang N, Pielak G, Mauk AG, Smith M, Hoffman BM (1987) Proc Natl Acad Sci 14: 1249
24. Cyt *c*: Smith M (1985) Ann Rev Genet 19: 423
    Hampsen D, Das G, Sherman F (1988) FEBS Lett 231: 275
    ccp: Cicarelli R et al (1988) Rev of Indust Microbiol 29: 161
25. Poulos T, Kraut J (1980) J Biol Chem 255: 10322
26. Northrup S, Boles J, Reynolds J (1988) Science 241: 67
27. Poulos T, Sherrif S, Howard A (1987) J Biol Chem 262: 13881
28. Hazzard J, McLendon G, Cusanovich M, Tollin G (1988) Biochem Biophys Res Comm 151: 429
29. Kang DS, Erman J (1982) J Biol Chem 257: 12775
30. a. Vitello L, Erman J (1980) J Biol Chem 255: 6224
    b. Vitello L, Erman J (1988) Biochem Biophys Acta
    c. Kane C, Ferguson Miller S, Margoliash E (1977) J Biol Chem 252: 919
31. Leonard J, Yonetani T (1973) Biochem 13: 1465
32. Billstone V, Pan Zhang Q, McLendon G, Hoffman BM, Biochem (in press)
33. Pardue K, McLendon G (1989) J Biol Chem, submitted
34. Das G et al (1988) J Biol Chem 263: 18290
35. Darwin C (1872) The Origin of Species. New York Earlton House

Plastocyanin and the Blue Copper Proteins

A.G. Sykes

Department of Chemistry, The University, Newcastle upon Tyne, NE1 7RU, UK

This review is concerned with the properties of the type 1 blue copper proteins [1–4], a class of protein[1] incorporating a single Cu atom and involved in redox processes. Within the class most information is available for plastocyanin, a component of photosynthetic electron transport, which has proved to be a particular focus of recent research. Relevant to the whole area of metalloprotein study is the more general question of how and over what distance electrons are transferred. Rapid, i.e. efficient, long-distance (>10 Å) electron transfer between metal centres is known to occur in biological systems, and attempts to better understand such processes is a subject of widespread current interest [7–14]. For reactions of two large biomolecules, specific relative orientations are required at the time of reaction, and questions relating to the docking process and association prior to electron transfer are highly relevant. For example, plastocyanin, which is a solute in the inner thylakoid of the chloroplast, is required to associate and electron transfer with its redox partners. In the case of single proteins having more than one active site [5], and membrane bound complexes made up of different protein molecules, intramolecular electron transfer is of prime concern, and the orientation of domains and molecules within these units is of considerable importance [15].

[1] The type 1–3 terminology to distinguish different Cu protein active sites remains extremely useful. Sub-groupings are appearing however in all three categories particularly in the case of the binuclear (EPR inactive) type 3 centers. Thus, in the recently determined X-ray crystal structure of ascorbate oxidase the type 3 and type 2 centers are present as a single trimer unit [5]. A discrete binuclear type 3 center is, however, retained in hemocyanin [6].

There is now a wealth of structural information on plastocyanin using a wide range of physical techniques which, together with studies on its reactivity in aqueous solution, make this one of the best documented metalloproteins. The protein makes use of two oxidation states here designated as PCu(II) and PCu(I). The structure of poplar plastocyanin from X-ray crystallography is similar in both oxidation states, Fig. 1 [12, 13]. Studies using inorganic complexes as probes for reactivity, through to the reactivity with other metalloproteins, and in particular the physiological partners cytochrome f and P700 in the photosynthetic electron transport chain, are considered here. Structural information on azurin in the Cu(II) state has also been obtained, and additional insight has been provided in the case of this protein by recent studies on mutants from site directed mutagenesis. A number of other type 1 Cu proteins have been isolated which require further study to add depth to the whole area.

1 Introduction

It is appropriate to begin on a cautionary note with some general comments on biological electron transfer. X-ray crystal structures of a number of electron transport metalloproteins have now been determined [18], from which it is clear

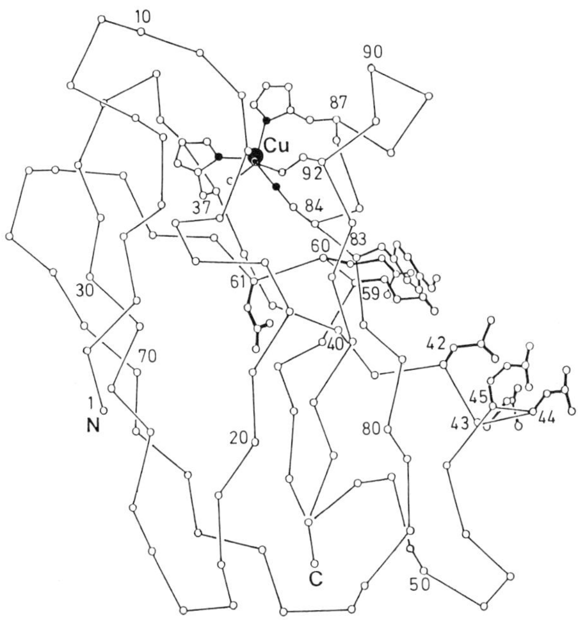

Fig. 1. The α-carbon chain structure of poplar plastocyanin [16] including details of the active site and the remote acidic patches 42–45 and 59–61. The first of these has been modified to include an acidic residue at position 45 as e.g. for spinach and French bean plastocyanins

that active sites are invariably close to the surface. For example, in the case of the *Clostridium plasteurianum* and *Spirulina platensis* ferredoxins (reduction potentials around −400 mV), the corners of $Fe_4S_4(SR)_4^{n-}$ and $Fe_2S_2(SR)_4^{n-}$ clusters, respectively (SR^- is a cysteinate residue; n=2, 3) are solvent exposed and accessible to redox partners [19, 20]. In the case of the *Chromatium vinosum* high potential Fe/S protein the Fe_4S_4 cluster is ~4.5 Å buried [21], a relatively small distance, possibly contributing to the different redox properties (350 mV; n=1, 2). Also with the cytochromes, e.g. cytochrome c [22], the edge of the protoporphyrin IX ring is known to be solvent exposed. Since the porphyrin has conjugate double bonds and available low lying molecular orbitals, delocalization of electron density with the central Fe can occur,[1] the result of which is that the exposed heme edge can serve as a donor/acceptor group for electron transfer. Furthermore, in the case of plastocyanin and azurin, it is known that a coordinated imidazole is solvent exposed, and electron delocalization over the Cu and imidazole could facilitate electron transfer. The need for long-distance electron transfer, and whether the description applies to any of these systems, depends very much on the definition used. An often quoted example is the reaction of cytochrome b_5 (reductant) with cytochrome c (oxidant) [24]. Association of the two ($K = 8.3 \times 10^4\ M^{-1}$ at 25 °C, pH 7.0, I=0.01 M) [25] is favorable because of the negative (cytochrome b_5) and positive (cytochrome c) charges surrounding the respective heme edges, and the intramolecular rate constant of $1.6 \times 10^3\ s^{-1}$ indicates favorable cytochrome b_5(II) to cytochrome c(III) electron transfer [26]. Docking has been computer modeled using energy minimization procedures, which indicate that an initial Fe–Fe separation of 17.8 Å relaxes to an average distance of 15.7 Å [24]. Also important is the observation that the extremities of the two heme rings, which in this treatment are co-planar, approach to within a distance 5–7 Å. The problem is therefore more sharply focused on how electrons are transferred over this shorter distance.[2]

The distance generally used in defining a long range electron transfer is that between the metal centers, which in the above cytochrome example is certainly long compared to that applicable in the case of two low molecular weight inorganic complexes. Just as some ligands of coordination complexes more readily transmit electrons (e.g. CN^- as compared to NH_3), it is reasonable that similar effects should be observed in the case of metalloproteins. Thus, the widespread occurrence of porphyrin, imidazole and *S*-donor ligands coordinated at metalloprotein active sites is believed to be beneficial to electron transfer.

There have been many studies on the reactivity of plastocyanin. Two sites (or regions) have been identified on the surface of the molecule as relevant to electron transfer. One is the *adjacent site* at or near His87, and the other a more

---

[1] One of the simplest illustrations of the effect a porphyrin can have is its effect on the substitution of axial aqua ligands of Co(III), which substitute some ~10^6 times faster than would normally be expected if the metal were 'pure' t_{2g}^6 [23].

[2] Recent experiments with genetically engineered cytochrome c [27] suggest that the aromatic ring of residue Phe82 is not involved in mediating electron transfer over this 5–7 Å distance, as has been suggested [24].

remote site at or near Tyr83, the α-carbon of which is ~10 Å from the Cu. The first of these is hydrophobic with no charge except that originating from the nearby Cu. The second, in a region of high negative charge arising largely from residues 42–45 and 59–61, is used by cationic reactants. At the time of electron transfer the two metal centers are separated by ligating groups and polypeptide, and are certainly >10 Å apart for reaction at the remote site, and in many cases also for reaction at the adjacent site.

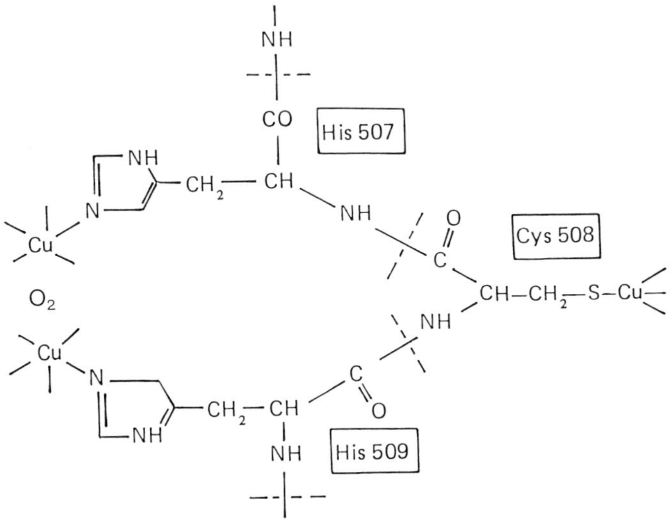

Fig. 2. The binding of Cu's in ascorbate oxidase. A section of the polypeptide (plus side chains), and position of the type 1 Cu in a plastocyanin like domain (to the right) are shown. The type 3 Cu's are attached at what corresponds to the remote site. Residues His507 and Cys508 are analogous to Tyr83 and Cys84 in plastocyanin

While there is at present no full understanding as to why plastocyanin should require two sites for reaction, there is now much evidence detailing this two-site reactivity. Moreover, the recent X-ray crystal structure of ascorbate oxidase (which has 4 Cu atoms per molecule) has indicated a plastocyanin-like domain, with the two type 3 Cu's (in close proximity with the type 2 Cu) located at the remote site, Fig. 2 [5]. Since electrons are transferred, from the type 1 Cu to O_2 bound at the type 3 center this structure defines two very similar through-bond routes for biological electron transfer.

2 Properties of Plastocyanin

2.1 Amino-Acid Sequences

Plastocyanins from the green leaves of higher plants and algae have ~99 amino-acid residues (M_r 10,500). Full sequences for 19 higher plant, 4 green algal, and 1 blue-green prokaryotic algal plastocyanin have been determined. These include spinach, French bean, potato, elder, marrow, broad bean, lettuce, dog's mercury, shepherd's purse, solanum, dock, poplar (two forms), cucumber, parsley, campion, barley, rice and carrot (19 higher plant) [28–30]; *Scenedesmus obliquus*, *Chlorella fusca*, *Enteromorpha prolifera*, *Ulva arasakii* (4 green algae) [31–32]; and *Anabaena variabilis* (a blue-green alga) [33], (references are to the most recent studies). Interestingly the latter is basic. An acidic blue-green algal plastocyanin has been reported in the thermophilic cyanobacterium *Phormidium laminosum*, and in recent work has been isolated from the photosynthetic chain of *Microcystis aeruginosa*, a unicellular cyanobacterium [34]. The isoelectric point for the latter is similar to values reported for other acidic plastocyanins. Nine of the first 15 amino-acids at the *N*-terminal have been shown to be identical to those of *A. variabilis* plastocyanin, and the full sequence is awaited with considerable interest. Other original sources for the sequences can be traced through Ref [1]. Those for the six plastocyanins most relevant to this review are shown in Fig. 3. Of the 19 higher plant sequences 47 of the 99 residues are invariant. Sequence homologies are not as strong if the four green algae are included (28 invariant residues), and with the inclusion of PCu from the blue-green alga *A. variabilis* only 23 remain invariant as indicated in Fig. 3. The latter include His37, Cys84, His87 and Met92 which coordinate to the Cu via N, S, N and S atoms, and constitute the active site. Conservation is most striking in the 31–44 and 78–92 regions which suggests functionality of these residues. Two deletions are observed at positions 57 and 58 in green algal and some higher plant (parsley, barley, rice and carrot) plastocyanins. *A. variabilis* plastocyanin has six additional residues which are inserted at the beginning and end of the sequence and close to position 75 [33]. Using the program, ALNED, sequences have been aligned to indicate homologies [28], when a different numbering system similar to that of *A. variabilis* and additional deletions are indicated. A phylogenetic tree of the blue proteins [3a], and of the plastocyanins has been reported [28].

Estimates of the overall charge on PCu(I) at pH's around or greater than 7 from amino-acid compositions (Asp and Glu as negative; Lys, uncoordinated His, and Arg as positive), indicate that this is conserved at $-9(\pm 1)$ for the higher plant and green algal sources, whereas for *A. variabilis* PCu(I) it is $+1$. Consistent with these estimates, from isoelectric focusing a pI of 4.2 has been reported for spinach PCu [35], and 3.97 for *S. obliquus*, whereas *A. variabilis* PCu gives 7.8. The number of charged residues on *A. variabilis* (21) is similar to those for many higher plant PCu's, but the distribution of + and − charge is different. In particular, only one of the acidic residues in the highly conserved 42–45 and 59–61 regions, which are part of the remote binding site, is retained (Asp42), and there are no two consecutive acidic residues in the sequence for *A. variabilis* plastocyanin. The quite different charge distribution with basic residues His and Lys at positions 59 and 60 have made it important to compare the reactivity of this basic plastocyanin with acidic forms [36]. There are other changes including that of the otherwise conserved residues Phe82 and Tyr83, which are interchanged in *S. obliquus* (and *C. fusca*) [1, 37], see Fig. 3.

```
                                        5                   10                  15                  20                  25
A.variabilis    Glu-Thr-Tyr-Thr-Val-Lys-Leu-Gly-Ser-Asp-Lys-Gly-Leu-Leu-Val-Phe-Glu-Pro-Ala-Lys-Leu-Thr-Ile-Lys-Pro-Gly-Asp
S.obliquus              Ala-Asn-Val-Lys-Leu-Gly-Ala-Asp-Ser-Gly-Ala-Leu-Val-Phe-Glu-Pro-Ala-Thr-Val-Thr-Ile-Lys-Ala-Gly-Asp
  Poplar                Ile-Asp-Val-Leu-Leu-Gly-Ala-Asp-Asp-Gly-Ser-Leu-Ala-Phe-Val-Pro-Ser-Glu-Phe-Ser-Ile-Ser-Pro-Gly-Glu
  Spinach               Val-Glu-Val-Leu-Leu-Gly-Gly-Gly-Asp-Gly-Ser-Leu-Ala-Phe-Leu-Pro-Gly-Asp-Phe-Ser-Val-Ala-Ser-Gly-Glu
  Fr. bean              Leu-Glu-Val-Leu-Leu-Gly-Ser-Gly-Asp-Gly-Ser-Leu-Val-Phe-Val-Pro-Ser-Glu-Phe-Ser-Val-Pro-Ser-Gly-Glu
  Parsley               Ala-Glu-Val-Lys-Leu-Gly-Ser-Asp-Asp-Gly-Gly-Leu-Val-Phe-Ser-Pro-Ser-Ser-Phe-Thr-Val-Ala-Ala-Gly-Glu
[Conservation]                          ●   ●               ●       ●       ●       ●                               ●   ○

                                        30                  35      ▽           40                  45                  50
A.variabilis            Thr-Val-Glu-Phe-Leu-Asn-Asn-Lys-Val-Pro-Pro-His-Asn-Val-Val-Phe-Asp-Ala-Ala-Leu-Asn-Pro-Ala-Lys-Ser
S.obliquus              Ser-Val-Thr-Trp-Thr-Asn-Asn-Ala-Gly-Phe-Pro-His-Asn-Ile-Val-Phe-Asp-Glu-Asp-Ala-Val-Pro-Ala-Gly-Val
  Poplar                Lys-Ile-Val-Phe-Lys-Asn-Asn-Ala-Gly-Phe-Pro-His-Asn-Ile-Val-Phe-Asp-Glu-Asp-Ser-Ile-Pro-Ser-Gly-Val
  Spinach               Glu-Ile-Val-Phe-Lys-Asn-Asn-Ala-Gly-Phe-Pro-His-Asn-Val-Val-Phe-Asp-Glu-Asp-Glu-Ile-Pro-Ser-Gly-Val
  Fr. bean              Lys-Ile-Val-Phe-Lys-Asn-Asn-Ala-Gly-Phe-Pro-His-Asn-Val-Val-Phe-Asp-Glu-Asp-Glu-Ile-Pro-Ala-Gly-Val
  Parsley               Lys-Ile-Thr-Phe-Lys-Asn-Asn-Ala-Gly-Phe-Pro-His-Asn-Ile-Val-Phe-Asp-Glu-Asp-Glu-Val-Pro-Ala-Gly-Val
[Conservation]              ○       ○   ○   ●   ●   ▲   ▲       ●   ●   ●           ●   ●   ▲   ▲           ●       ▲   ○

                                        55                  60                  65                  70                  75
A.variabilis            Ala-Asp-Leu-Ala-Lys-Ser-Leu-Ser-His-Lys-Gln-Leu-Leu-Met-Ser-Pro-Gly-Gln-Ser-Thr-Ser-Thr-Thr-Phe-Asp
S.obliquus              Asn-Ala-Asp-Ala-Leu-Ser- * - * -His-Asp-Asp-Tyr-Leu-Asn-Ala-Pro-Gly-Glu-Ser-Tyr-Thr-Ala-Lys-Phe-Asp
  Poplar                Asp-Ala-Ser-Lys-Ile-Ser-Met-Ser-Glu-Glu-Asp-Leu-Leu-Asn-Ala-Lys-Gly-Glu-Thr-Phe-Glu-Val-Ala-Leu-Ser
  Spinach               Asp-Ala-Ala-Lys-Ile-Ser-Met-Ser-Glu-Glu-Asp-Leu-Leu-Asn-Ala-Pro-Gly-Glu-Thr-Tyr-Lys-Val-Thr-Leu-Thr
  Fr. bean              Asp-Ala-Val-Lys-Ile-Ser-Met-Pro-Glu-Glu-Glu-Leu-Leu-Asn-Ala-Pro-Gly-Glu-Thr-Tyr-Val-Val-Thr-Leu-Asp
  Parsley               Asn-Ala-Glu-Lys-Ile-Ser- * -  *-Gln-Pro-Glu-Tyr-Leu-Asn-Gly-Ala-Gly-Glu-Thr-Tyr-Glu-Val-Thr-Leu-Thr
[Conservation]                      ○   ○   ●                           ●               ●   ○               ○       ○

                                        80              ▽   85      ▽           90      ▽           95
A.variabilis            Ala-Ala-Gly-Glu-Tyr-Thr-Phe-Tyr-Cys-Glu-Pro-His-Arg-Gly-Ala-Gly-Met-Val-Gly-Lys-Ile-Thr-Val-Ala-Gly
S.obliquus              Thr-Ala-Gly-Glu-Tyr-Gly-Tyr-Phe-Cys-Glu-Pro-His-Gln-Gly-Ala-Gly-Met-Val-Gly-Lys-Val-Ile-Val-Gln
  Poplar                Asn-Lys-Gly-Glu-Tyr-Ser-Phe-Tyr-Cys-Ser-Pro-His-Gln-Gly-Ala-Gly-Met-Val-Gly-Lys-Val-Thr-Val-Asn
  Spinach               Glu-Lys-Gly-Thr-Tyr-Lys-Phe-Tyr-Cys-Ser-Pro-His-Gln-Gly-Ala-Gly-Met-Val-Gly-Lys-Val-Thr-Val-Asn
  Fr. bean              Thr-Lys-Gly-Thr-Tyr-Ser-Phe-Tyr-Cys-Ser-Pro-His-Gln-Gly-Ala-Gly-Met-Val-Gly-Lys-Val-Thr-Val-Asn
  Parsley               Glu-Lys-Gly-Thr-Tyr-Lys-Phe-Tyr-Cys-Glu-Pro-His-Ala-Gly-Ala-Gly-Met-Lys-Gly-Glu-Val-Thr-Val-Asn
[Conservation]                  ●       ●       ○   ○   ●       ●   ●       ●   ●   ●   ●       ○       ○       ●
```

Fig. 3. Plastocyanin amino-acid sequences relevant to this review. Other sequence information is included. There are now 24 known sequences (18 formerly in Ref. [1]). The 23 residues invariant throughout all 24 known higher plant and algal plastocyanin sequences are indicated (●), 5 others (▲) are invariant if *A. variabilis* is excluded, and a further 19 (o) if only the higher plant sequences are considered. Deletions are indicated (*), and residues which coordinate the Cu(▽)

A second plastocyanin (PCb) has been isolated from poplar leaves, *Populus nigra italica* [29]. As compared to the PCa form which was used in the crystal structure determinations, the sequence has 12 replacements, there being one extra carboxylate to give a PCu(I) charge of -10. Two type 1 Cu proteins have also been isolated from cucumber. One is a plastocyanin, whereas the other obtained from seedlings is referred to as the basic blue protein or plantacyanin. More generally, if only a single type 1 protein is present, the possible existence of different plastocyanins in different species and sub-species (for example the different types of spinach *Spinachia oleracea* and *Beta vulgaris*) suggests that greater care is required in reporting the source with full generic name. In this respect it would be helpful to have more information on sequences from different types, e.g. of spinach, to assess the extent of such variations.

2.2 Structure of Plastocyanin

Plastocyanin has the shape of a slightly flattened cylinder $40 \times 32 \times 28$ Å, the 40 Å axis corresponding to what is sometimes referred to as the north-south direction, Fig. 1 [16]. Based on this view the top of the molecule in Fig. 1 is sometimes known as the "north", and the remote (acidic patch) around Tyr83 as the "east". Although the compass directions are not ideal terminology, they do have the advantage of easy reference to what are distinctive regions. There are eight strands of polypeptide chain which are connected by seven loops at the ends of the cylinder. Seven of the strands have substantial β-structure and strand five is irregular and contains the only helical structure (about 1.5 turns). From the PCu(II) structure (now to 1.6 Å resolution) the Cu is coordinated to the N^{δ} of His37 and His87, and the S-atoms of Cys84 and Met92 in an irregular tetrahedral geometry. Two of the bond angles at the Cu differ by more than 20° from those of a regular tetrahedron. The two imidazole Cu–N bonds at 2.10 and 2.04 Å, and the Cu–S(Cys) bond at 2.13 Å may be regarded as normal, but the thio-ether Cu–S(Met) bond (2.90 Å) is unusually long, and is not detected by EXAFS [38]. As already indicated, the Cu atom is buried ~ 6 Å in the interior at the north end, the exposed edge of the imidazole ring of His87 being level with the molecular surface.

There is a striking imbalance in the charge on the surface of PCu. None of the ten charged residues which are most conserved in plant PCu's (eight Glu/Asp and two Lys) occur in the upper quarter of the molecule, Fig. 1 [16]. The partly exposed His87 is surrounded by conserved non-polar groups, constituting the so-called hydrophobic patch. It is important to note that the highly conserved Asp and Glu residues at 42–45 and 59–61 are concentrated on two kinks in the protein backbone, Fig. 1, giving two prominent bulges in the surface [37a]. These occur on either side of the conserved Tyr83, and (at the higher pH's) negatively charged carboxylates are directed into the solvent and form an elongated region of negative charge. The crystal structure of the PCu(II) active site is not affected by pH (4.2 and 6.0). Also the structure of PCu(I) at pH 7.8 (to

Table 1. X-ray crystal structure dimensions (Å) of the Cu site in reduced and oxidized poplar plastocyanin [16, 17]. The Cu–N (His87) distance for PCu(I) at pH3.8 is with the imidazole in the same orientation as in the high pH form, i.e. with no rotation

	PCu(I)(pH 3.8)	PCu(I)(pH 7.8)	PCu(II)(pH 6.0)
Cu–N(His37)	2.11	2.12	2.04
Cu–N(His87)	3.15	2.25	2.10
Cu–S(Cys84)	2.13	2.11	2.13
Cu–S(Met92)	2.52	2.90	2.90
Resolution	1.9	2.15	1.6

1.9 Å) is substantially the same as PCu(II) [17], with the two Cu–N bonds ~0.1 Å longer (Table 1), and small changes only in two of the bond angles involving Cu–S(Cys). Reorganization energy requirements are minimized therefore, and the active site meets one of the most important requirements for efficient electron transfer. However, when the structure was determined at five other pH's down to 3.8 (the pH was measured in 2.7 M$(NH_4)_2SO_4$) the Cu-coordination was seen to change substantially, and at pH 3.8 the Cu(I) becomes trigonally coordinated by His37, Cys84 and Met92. The bond to His87 is broken, the Cu making only van der Waal's contact with the imidazole ring of His87. Because the trigonal(planar) form of Cu(I) is less readily oxidized, this confirms the original explanation in terms of 'protonation at or near to the Cu(I)", for the decrease in kinetic redox activity [39]. Protonation at the N^{δ} atom of His87 (formerly bonded to the Cu) is responsible for this change, a particularly interesting feature being the existence of a second conformer in which the imidazole ring is rotated by 180° about the C^{β}–C^{γ} bond, the position of which remains unchanged, Fig. 4. At intermediate pH's in solution a dynamic equilibrium exists between the two forms. The most striking pH dependent structural change apart from those at the Cu site center around the flip of the Pro36 side chain from a C^{δ}-*exo* conformation in PCu(I) to a C^{δ}-*endo* conformation in H^+PCu(I) [17]. There is a high degree of flexibility at Pro36 which is part of the hydrophobic cavity around the Cu. This may help to explain how the tightly packed hydrophobic side chains at the northern surface can open up a pathway whereby the imidazole ring can rotate [17].

The solution conformation of plastocyanin from French bean, spinach, and *S. obliquus* has now been determined from distance and dihedral angle constraints derived by NMR spectroscopy [37, 40]. These two-dimensional NMR studies have indicated a well defined backbone conformation, which is very similar to that of poplar PCu in the crystalline state. However, in the case of *S. obliquus* there are deletions at positions 57 and 58 which influence the shape in the acidic region and in particular close to residues 59–61. The "gap" which is created is in effect repaired with consequent tightening of the loop 57–62 as indicated in Fig. 5. One of the pronounced bulges at the remote site of poplar and presumably other higher plant plastocyanins is not therefore present in *S. obliquus* (or plastocyanin from other green algae) [31, 32], as well as parsley

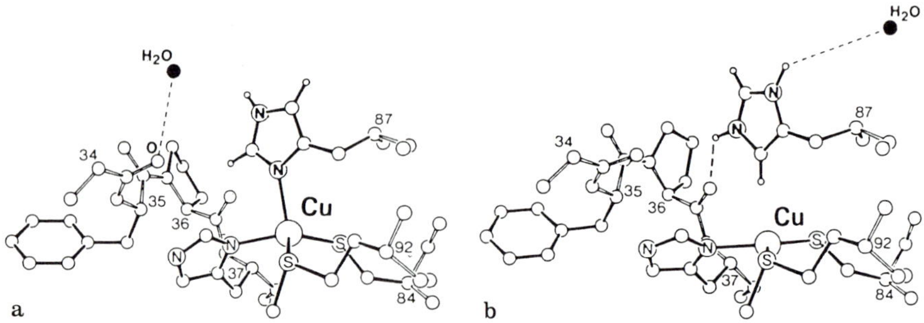

Fig. 4a,b. The Cu site in poplar PCu(I) (**a**) at high and (**b**) at low pH, here 7.8 and 3.8 respectively, indicating the effect of protonation of His87. At pH 3.8 the position of the water molecule indicated requires that the imidazole ring is rotated by 180° as compared to the pH 7.8 orientation [17]

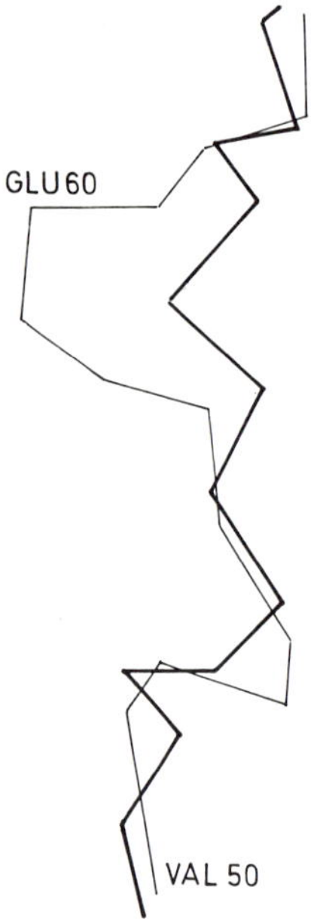

Fig. 5. The α-carbon chain of poplar plastocyanin residues 50–64 from the X-ray crystal structure (*thin line*), compared with that of *S. obliquus* plastocyanin obtained from a 2D NMR study [37a], showing the effect of deletions of residues 57 and 58

barley, rice and carrot plastocyanins [1, 28, 30], all of which have the deletions at 57 and 58. Here we retain the aligned sequence numbers with positions 57 and 58 vacant. It should be noted that 2D NMR studies have reported discrepancies in the sequences of plastocyanin from poplar (for Ile39 read Val39) [41] and *S. obliquus* (the C-terminus now has an extra Val as shown in Fig. 3) [37].

The crystal structure of poplar apoplastocyanin has been reported at 1.8 Å resolution [42]. The structure closely resembles that of the holoprotein, and positions of the Cu-binding residues are different by only 0.1–0.3 Å. Tetrahedral coordination, with proportionately larger metal–ligand bond distances, is also observed on replacing the Cu by Hg(II) [43]. This suggests that the irregular geometry of the active site is imposed on the metal atom by the polypeptide chain. The structure of green algal plastocyanin *E. prolifera*, has also been determined by Freeman and colleagues [44].

The long Cu–S(Met) bond (2.9 Å) does not contribute to the EXAFS fit for PCu(II) under a variety of conditions, including studies on orientated single crystals at liquid He temperatures [38]. From fitting procedures reported the two Cu–N(His) bonds at 1.97 Å are ~0.1 Å less and the Cu–S(Cys) distance 2.11 Å (0.02 Å smaller) than those obtained from the X-ray crystal structure determination [16].

Because of its distorted tetrahedral shape, and the long Cu–S(Met) bond, the type 1 Cu active site has proved difficult to model in low M_r complexes. The unusual EPR spectra with characteristically small hyperfine splitting constants are a consequence of the asymmetry of the copper environment.

2.3 *Other Properties of Plastocyanin*

PCu(II) has a dominant 597 nm band, which from studies on Co(II) substituted protein is assigned to S(Cys84)→Cu charge transfer. There are other satellite bands at 427, 468, 535, 717, 781 and 926 nm, which involve the cysteine and histidine ligands [45]. It has been proposed that transitions involving the S(Met) ligand contribute to the spectrum. The reduced PCu(I) form does not absorb in the visible, but both PCu(I) and PCu(II) have peaks and shoulders in the 240–300 nm range, and assignments to tyrosine and phenylalanine residues have been made [46]. In this region the absorbance for spinach PCu(II) is ~70% that for the PCu(I) form.

From ^1H NMR studies protonation and dissociation of the two histidine ligands at the active site of spinach PCu(I) has been reported [47]. The acid dissociation pK_a value for the reversible process (1),

$$H^+PCu(I) \rightleftharpoons H^+ + PCu(I) \quad (1)$$

involving His87 is 4.9. An upper limit for His37 is 4.5. Protonation of His87 results in a loss in redox activity [39], an effect which has (as already discussed) been fully documented by X-ray crystallography [17]. Using ^1H NMR His87 pK_a values for other plastocyanins have been determined, Table 2 [48]. Higher pK_a values are noted for parsley and *S. obliquus* plastocyanin which have deletions at positions 57 and 58. From kinetic studies with $[Fe(CN)_6]^{3-}$ as oxidants similar pK_a values are obtained [36, 49], but with $[Co(phen)_3]^{3+}$ somewhat higher "apparent" pK_a's are obtained. The latter are believed to be composite, arising from the use $[Co(phen)_3]^{3+}$ makes of the remote site. This will be returned to later.

Reduction potentials (E_0) for different plastocyanins, the PCu(II)/PCu(I) couple, have been determined by spectrophotometric titration against, e.g. $[Fe(CN)_6]^{4-}$. At pH 7.5 for higher plant and green algal plastocyanins values are close to 370 mV at 25 °C, I=0.10 M(NaCl) [1]. Thus French bean gives a value 360 mV and *S. obliquus* 363 mV [50]. However, *A. variabilis* gives an

Table 2. Acid dissociation pK_a values, I = 0.10 M(NaCl), relating to the active site protonation of different plastocyanins, PCu(I), as determined by (a) proton NMR (b) the variation of rate constants (25 °C) with pH for the $[Fe(CN)_6]^{3-}$ oxidation of PCu(I), I = 0.10 M(NaCl), and (c) similar experiments with $[Co(phen)_3]^{3+}$ as oxidant. The latter is an 'apparent' value only, and is believed to be composite due to reaction occurring at the remote site

	pK_a		
	NMR	$[Fe(CN)_6]^{3-}$	$[Co(phen)_3]^{3+}$
Parsley	5.7	5.5	6.1
S. obliquus	5.4	5.0	5.65
Spinach	4.9	4.8	5.5
French bean	4.7	4.6	5.4
Poplar		4.7	5.2
A. variabilis	5.1	5.0	5.4[a]

Data from Refs. [1], [48], [49], [101], [102] and Refs. [10], [70], [71].
[a] May require revision.

appreciably lower value of 340 mV [36, 50]. From kinetic data the E^0 values of all plastocyanins increase with decreasing pH, due to the protonation of His87 of PCu(I). Typical values are >390 mV at pH 5 and continue to increase as the pH is further decreased. Most plastocyanins denature at or around pH 4, a process which is probably related to the protonation of His37 in addition to His87.

Recently, unmediated electrochemistry of plastocyanin has been achieved at an edge-orientated pyrolytic graphite electrode over the pH range 4–8 [51]. Promotion of rapid electron transfer is observed on addition of redox inert multivalent cations such as Mg^{2+} and $[Cr(NH_3)_6]^{3+}$ (<5 mM), or aminoglycosides (e.g. neomycin), or by mild acidification. The edge face of the pyrolytic graphite, subjected to standard polishing procedures in air, is believed to contain a variety of hydrophillic CO functional groups, which result in an excess of negative charge at the electrode surface. The presence of cations or acid is necessary for favorable interaction with the negatively charged plastocyanin.

A transit peptide consisting of a hydrophobic 66 amino acid long peptide interspersed with positively charged residues has been identified and sequenced [52]. This is initially attached to the 99 amino acids of the mature plastocyanin, and is responsible for taking the plastocyanin across membranes into the thylakoid region of the chloroplast.

3 Properties of Azurin

3.1 General

Studies on the bacterial type 1 protein azurin have been extensive. Ten different azurin amino-acid sequences have been determined with 47 out of 129 residues (M_r 14,000) conserved. Reduction potentials are in the range 280–339 mV at pH 7.5. The crystal structures of azurin ACu(II) from *Pseudomonas aeruginosa* (2.7 Å resolution) [53] and *Alcaligenes denitrificans* (1.8 Å resolution) have been determined [54]. Eight extended strands of polypeptide chain form a β barrel structure with topology like that of plastocyanin. Only 9 residues can be said to be invariant between the two families, which includes the Cu ligands His46, Cys112, His117 and Met121. As in the case of plastocyanin the Cu to His and Cys bond lengths are close to 2.0 Å, and the Cu–S (Met) distance longer (~2.6 Å). From EXAFS studies the Cu–S (Met) distance cannot be determined unambiguously in the case of ACu(II), but for ACu(I) from *P. aeruginosa* a value 2.7 Å has been obtained [55].

3.2 Structure

An additional interaction between the Cu and the carbonyl O-atom of Gly45 is short enough (3.1 Å) in the case of *A. denitrificans* azurin to imply a weak bond. A similar distance (~3.5 Å) has been detected in *P. aeruginosa.* The extra loop in azurin, ten residues linking His35 to His46, is to be compared with plastocyanin in which only three residues join the analogous Ala33 to His37. This is probably responsible for the closer approach of Gly45 to the Cu, the comparable distance in plastocyanin being 3.8 Å. A major difference in primary structure is between residues 52 and 90 (azurin numbering), where there is no structural homology at all with plastocyanin. Residues 52 to 81 on azurin constitute a prominent flap on the outside of the β-strand 5. This is inserted at position 42 in the plastocyanin structure and replaces the section including the negative patch 42–45. Furthermore, the section including residues 59–61 in plastocyanin is replaced in the azurin structure. Azurin has no highly charged regions, and most charged residues are neutralized by a nearby residue of opposite charge. The pI for *P. aeruginosa* azurin is 5.6, consistent with a small negative charge at pH ~7. The hydrophobic patch around the exposed His87 of plastocyanin, His117 in azurin, is retained, and is a likely site for electron transfer. The Cu appears to be slightly more buried (~7 Å) than in the case of plastocyanin (~6 Å).

3.3 Influence of His35

Earlier suggestions that the two uncoordinated and invariant residues His35 (inaccessible to solvent and covered by polypeptide) and His83 (remote and 13 Å from Cu) are, from effects of [H^+] on rate constants (and related pK_a values), sites for electron transfer may require some re-examination. Thus, it has been demonstrated in plastocyanin studies [50] that a surface protonation can influence the reduction potential at the active site, in which case its effect is transmitted to all reaction sites. In other words, an effect of protonation on rate constants need not necessarily imply that the reaction occurs at the site of protonation. His35 is thought to be involved in pH-dependent transitions between active and inactive forms of reduced azurin [53]. The proximity of

His35 to ligated His46 may be important in an electron transfer role or in His35 exercising some conformational control of the active site. The reduction potential of *P. aeruginosa* azurin increases from 300 mV (pH 8) to 360 mV (pH 5), which is believed to be related to His35 protonation. A pK_a of 6.6 is observed for this process, or alternatively pK_a's for azurin in the oxidized (6.1) and reduced (7.2) forms can be obtained [56].

The His35, with coordinated His46 in close proximity, has frequently been suggested as a site for electron transfer reactivity of azurin. Two processes have been detected in a temperature-jump study on the equilibration of azurin with cytochrome c_{551}, its physiological partner [57]. The fast process is assigned to electron transfer, and the slower process to a conversion between inactive and active forms of reduced azurin. It has been concluded that the active form is protonated. A second H-bonded form of His35 is believed to result from the protonation [2].

3.4 Site Directed Mutagenesis

Azurin from *Pseudomonas aeruginosa* has been expressed in *Escherichia coli* [58]. Site-directed mutagenesis has been used to obtain the Met121→Leu replacement [59]. The strong blue color of the Cu(II) form and characteristic EPR signal are retained, showing that this Met residue is not an obligatory component of the active site. The visible absorbance band at 625 mm is shifted 5 nm to longer wavelengths, and ε increased by 10% as compared to wild type azurin [59]. The reduction potential is increased by 70 mV. In work from the Canters' group the replacement Met44→Lys has little or no effect on structure from a comparison of ^{1}H NMR spectra [60]. The mutation, located close to the Cu ligated His117 on the hydrophobic north surface, produces a significant effect on reactivity. Thus, the self-exchange rate constant (37 °C) at pH 9 (1.0×10^5 M^{-1} s^{-1}) decreases to a value $< 1.0 \times 10^3$ M^{-1} s^{-1} at pH 5. In the case of wild type azurin the rate constant of 1.3×10^6 M^{-1} s^{-1} is independent of pH. It is concluded that protonation of Lys44 decreases the self-exchange process, and that the hydrophobic patch is involved in the electron-exchange reaction. Preliminary results obtained on replacing His35→Gln indicate that neither the exchange rate constant nor the pH dependence differ from wild type azurin.

4 Other Type 1 Cu Proteins

There are four other proteins – stellacyanin, rusticyanin, umecyanin and amicyanin (Table 3) which have been fairly extensively studied. A crystal structure determination for amicyanin from *Thiobacillus versutus* is now under way [61]. A number of other type 1 proteins have been identified. These include pseudo-

Table 3. Properties of single type blue Cu proteins

Protein	Source	M_r	Amino acids	pI	E^0/mV	(Absorbance)[a]
Plastocyanin	Plants/algae	10,500	99	4.2	370	597 (4500)
Azurin	Bacteria	14,000	128	5.4	305[b]	625 (5200)
Stellacyanin	Lacquer tree[c]	20,00[d]	107	9.8	184	608 (4080)
Rusticyanin	*T. ferroxidans*	16,000	159	9.1	680[e]	597 (2240)
Umecyanin	Horse-radish roots	14,600	125	5.8	283	610 (3400)
Amicyanin	Methylotrophs[f]	11,000	106	4.7	260	596 (3900)

[a] Peak position λ/nm(ε/$M^{-1}cm^{-1}$); [b] At pH 7.5; Value for *Alcaligenes denitrificans* 276 mV (Ainscough EW, Bingham AG, Brodie AM, Ellis WR, Gray HB, Loehr TM, Plowman JE, Norris GE, Baker EN (1987) Biochemistry 26: 71). [c] *Rhus vernicifera*; [d] 40% carbohydrate; [e] pH2; [f] e.g. *T. versutus*.

azurin [62], cucumber basic protein [63], mavicyanin [64], mung-bean protein [65], rice bran [66], and auracyanin [67]. The latter, (pI = 4.0) from a green photosynthetic bacterium *Choloroflexus aurantiacus*, was isolated as a disulfide-bridged dimer (monomer M_r 12,800). It appears to be peripherally associated with the cytoplasmic membrane. It's role is uncertain, but it has a reduction potential (240 mV) compatible with a role in photosynthesis. Two type 1 copper proteins have been isolated from the nitrite-reducing system in *Pseudomonas aureofaciens* [68]. The nitrite reductase protein was found to be dimeric (monomer M_r 40,000), pI = 6.05. The pseudoazurin protein (M_r 15,000) has been expressed in *E. coli* and the correctly processed blue protein isolated [69]. The methylotropic bacteria *Pseudomonas* strain AM1 and organism 4025 also appears to contain two type 1 Cu proteins, an amicyanin and an azurin [70]. Type 1 sites are also present in the multi-Cu proteins ascorbate oxidase [5], laccase [71, 72] and ceruloplasmin [73]. There is now evidence that all three have trinuclear Cu sites as a result of a merging of the type 2 and 3 sites. Stellacyanin and laccase are both isolated from the Japanese lacquer tree *Rhus vernicifera*. Ascorbate oxidase is one of a number of blue Cu proteins obtained from cucumber peelings. Laccase is obtained from both tree and fungal sources, but has different reduction potentials for the type 1 site (395 and 785 mV, respectively) in the two cases.

The crystal structure of the pseudoazurin from *Alcaligenes faecalis* S-6 sometimes referred to as the 'blue protein' (also as cupredoxin), has been reported to 2.0 Å [74]. The protein folds in β-sandwich which is described as being similar to plastocyanin and azurin.

The structure of the blue basic cucumber protein (pI ~ 10.5), sometimes referred to as plantacyanin, has also been determined [63]. The same coordinating ligands are observed in this case His39, Cys79, His84, Met89, as in plastocyanin, azurin and pseudoazurin (His40, Cys78, His81, Met86). The reduction potential is 314 mV. A disulfide bridge joins Cys52 to Cys85. The folds of plastocyanin, azurin and cucumber basic protein (CBP) are distinctly different. In azurin strands 4 and 5 of the polypeptide backbone are part of the β sandwich, and connecting these ends is a flap comprising of 30 residues and including three turns of α-helix hanging off the main body of the molecule. In plastocyanin strand 5 is too irregular to be part of the β sandwich. In CBP the β-sandwich structure is further depleted by a bend and a twist in strands 4 and 5 that place these strands at a large angle from the others. There are, in other words, quite striking differences between CBP and plastocyanin.

Mavicyanin ($M_r = 18{,}000$) is obtained from green squash (*Cucurbito pepo medullosa*), where it occurs alongside ascorbate oxidase [64]. It has a peak at 600 nm ($\varepsilon \sim 5000\ M^{-1}\ cm^{-1}$) and reduction potential of 285 mV. Further studies on this and the mung bean and rice bran proteins [65, 66] would be of interest. All the above type 1 Cu proteins have an intense blue color and characteristic narrow hyperfine EPR spectrum for the Cu(II) state. Table 3 summarizes the properties of those most studied. There is some variation in reduction potential and position of the main visible absorbance peak. In the case of azurin, for example, the latter is shifted from 597 to 625 nm. Stellacyanin has no methionine and the identity of the fourth ligand is therefore different [75]. The possibility that this is the *O*(amide) of Gln97 has been suggested [63b]. It now seems unlikely that the disulfide is involved in coordination. Stellacyanin has 107 amino acids, with carbohydrate attached at three points giving a 40% contribution to the M_r of 20,000 [75].

Rusticyanin from *Thiobacillus ferrooxidans* [76] is unusual in having a high reduction potential (680 mV), and in being stable at pH 2, which is close to the working pH. From studies on the Fe^{2+} reduction of rusticyanin in sulfate, it has been concluded that the reaction is too slow to be directly implicated in the Fe-dependent respiratory electron transport chain of *T. ferrooxidans* [77]. Instead, an acid stable cytochrome c is believed to be involved as a catalyst, and the electron transport chain is $Fe^{2+} \rightarrow$ cytochrome c $\rightarrow$ rusticyanin. It has been suggested that reduction potentials can be fine tuned by effects of polypeptide backbone structure on the Cu–S(Cys) and Cu–S(Met) bond distances (possibly also angles) and Cu ligand field. A further control may be by hydrogen-bonding of the coordinated cysteine thiolate, which has been detected by resonance Raman spectroscopy [79]. The sequence of rusticyanin has a sufficiently 'normal' distribution of His, Cys, His and Met residues for these to be coordinated to the Cu [80]. One variant which may determine the different properties is the length of peptide separating the coordinating group Cys, His, Met which are positioned 84, 87, 92 (plastocyanin), 112, 117, 121 (azurin), and 127, 132, 137 for rusticyanin. Umecyanin [81], which has 38% homology (first 88 residues) with stellacyanin [82], has three methionines but none after residue 74 in the sequence, which again indicates a change in ligation. An EXAFS fit for umecyanin is consistent with His Cys His coordination, but again it was not possible to identify the fourth ligand [83].

Two amicyanins have been a recent focus of attention [84, 85]. These are from the methylotrophic bacteria *Pseudomonas* AMl and *Thiobacillus versutus*, where the function is to mediate electron transport between bacterial cytochrome c and methylamine dehydrogenase in a relatively short electron transport chain. The sequence of *Pseudomonas* AMI (99 residues) has been determined, and residues His, Cys, His, Met are available and the likely ligating groups [86]. From ^{1}H NMR studies on the Cu(I) form one of the His ligands protonates, pK_a 6.74 [87]. This is the only other example so far of a protein other than plastocyanin exhibiting active-site protonation. In the case of amicyanin the kinetics of protonation and rotation as reported for plastocyanin have been determined by the Canters' group [87].

Kinetic studies on the reactions of azurin [88, 89], stellacyanin [90], rusticyanin [91], and umecyanin [83] with inorganic redox partners have been carried out. Self-exchange rate constants/$M^{-1} s^{-1}$ have been determined for azurin (9.6×10^5) [92], stellacyanin (1.2×10^5) [93], amicyanin (1.3×10^5) [87] and plastocyanin ($<3 \times 10^3$) [94], using ^{1}H NMR and EPR methods. The plastocyanin limit requires some comment. Studies at 25 °C and pH 6 on spinach plastocyanin (1.6 mM) have been carried out using ^{1}H NMR with (a) $[Co(NH_3)_6]^{3+}$, (b) Mg^{2+}, and (c) KCl added. Typical rate constants are 7×10^4 $M^{-1} s^{-1}$ (8 mM$[Co(NH_3)_6]^{3+}$), 2.7×10^4 $M^{-1} s^{-1}$ (20 mM $MgCl_2$) and 4×10^3 $M^{-1} s^{-1}$ (0.1 M KCl). With the increased screening provided by different cations, which associate more strongly at the lower ionic strengths, the self-exchange rate constant is seen to increase. Since cations are expected to associate at the acidic patch, this might be a part of the exchange process observed. Possible mechanisms are therefore the interaction of two negatively charged (remote) plastocyanin sites, one remote and one adjacent site, or two hydrophobic adjacent sites. Certainly in the case of azurin the latter would seem to be the more likely mechanism, and it is of interest that in the crystal structure of *A. denitrificans* azurin [54], two molecules are orientated and in contact in this manner (with the exclusion of solvent). The distance between the Cu's is 14.3 Å, and between the two His117 edges 7.1 Å.

5 Kinetic Studies

Inorganic complexes have been used extensively over the last decade in the study of electron-transfer reactions of metalloproteins. The reactivities of six different plastocyanins (sequences Fig. 3), including *S. obliquus* (green alga) and *A. variabilis* (blue-green alga), have been investigated. Most extensive studies have been with the parsley and spinach plastocyanins. Spinach and French bean are the more typical higher plant plastocyanins, since parsley has deletions at 57 and 58, which is more a feature of green algal plastocyanins. Also poplar plastocyanin has only three acidic residues at 42–45. References to studies with plastocyanin from different sources are parsley [39, 95–99], spinach [100], poplar [101], *S. obliquus* [49], *A. variabilis* [102], and French bean [103, 104].

Table 4. Comparison of rate constants (25 °C) for the reactions of parsley and spinach plastocyanins, pH 7.5, I = 0.10 M(NaCl) [95, 99, 100]

Reaction	$E^{0\,a}$ mV	k (parsley) $M^{-1}s^{-1}$	k (spinach) $M^{-1}s^{-1}$	Ratio
PCu(I) + [5−][b]	460	3.4×10^4	—	
PCu(I) + $[Fe(CN)_6]^{3-}$	410	9.4×10^4	8.0×10^4	1.18
PCu(I) + $[Co(dipic)_2]^-$	747	4.9×10^2	4.0×10^2	1.22
PCu(I) + $[Ru(NH_3)_5(CH_3CN)]^{3+}$	462	—	1.8×10^5	
PCu(I) + $[Co(phen)_3]^{3+}$	370	3.0×10^3	2.5×10^3	1.20
$[Ru(NH_3)_5py]^{2+}$ + PCu(II)	273	4.4×10^5	5.8×10^5	0.76
$[Co(terpy)_2]^{2+}$ + PCu(II)	260	7.3×10^4	8.2×10^4	0.89
$[Co(phen)_3]^{2+}$ + PCu(II)	370	2.5×10^3	2.8×10^3	0.89
$[Fe(CN)_6]^{4-}$ + PCu(II)	410	1.7×10^4	1.8×10^4	0.94

[a] Reduction potential for inorganic redox partner.
[b] [5−] is the one-equivalent oxidant $[(NC)_5FeCNCo(CN)_5]^{5-}$.

Table 5. Rate constants/$M^{-1}s^{-1}$ (25 °C) [49, 102] for the oxidation of different plastocyanins, PCu(I), by $[Fe(CN)_6]^{3-}$ and $[Co(phen)_3]^{3+}$ at pH ~7.5, I = 0.10 M(NaCl)

Source	$10^{-4}k_{Fe}$	$10^{-3}k_{Co}$[a]	Ratio
Parsley	9.4	3.0	31
S. obliquus	8.5	1.85	49
Spinach	8.5	2.50	34
French Bean	5.8	4.7	12
Poplar	6.9	2.90	25
A. variabilis	67.0	0.63	1060

[a] Determined at relatively low (~0.4 mM) $[Co(phen)_3]^{3+}$ so that saturation kinetics do not apply.

5.1 Rate Constant Patterns

Rate constants have been reported for the reactions of parsley and spinach plastocyanins with inorganic redox partners, charges in the range −5 to 3+, Table 4. A self-consistent pattern is observed with parsley PCu(II), some 20% more reactive, and spinach PCu(I) 14% more reactive, which suggests that the reduction potential for spinach plastocyanin is slightly lower than that for parsley plastocyanin. For all six plastocyanins, the ratio of $[Fe(CN)_6]^{3-}$ to $[Co(phen)_3]^{3+}$ rate constants, Table 5, shows a consistent pattern, except in the case of *A. variabilis* plastocyanin. Here the less prominent/non-existent acidic patch and overall positive charge result in significant changes, with an increase in rate constant for the reaction with $[Fe(CN)_6]^{3-}$, and a decrease with $[Co(phen)_3]^{3+}$. The increase with $[Fe(CN)_6]^{3-}$ may in part be a result of the bigger driving force, since *A. variabilis* has a reduction potential some 30 mV less than for other plastocyanins. However, the same argument would be expected to

Table 6. Charge distribution on different plastocyanins PCu(I) at pH ~7.5, based on a count of acidic (Asp, Glu) and basic (Lys, His, Arg) residues with no allowance for H-bonding. Information obtained from sequences in Fig. 3

Source	Overall Charge	42–45	59–61
Parsley	8−	4−	1−
S. obliquus	8−	3−	2−
Spinach	9−	4−	3−
French bean	9−	4−	3−
Poplar	9−	3−	3−
A. variabilis	1+	1−	2+

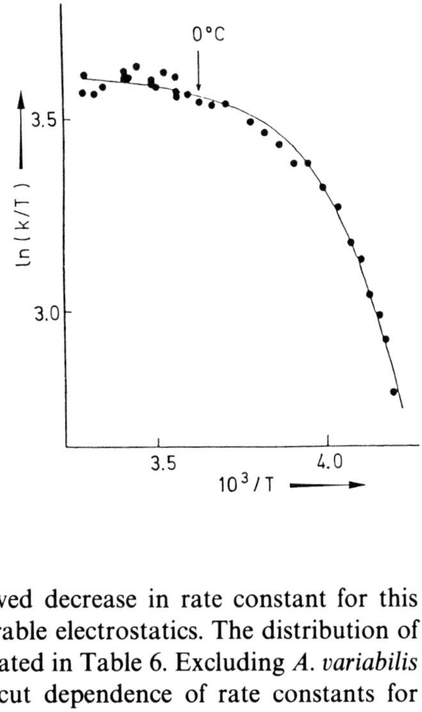

Fig. 6. Curved Eyring plot of second-order rate constants k for the $[Fe(CN)_6]^{3-}$ oxidation of spinach PCu(I) in 50:50 (v/v) MeOH–H_2O solvent at temperatures 25 to −35 °C (reproduced with permission from Ref. [105])

hold for $[Co(phen)_3]^{3+}$, and the observed decrease in rate constant for this oxidant draws attention to the less favorable electrostatics. The distribution of charge on different plastocyanins is indicated in Table 6. Excluding *A. variabilis* there does not appear to be any clearcut dependence of rate constants for $[Co(phen)_3]^{3+}$ (Table 5) on charge at 59–61, and the less variable charge at 42–45 would seem to be more important.

5.2 *Activation Parameters*

These have been determined for the $[Fe(CN)_6]^{3-}$ oxidation of parsley plastocyanin, and at pH 7.5, I = 0.10 M(NaCl), are $\Delta H\ddagger = -3.3$ kcal mol^{-1} and $\Delta S\ddagger = -47$ cal K^{-1} mol^{-1} [39]. To account for the negative enthalpy term it has been suggested that the second-order rate constant is composite incorporating an equilibrium process. Although no saturation (sometimes referred to as limiting) kinetic behavior is observed, it is likely that there is association (with K small) prior to electron transfer as discussed below for $[Co(phen)_3]^{3+}$. Thus, the second-order rate constant k corresponds to $k_{et}K$, where K for association has a negative enthalpy term which is sufficient to outweigh a positive value from k_{et}. Activation parameters for the one-equivalent oxidant $[(NC)_5FeCNCo(CN)_5]^{5-}$ are also negative, $\Delta H\ddagger = -2.9$ kcal mol^{-1} and $\Delta S\ddagger = -48$ cal K^{-1} mol^{-1}, as are $\Delta H\ddagger$ and $\Delta S\ddagger$ for the same two oxidants with azurin [104].

Activation parameters for the $[Fe(CN)_6]^{4-}$ reduction of parsley plastocyanin in H_2O, I=0.10 M(NaCl), $\Delta H\ddagger = 6.3$ kcal mol^{-1} and $\Delta S\ddagger = -17.5$ cal K^{-1} mol^{-1} [39], are essentially the same as values 8.8 and 8.4 ($\Delta H\ddagger$/kcal mol^{-1}), -9.1 and -10.6 ($\Delta S\ddagger$ cal K^{-1} mol^{-1}) for spinach and French bean respectively [106].

To summarise, although the driving force is small, the reactions of plastocyanin PCu(I) (370 mV) with $[Fe(CN)_6]^{3-}$ (410 mV) and $[Co(phen)_3]^{3+}$ (370 mV) are fast. All the reactants in Table 4 are in the stopped-flow range.

5.3 *Effect of Mixed Solvent*

The Eyring plot for the monophasic $[Fe(CN)_6]^{3-}$ oxidation of spinach plastocyanin in 50:50 (v/v) MeOH–H_2O mixed solvent over the extended temperature range 25 to -35 °C, pH 7.0, I=0.10 M(NaCl), exhibits a marked downward curvature, Fig. 6 [105]. The rate constant at 25 °C of 1.04×10^4 M^{-1} s^{-1} compares with the value in Table 4 of 8.0×10^4 M^{-1} s^{-1}. A plot of first-order rate constants against $[Fe(CN)_6^{3-}]$ (reactant in excess) is linear over the range up to 3 mM at all temperatures. There appear to be two resolvable processes with $\Delta H\ddagger = 0.18$ kcal mol^{-1}, and $\Delta S\ddagger = 21.4$ cal K^{-1} mol^{-1} (low temperature), and $\Delta H\ddagger = 14.7$ kcal mol^{-1}, $\Delta S\ddagger = -39.3$ cal K^{-1} mol^{-1} (high temperature). The contrast in these is striking. From a similar study on the ferrocenium oxidant $[Fe(Cp)_2]^+$ (Cp is the η^5-cyclopentadienide ligand) the Eyring plot is linear 25 to -30 °C, with $k = 1.9 \times 10^6$ (1.06×10^6) M^{-1} s^{-1} (results in H_2O solvent in brackets). Further work is called for to better understand these effects.

5.4 *Effect of Optical Isomers*

The kinetics of the reduction of spinach plastocyanin PCu(II) by the optically active complexes 2,6-bis[3-(*S*)- or 3-(*R*)-carboxyl-2-azabutyl] pyridine, here abbreviated to (*S,S*)- or (*R,R*)-ALAMP have been studied [107]. The latter enantiomer (Δ-configuration) reacts 1.6–2.0 times faster at different values of pH and temperature than the *S,S* form. Activation parameters have shown that the observed stereoselectivity is a consequence of the difference in activation entropies (36 cal K^{-1} mol^{-1}), which overcompensates the enthalpy change favoring the Λ configuration (0.72 kcal mol^{-1}).

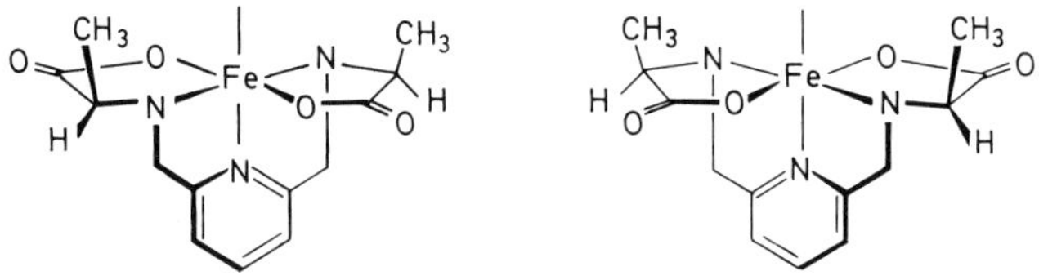

5.5 *Saturation Kinetics*

There are surprisingly few examples and considerable care is required. A number of earlier reports have been checked out [90, 95, 108], and shown to be incorrect with effects certainly not as extensive as claimed. The first plastocyanin example (1978) was the $[Co(phen)_3]^{3+}$ oxidation of parsley PCu(I) [39]. In this study first-order rate constants (k_{obs}) obtained with $[Co(phen)_3]^{3+}$ in >10-fold excess of PCu(I) ($\sim 10^{-5}$ M) give a non-linear dependence on $[Co(phen)_3^{3+}]$ (concentrations of oxidant to 3×10^{-3} M), Fig. 7. Such behavior can be accounted for by the reaction sequence (2)–(3),

$$\mathrm{PCu(I)} + [\mathrm{Co(phen)_3}]^{3+} \overset{K}{\rightleftharpoons} \mathrm{PCu(I)}, [\mathrm{Co(phen)_3}]^{3+} \tag{2}$$

$$\mathrm{PCu(I)}, [\mathrm{Co(phen)_3}]^{3+} \xrightarrow{k_{et}} \mathrm{PCu(II)} + [\mathrm{Co(phen)_3}]^{2+} \tag{3}$$

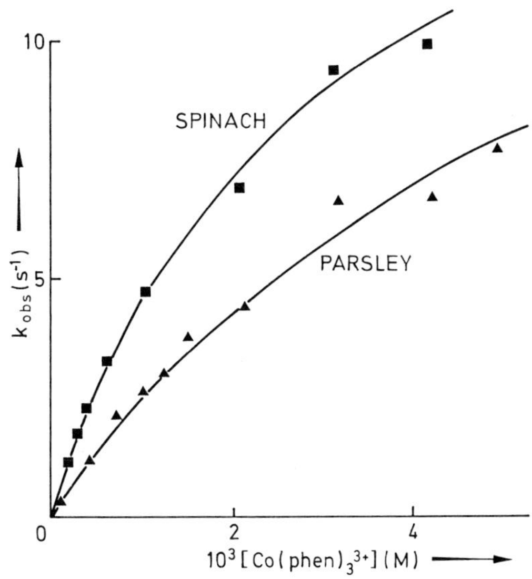

Fig. 7. The variation of first-order rate constants k_{obs} (25 °C) with oxidant for the $[Co(phen)_3]^{3+}$ (reactant in large excess) oxidation of parsley (▲) and spinach (■) plastocyanin, PCu(I), I = 0.10 M(NaCl) [39]

which gives (4).

$$k_{obs}=\frac{k_{et}K[Co(phen)_3^{3+}]}{1+K[Co(phen)_3^{3+}]} \tag{4}$$

This can be tested for by plotting $1/k_{obs}$ against $[Co(phen)_3^{3+}]$, to correspond with the linearized form of (4), and K and k_{et} are readily obtained. Two alternative mechanisms, have been considered [1]. Of these the so-called 'dead-end' mechanism is difficult to rule out. There is, however, at present no evidence for it applying, and until such evidence is forthcoming discussion has proceeded in terms of (2)–(3). Were it possible to use sufficiently high $[Co(phen)_3^{3+}]$, a plot of k_{obs} against $[Co(phen)_3^{3+}]$ would level off at k_{et}. The use of $[Co(phen)_3^{3+}] > 3 \times 10^{-3}$ M, I = 0.10 M(NaCl), would introduce significant variations in anion and cation contributions. Even at 3×10^{-3} M such effects may contribute. However, more extensive association of redox inactive analog complexes of higher charge is observed in inhibition studies (see later). No similar effects are observed for reactions of plastocyanin with negatively charged reactants. In other words, a self-consistent picture has emerged, in which electrostatics involving interaction of the remote site with catonic species is a prime controlling feature of reactivity. From the simple treatment for $[Co(phen)_3]^{3+}$ involving (4), $K = 167\ M^{-1}$ and $k_{et} = 17.9\ s^{-1}$ at 25 °C. A further elaboration is considered in the next but one section, which results in a modification of these values.

The above results are consistent with cationic association at the acidic patch 42–45. Saturation (or limiting) kinetic behavior is observed also for $[Co(phen)_3]^{3+}$ with spinach (Fig. 7) [39], and French bean PCu(I) [109]. It is not detected however in the corresponding reactions of poplar [101] and *S. obliquus* [109] PCu(I) at I = 0.10 M(NaCl), which have only three negative residues at 42–44, or with *A. variabilis* plastocyanin [102], which retains only Asp42 (with Glu85 nearby). On decreasing the ionic strength to < 0.10 M, increased association has been demonstrated in the case of French bean PCu(I) [109]. At I = 0.50 M(NaCl) on the other hand, no saturation kinetics is observed, and K is less [110]. These trends are explained by a shielding of the negative charge on the protein by the ionic medium. Although $[Fe(CN)_6]^{3-}$ (410 mV) and $[Co(phen)_3]^{3+}$ (370 mV) have similar reduction potentials, the $[Fe(CN)_6]^{4-/3-}$ couple has a more favorable self-exchange rate constant ($1.9 \times 10^4\ M^{-1}\ s^{-1}$)[111] as compared to $[Co(phen)_3]^{2+/3+}$ ($41.7\ M^{-1}\ s^{-1}$) [112]. A comparison of kinetic data for $[Fe(CN)_6]^{3-}$ and $[Co(phen)_3]^{3+}$ is given in Table 5. In the case of *A. variabilis* PCu(I) it is not at present clear whether there are contributions from reaction at Asp42 [102].

The observation that there is significant association of $[Co(phen)_3]^{3+}$ with parsley PCu(I) at I = 0.10 M(NaCl), has led to competitive inhibition studies using redox inactive complexes of higher positive charge. Thus, on adding increasing amounts of $[Cr(phen)_3]^{3+}$, competitive inhibition (or blocking) of the $[Co(phen)_3]^{3+}$ redox process is observed. Other redox inactive complexes of charge 4+ through to tetranuclear 6+ and 7+ complexes have been studied [96, 98]. This effect is considered further in section 7.

6 NMR Line Broadening

The suggestion that two different binding sites on plastocyanin are used according to charge has been further established using high resolution ^1H NMR spectroscopy in D_2O solvent [104]. Thus effects of redox inactive analog complexes $[Cr(CN)_6]^{3-}$ and $[Cr(phen)_3]^{3+}$, as well as $[Cr(NH_3)_6]^{3+}$ (0.2–1.2 mM) on the NMR spectrum of French bean, cucumber, and parsley PCu(I) (3.5 mM) have been reported [104, 113]. The paramagnetic Cr(III) complexes bring about line-broadening at sites where a favorable interaction occurs. Aromatic residues were among the first to be assigned in NMR spectra and have proved most useful. The evidence obtained indicates preferential binding of $[Cr(CN)_6]^{3-}$ close to His87, whereas with $[Cr(phen)_3]^{3+}$ and $[Cr(NH_3)_6]^{3+}$ line broadening of Tyr83 occurs. The Tyr83 is flanked either side by acidic residues 42–45 and 59–61. Since line broadening is dependent on the inverse sixth power of the distance of the amino acid from the paramagnetic center, the His87 and Tyr83 residues are likely to be a part of or close to the respective binding sites. No benefit appears to result from the presence of aromatic (phen) as opposed to ammine complexes, and although some interaction of phenanthroline with Tyr83 is possible [103], the evidence obtained at least with different inorganic inhibitors suggests that charge is particularly important. Stacking effects are not excluded in considering the interaction of two proteins.

More recent studies on the binding of $[Cr(NH_3)_6]^{3+}$ to spinach PCu(I) by both one and two-dimensional NMR techniques have indicated a broadening of several resonances in the aliphatic region corresponding to spatially separated groups, and therefore reflecting a large $[Cr(NH_3)_6]^{3+}$ accessible surface [114]. Three groups of negatively charged residues are tentatively assigned as cation binding sites on the protein surface. Thus, binding of $[Cr(NH_3)_6]^{3+}$ at the acidic groups of residues 59–61 (and possibly Glu68) can account for the observed relaxation of the methyl group resonances of Met57, Leu62 and Leu63. The broadening of Val39, Val40, Ile46, Thr79, Phe82, Tyr83 and Val93 cross-peaks is explained by binding of the probe to the acidic residues 42–45 and 59. Binding at Asp51 is invoked to explain the observed strong effects at Val50, Ala52 and Ala53. An interaction at Glu25 and Glu26 is a further possibility. Of these, the area incorporating Tyr83 and Val40 is seen to have the highest affinity for $[Cr(NH_3)_6]^{3+}$.

Similar NMR studies have been reported for *S. obliquus* and *A. variabilis* plastocyanin [37]. With $[Cr(phen)_3]^{3+}$ binding to *S. obliquus* PCu(I) occurs over a delocalized region of protein surface which includes the acidic residues 42–44 and 60–61. The resonances most affected are those for the C4H and C3, 5H protons of Phe83, and the C3, 5H protons of Tyr62, with somewhat less broadening of resonances from His59 C4H and Phe83 C2, 6H protons. Other slight but significant broadening effects have been noted. It is of interest here that the highly conserved Phe82 and Tyr83 residues are interchanged in *S. obliquus*. No specific broadening was observed with $[Cr(CN)_6]^{3-}$. Similar experiments have been carried out with *A. variabilis* PCu(I) [102], which has fewer acidic residues at the remote site (Glu85 is in close proximity to Asp42). No specific broadening effects were observed with $[Cr(phen)_3]^{3+}$. With $[Cr(CN)_6]^{3-}$ interaction at sites in the vicinity of His59/Lys60 and Lys9/Lys33 in the northern part of the molecule are observed with only a slight shift in His87 resonances.

To summarize, in all cases so far studied (except *A. variabilis*) there is evidence for $[Cr(phen)_3]^{3+}$ association at the acidic patch close to residue 83, here referred to as the remote site. There is also a strong case for a favorable $[Cr(CN)_6]^{3-}$ interaction close to His87 at the hydrophobic adjacent site. There is no compelling reason why an inorganic complex should be rigidly held and some movement around a specific region of surface may well occur. Redox active $[Fe(CN)_6]^{3-}$ and $[Co(phen)_3]^{3+}$ are expected to interact at the same sites as the Cr(III) analogs. Discussion will proceed, therefore, in terms of fairly broad regions of interaction at remote and adjacent sites. Whether electron transfer can take place from all positions within these regions, or is a much more specific process, has still to be considered.

The interaction of positively charged (9+) cytochrome c(III) with PCu(II) has been studied by the same NMR procedures [115]. The results obtained again indicate that there is a favorable interaction at the remote site. The corresponding interaction of PCu(II) with cytochrome f(M_r 33,000) is at present more difficult to interpret without knowledge of the cytochrome f structure.

7 Effect of Redox Inactive Complexes

Not only does $[Cr(phen)_3]^{3+}$ inhibit the $[Co(phen)_3]^{3+}$ oxidation of PCu(I) [116], but other complexes (I) of high positive charge including $[Co(NH_3)_6]^{3+}$, $[Pt(NH_3)_6]^{4+}$ and $[(NH_3)_5CoNH_2Co(NH_3)_5]^{5+}$ are also effective in this way [96]. The association step (5) has to be included in the reaction scheme alongside (2) and (3).

$$\mathrm{PCu(I)} + \mathrm{I} \overset{K_I}{\rightleftharpoons} \mathrm{PCu(I), I} \tag{5}$$

If the amount of inhibitor is restricted to an upper limit of 10^{-3} M, the effect of the 3+ complexes is not sufficiently big to give very precise values of K_I. A further point is that $[Pt(NH_3)_6]^{4+}$ undergoes acid dissociation to give the amido product $[Pt(NH_3)_5(NH_2)]^{3+}$ (pK_a 7.1) [96], and for maximum effectiveness has to be used at lower pH (5.8). As expected, the 4+ and 5+ complexes give more extensive inhibition than the 3+ complex.

An alternative explanation of inhibition which has features in common with the 'dead-end' mechanism, may be considered but is difficult to accept. In this mechanism, peptide conformational changes occur as a result of the association (5), and these have the effect of changing the reduction potential of the protein. Thus, in the present case association of a redox inactive complex would have the effect of increasing the reduction potential, and rate constants for the oxidation of PCu(I) would decrease with increasing amounts of inhibitor. However, any one inhibitor would be expected to give the same relative effect with all oxidants, e.g. $[Co(phen)_3]^{3+}$ and $[Fe(Cp)_2]^+$, as well as $[Fe(CN)_6]^{3-}$ whether reaction is at the remote or adjacent site. This is not observed. Moreover, a decrease in rate is observed when inhibitor is used in a reduction e.g. the reaction of PCu(II) with $[Ru(NH_3)_5py]^{2+}$. A decrease in reactivity of PCu(I), I adduct with positively charged reactants might also be explained in terms of overall electrostatics, since the net charge on plastocyanin is decreased by the presence of say $[(NH_3)_5CoNH_2Co(NH_3)_5]^{5+}$ regardless of the site of binding, and for this reason could lead to a slower reaction with positively charged oxidants. Such an explanation would not, however, account for the significantly different inhibition of the like-charged oxidants $[Fe(Cp)_2]^+$ and $[Fe(Cp(CpCH_2OH)]^+$ by $[(NH_3)_5CoNH_2Co(NH_3)_5]^{5+}$ [117].

In all the plastocyanin studies with inorganic redox partners to date, inhibition is incomplete, Fig. 8. The possibility that because of the large protein surface area involved, the binding of inhibitor at the acidic patch decreases but does not entirely eliminate reaction at this site has been considered. It is not, however, consistent with further experiments in which the effect of the larger 6+ and 7+ tetranuclear inhibitors (formulae as shown) are compared with $[Pt(NH_3)_6]^{4+}$ and $[(NH_3)_5CoNH_2Co(NH_3)_5]^{5+}$ at pH 5.8. For any one plastocyanin the extent of maximum percentage inhibition is the same in all cases, Fig. 8. This observation is difficult to reconcile with the partial inhibition at one site, since it is unlikely that the bulkier inhibitors and oxidant could associate with the protein at the same time and to the same extent. Therefore it is now believed that electron transfer with $[Co(phen)_3]^{3+}$ takes place at, at least, one other site. Although the second site has not been identified, it clearly has to be of low or zero charge and the strongest contender would seem to be the adjacent site at or close to His87. Therefore, on the basis of these inhibition studies, reaction is partitioned between the adjacent (A) and remote (B) sites, Table 7.

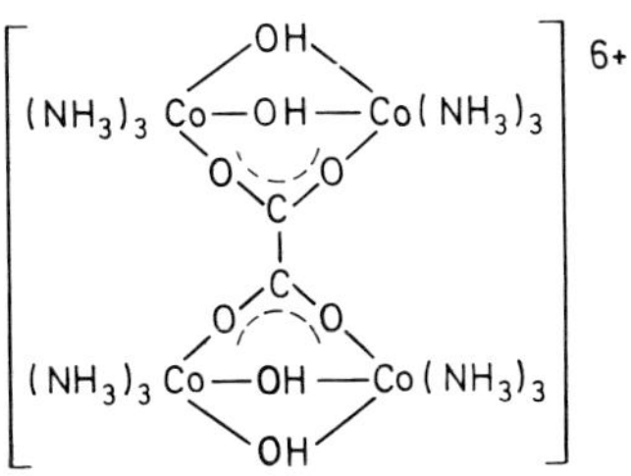

$[Co(III)_4]_A$

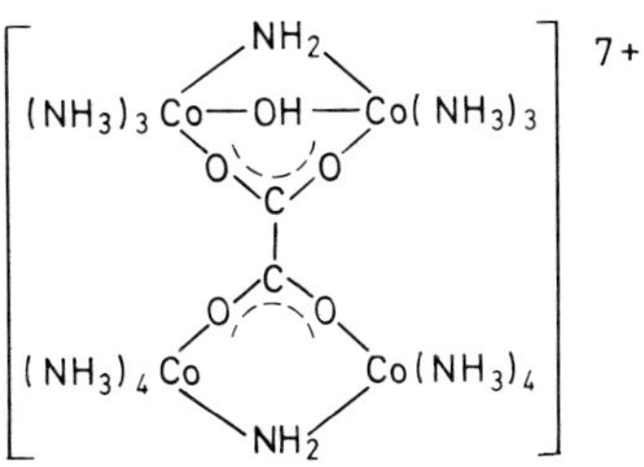

$[Co(III)_4]_B$

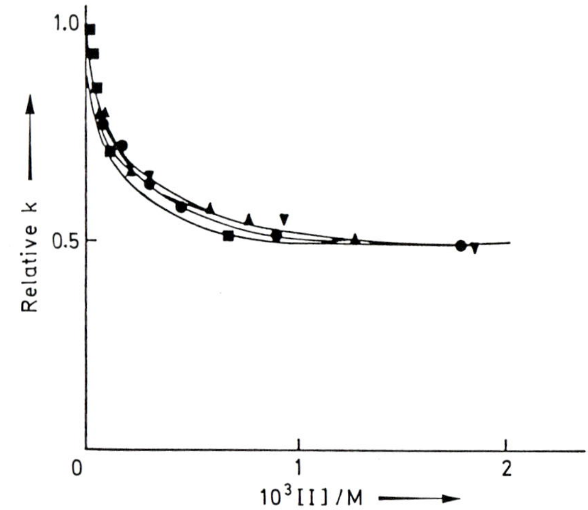

Fig. 8. Competitive inhibition of redox inactive complexes (I) on the $[Co(phen)_3]^{3+}$ oxidation of parsley plastocyanin PCu(I) [98]. Second-order rate constants (25 °C), shown as relative values, were determined at pH 5.8 (Mes), I = 0.10 M (NaCl) with $[Pt(NH_3)_6]^{4+}$ (▼), $[(NH_3)_5CoNH_2(NH_3)_5]_5]^{5+}$ $[Co(III)_4]_A$ (▲) and $[Co(III)_4]_B$ (■). Full formulae of the latter two complexes are as indicated above

Table 7. Partitioning of electron transfer between adjacent (k_A) and remote (k_B) binding sites on spinach plastocyanin PCu(I) at 25 °C, pH 7.5, I = 0.10 M (NaCl), using redox inactive $[(NH_3)_5CoNH_2Co(NH_3)_5]^{5+}$ [100, 117]

Oxidant	k_A (%)	k_B (%)
$[Fe(CN)_6]^{3-}$	100	0
$[Fe(Cp)_2]^+$	72	28
$[Fe(Cp)(CpHgCl)]^+$	60	40
$[Fe(Cp)(CpCH_2OH)]^+$	40	60
$[Co(phen)_3]^{3+}$	25	75[a]
$[Ru(NH_3)_5(CH_3CN)]^{3+}$	9	91

[a] Recalculated, 80% in original.

Returning to the saturation kinetic effect with $[Co(phen)_3]^{3+}$ as oxidant, it is necessary now to refit data to a scheme in which reaction takes place at two sites (A and B), only one of which (B) becomes saturated [98]. The scheme is shown in (6)–(8),

$$PCu(I) + [Co(phen)_3]^{3+} \overset{K}{\rightleftharpoons} PCu(I), [Co(phen)_3]^{3+} \tag{6}$$

$$PCu(I), [Co(phen)_3]^{3+} \xrightarrow{k_{et}} PCu(II) + [Co(phen)_3]^{2+} \tag{7}$$

$$PCu(I)_T + [Co(phen)_3]^{3+} \xrightarrow{k_2} PCu(II) + [Co(phen)_3]^{2+} \tag{8}$$

where $PCu(I)_T$ represents the total Cu(I) protein, and k_2 in (8) is for reaction at site A. This gives the expression (9)

$$k_{obs} = \frac{Kk_{et}[Co(phen)_3^{3+}]}{1 + K[Co(phen)_3^{3+}]} + k_2[Co(phen)_3^{3+}] \tag{9}$$

from which K is now revised to 340 M^{-1} (instead of 167 M^{-1}), and k_{et} to 5.7 s^{-1} (previously 17.9 s^{-1}). Similarly, with the addition of inhibitor, the additional step (10) has to be included in the scheme.

$$PCu(I) + I \overset{K_I}{\rightleftharpoons} PCu(I), I \tag{10}$$

Any reaction of PCu(I), I with $[Co(phen)_3]^{3+}$ at site A is included in (8). For a series of experiments with $K[Co(phen)_3^{3+}] \ll 1$ the relationship (11) is obtained.

$$k_{obs} = \frac{Kk_{et}[Co(phen)_3^{3+}]}{1 + K_I(I)} + k_2[Co(phen)_3^{3+}] \tag{11}$$

Values of K_I at pH 5.8 and 7.5 are listed in Table 8.

It has been shown that the association of $[Pt(NH_3)_6]^{4+}$ with parsley PCu(I) decreases with increasing protonation at the remote site [96]. Rate constants for oxidation of the adduct PCu(I), $[Pt(NH_3)_6]^{4+}$ with $[Co(phen)_3]^{3+}$ also decrease with pH, an effect which stems from protonation at the active site. The pK_a's derived are both close to 5.8.

From the data in Table 8, K_I for the 3+ ammine complex appears to be greater than values for the phenanthroline complexes. A possible explanation is

Table 8. Equilibrium constants K_I (25 °C) for the association of inorganic complexes with parsley plastocyanin PCu(I), I = 0.10 M(NaCl) [96, 98]

Complex	K_I (pH 5.8)	K_I (pH 7.5)
$[Co(phen)_3]^{3+}$ [a]		340[b]
$[Cr(phen)_3]^{3+}$		360[b,c]
$[Co(NH_3)_6]^{4+}$	1.6×10^4	
$[Co(III)_2]^{5+}$ [e]	0.79×10^4	1.6×10^4 [d]
$[Co(III)_4]_A^{6+}$ [e]	0.79×10^4	
$[Co(III)_4]_B^{7+}$ [e]	1.01×10^4	

[a] Redox active;
[b] Recalculated for reaction at two sites using Eq. 11;
[c] Further data required for a more precise fit;
[d] A similar value is obtained for spinach PCu(I), Ref. 117. Lower values which have been reported may be due to some decomposition of the 5+ complex;
[e] Full formula on p 25.

that there is significant H-bonding of the ammine ligands to carboxylate residues at the remote site.

The 4− complexes $[Zr(C_2O_4)_4]^{4-}$ and $[Mo(CN)_8]^{4-}$ (2×10^{-3} M) have no effect on rate constants for the $[Fe(CN)_6]^{3-}$ oxidation of PCu(I) [96]. The aqua ion Mg^{2+}, which is known to be present in large amounts in the interior of the thylakoid, produces only a mild inhibition on the reaction of $[Co(phen)_3]^{3+}$ with spinach PCu(II) ($K \sim 38\ M^{-1}$) [100]. The aqua ion Gd^{3+} produces a larger effect [108] consistent with its higher charge. Modification of plastocyanin by attachment of ethylenediamine as the monoprotonated form enH^+ to the acidic patch using a carbodiimide procedure [118], results in an inhibition of the reaction with $[Co(phen)_3]^{3+}$ [109]. It has no effect, however, on the reaction with $[Fe(CN)_6]^{3-}$, consistent again with the use of the remote and adjacent sites respectively by these two reactants.

The effect of $[Pt(NH_3)_6]^{4+}$ on the $[Fe(CN)_6]^{3-}$ oxidation of parsley PCu(I) is more difficult to interpret, but merits further comment. At pH 7.5, with $[Pt(NH_3)_6]^{4+}$ (and its conjugate base) present at concentrations up to 1.31 mM, rate constants increase by a factor of 2.5 [96]. In the case of the $[Co(dipic)_2]^-$ oxidation (dipic is pyridine-2,6-dicarboxylate) [95], a smaller 8% increase is observed, drawing attention to the importance of size of charge. There are a number of possible explanations. These include the effect of association of $[Pt(NH_3)_6]^{4+}$ on the net protein charge and hence its interaction with $[Fe(CN)_6]^{3-}$. Association of $[Pt(NH_3)_6]^{4+}$ at the remote acidic patch may lead to $[Fe(CN)_6]^{3-}$ using this site to enhance the rate. A further contributing factor may be the quite separate association of $[Pt(NH_3)_6]^{4+}$ and $[Fe(CN)_6]^{3-}$ to give an adduct which is more redox active.

Inhibition experiments of the kind already discussed, and electrostatic interactions in general, touch on a number of fundamental questions. It seems appropriate in considering long range interactions that the overall charge on the protein should be relevant. However, on closer approach local charges are likely to take over and become dominant. In a recent paper in which the generation of a dipole axis was considered, reaction at the adjacent site only was rationalized [119].

Binding at the remote site has also been detected in studies on the quenching of the excited states $^*[Cr(phen)_3]^{3+}$ and $^*[Ru(bipy)_3]^{2+}$ by French bean plastocyanin [103]. The model adopted allows for electron transfer from the remote and adjacent sites, where at low protein concentrations the adjacent pathway is ~10 times faster. At the higher concentrations of protein, up to 4×10^{-3} M, an interesting feature is the evidence for an adduct in which two PCu(I) molecules are associated with one inorganic complex. The oxidant is believed to be sandwiched between two PCu(I)'s.

By a quite independent method involving an Amberlite XAD-2 gel column the constant for association of $[Fe(phen)_3]^{2+}$ (guest) with parsley PCu(II) (host) has been determined as $1.8 \times 10^3\ M^{-1}$ at $I = 0.01$ M, with phosphate buffer at pH 7.5 [120]. The high value for a 2+ ion, is accounted for by the lower ionic strength used in this study.

8 pH Effects at the Remote Site

The $[Fe(CN)_6]^{4-}$ reduction of parsley PCu(II) at the adjacent site gives a gradual ~20% increase in rate constants as the pH is decreased from 8.0 to 4.5, and no well defined pK_a is observed. It appears, therefore, that reaction at the adjacent site is not affected in a direct way by pH. In contrast, rate constants for the positively charged reductants $[Co(phen)_3]^{2+}$, $[Co(terpy)_2]^{2+}$ and $[Ru(NH_3)_5py]^{2+}$, with parsley PCu(II) decrease on decreasing the pH from 7 to 4, Fig. 9 [99]. All three reductants are influenced by a well defined acid dissociation process with pK_a 5.1. Moreover, since $[Co(phen)_3]^{2+}$ and $[Co(phen)_3]^{3+}$ are expected to use the same site or sites on plastocyanin, the pK_a can be assigned to an acidic residue or residues at the remote site. There is no variation of the pK_a with identity of reductant, and the low pH rate constant is 52% of the value at pH 7. The decrease observed here is very similar to the maximum effectiveness of redox inactive complexes, e.g. $[(NH_3)_5CoNH_2Co(NH_3)_5]^{5+}$ on the $[Co(phen)_3]^{3+}$ oxidation of parsley PCu(I) at pH 7.5 (61%), suggesting that protonation and blocking give the same net effect. Again there is evidence consistent with a partitioning of reaction between the remote and adjacent sites.

It is appropriate now to return to the effect of pH on the $[Co(phen)_3]^{3+}$ oxidation of PCu(I). If protonation at the remote site influences the reaction of PCu(II), then a similar effect might be expected for the reaction of PCu(I) with positively charged complexes. In the case of PCu(I) the kinetics are dominated by the inactivation resulting from the active site protonation. Whereas the pK_a for the $[Fe(CN)_6]^{3-}$ oxidation is in good agreement with the ^1H NMR independently measured value, the apparent pK_a obtained with $[Co(phen)_3]^{3+}$ is significantly higher, an effect which is clear from an inspection of Fig. 10. A two pK_a fit is possible in the case of $[Co(phen)_3]^{3+}$, as has been illustrated [1,100],

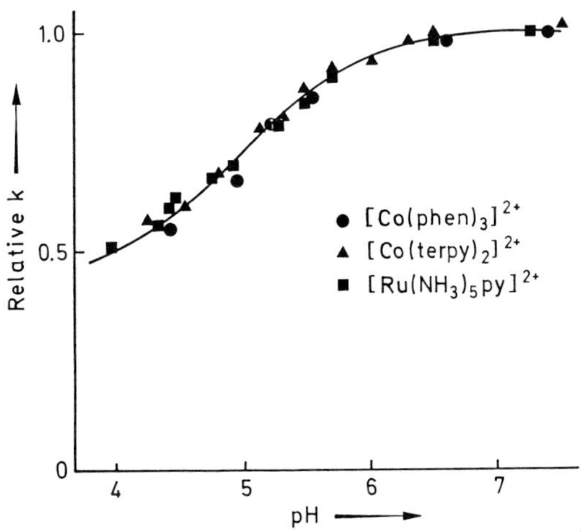

Fig. 9. Variation of second-order rate constants (relative values) with pH for the reduction of parsley plastocyanin PCu(II) with $[Co(phen)_3]^{2+}$, $[Co(terpy)_2]^{2+}$ and $[Ru(NH_3)_5(py)]^{2+}$ at 25 °C, I = 0.10 M(NaCl) [99]

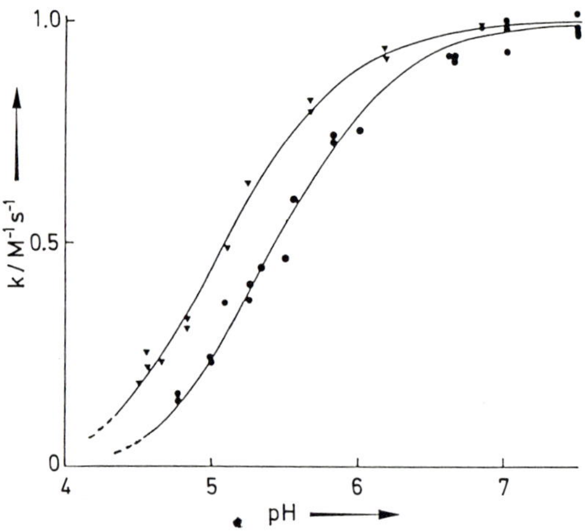

Fig. 10. The variation of second-order rate constants (relative scale) with pH for the $[Fe(CN)_6]^{3-}$ (▼) and $[Co(phen)_3]^{3+}$ (●) oxidations for *A. variabilis* PCu(I) at 25 °C, I = 0.10 M (NaCl) [102].

where one of the pK_a's used is the NMR value for active site protonation. The second pK_a is then assigned to acid dissociation at the remote site.

The pK_a values obtained for PCu(II) and PCu(I) are summarized in Table 9. Before discussing what appears to be a clear cut trend, the precision of individual pK_a's should first be considered. It has for example been necessary to revise (downwards) the spinach PCu(II) pK_a from 5.3 to 4.8 [121]. Plastocyanin is known to denature at or around pH 4.0 and in earlier work the lowest pH used was 4.5, which does not allow as accurate a fit to pK_a values. By using a pH-jump method in which the final low pH is attained at the time of stopped-flow mixing (one reactant solution carrying substantially more buffer is allowed to control the pH), it is now possible with some confidence to include data down to pH 4.0. The other uncertainty is in the precision of the pK_a for the PCu(I) remote site from a two pK_a fit. Again the data has to be free from artifacts introduced at

Table 9. Acid dissociation pK_a values relating to the protonation at the remote binding site of different plastocyanins [49, 101] as obtained from the variation of rate constants with pH 25 °C, I = 0.10 M(NaCl)

Source	PCu(I)[a]	PCu(II)
Parsley	5.8	5.1
S. obliquus	5.5	4.8
Spinach	5.6	4.8[b]
French bean	5.5	—
Poplar	5.3	5.1

[a] After allowance for active site pK_a in studies with $[Co(phen)_3]^{3+}$.
[b] Revised value, Ref. [121].

one or other extremity of the pH range. One important point is that active site pK_a's can only be shifted to the higher 'apparent' pK_a's, as observed for $[Co(phen)_3]^{3+}$, by inclusion of a larger remote site pK_a. The trend in remote site pK_a's, with values for PCu(I) consistently larger than those for PCu(II), is seen to repeat itself for other plastocyanins, Table 9.

Remote site pK_a's as high as 5.8 for PCu(I) require some comment and may be important. One possible explanation is that two carboxylates at or involving residues 42–45 share a proton [99]. A pK_a of up to or around 5.0 for PCu(II), on the other hand, is more likely to stem from protonation at a single carboxylate. The question then is how does the oxidation state of the Cu control this pK_a? Two structural features, namely that coordinated His37 is linked directly to Asp42 by a chain of highly conserved residues, and coordinated Cys84 is bonded directly to Tyr83, may be important. Fluorescence experiments [46b], have indicated a sensitivity of Tyr83 to the oxidation state of the Cu. However, the implied conformational change does not appear to be supported by crystallographic evidence for poplar PCu(I) and PCu(II) [16, 17]. It has been suggested that the conformational change is masked by the high (2.7 M) concentration of ammonium sulfate from which the crystals are grown. While this suggestion remains to be tested experimentally, it has been noted that ammonium sulfate does not prevent the crystallographic detection of other side chain conformational changes as a function of pH.

There are two other explanations to consider [101]. Firstly, the effect may originate from the protonation occurring at the active site of PCu(I). It is clear from the crystal structure that the change in Cu site geometry is transmitted to the acidic patch since there are small but significant positional changes at residues 42–44 and 59–61. The weakness in this explanation is that (except for parsley and *S. obliquus*), the remote site pK_a for PCu(I) is larger than that for the active site, and it is difficult therefore to see how the latter shifts the remote site pK_a for PCu(II) to a higher value in the case of PCu(I). Alternatively, the reason for the different pK_a's for PCu(I) and PCu(II) may be electrostatic in origin [101]. It can, for example, be argued that a change in charge of -1 in formal charge of the Cu accounts for a pK_a shift of 0.25–0.40 at a distance of ~ 16 Å if the intervening material has a dielectric constant of ~ 40. The argument used here is similar to that indicated by Rees [122] to estimate the electrostatic effect of change in charge at a remote position on the reduction potential at the metal center. In the case of PCu(I), at low pH the electrostatic effect of the -1 change in charge of the Cu must be partly balanced by the effect of protonation at His 87.

Whatever the explanation, the sensitivity of the remote pK_a to oxidation state of the Cu is of potential importance in relation to the functional role of plastocyanin. Plastocyanin and its physiological electron transport partner cytochrome f are believed to have complementary surfaces which lead to efficient interaction prior to electron transfer. As will be seen below there is substantial evidence for cytochrome f(II) (as reductant) reacting at the remote site of PCu(II). One problem which may be anticipated here is how dissociation of the product pair might be induced after electron transfer. A conformational change originating from the Cu, and bringing about a change in pK_a at the remote binding site, is one way in which this might be achieved.

In the case of the higher plant and *S. obliquus* plastocyanins, protonation of carboxylates at and around positions 42–45 provides the most acceptable explanation of the remote site pK_a's. This explanation carries less weight in the case of *A. variabilis* PCu(I) (estimated charge +1) because there are fewer carboxylates, and negative charge at 42–45 and 59–61 is no longer conserved, Table 6. For the $[Co(phen)_3]^{3+}$ oxidation of *A. variabilis* PCu(I) the "apparent" pK_a is again shifted to a higher value, in this case 5.4 as compared to the $[Fe(CN)_6]^{3-}$ kinetic (5.0) and NMR (5.1) values [102, 47b]. From the same two pK_a fitting procedure the remote site pK_a is 5.7. The proximity of Glu85 to Asp42 could provide the necessary site for protonation, and the two carboxylate involvement an explanation of the higher pK_a. An effect of pH is apparent therefore for reaction at the remote site of PCu(I). However, some caution is clearly required, since it has been noted that not only is the $[Fe(CN)_6]^{4-}$ reduction of *A. variabilis* PCu(II) independent ($\pm 5\%$) of pH in the range 6.1–8.8, but the $[Co(phen)_3]^{2+}$ and $[Co(terpy)_2]^{2+}$ reductions likewise give little or no dependence on pH over the range 4.6–7.5 [101, 121]. This behavior seems contrary to there being a remote site pK_a. Clearly the last word has not been said on *A. variabilis* plastocyanin reactivity. In the case of the cabbage (*B. oleracea*) cytochrome f(II) reduction of *A. variabilis* PCu(II) a pK_a of 5.19 which is obtained from studies (10 °C) at pH 4.5–8.5, I = 0.20 M(NaCl) [123], may arise in part from protonation on the cytochrome f.

Important to address in the case of the two algal plastocyanins is the effect of changes in amino-acid sequence at the remote site. *S. obliquus* plastocyanin has Asp, Glu, Asp, Ala residues at 42–45, and in addition to deletions at 57 and 58 has His, Asp, Asp at 59–61. *A. variabilis* has Asp, Ala, Ala, Leu at 42–45, with no deletions at 57 and 58, and with His, Lys, Glu at 59–61. Both proteins have a His residue at position 59. From NMR titrations the His59 residues have pK_a's of 7.8 (*S.o.*) [49] and 7.3 (*A.v.*) [47b]. The trend here is as expected if proton sharing occurs in the latter case since residue 60 is Asp (*S.o.*) and Lys (*A.v.*), Fig. 3. The more extensive studies on *S. obliquus* have indicated an interesting long-range influence of H^+. Thus, from ^1H NMR the active site His37 appears to be sensitive to protonation at His59, and His59 to protonation at the active site, Fig. 11 [49]. This is an unusual effect since there is no direct link between the two, and the imidazole ring of His59 is 10–12Å from the Cu site.

Rate constants for the oxidation of *S. obliquus* PCu(I) with $[Fe(CN)_6]^{3-}$ and $[Co(phen)_3]^{3+}$ respond to the His59 acid dissociation as well as the active site protonation of His87, Fig. 12 [49]. At the higher pH's the trend in rate constants corresponding to the His59 change is observed for both the 3− and 3+ oxidants, and is in the same direction for both oxidants. Since the two reactions are at different sites, an explanation based on local charge cannot be the sole explanation. The effect of protonation is believed to be transmitted to the active site (i.e. it changes E^0), and in this way influences reactions at both the remote and adjacent sites [49]. The amplitude of the effect is, however, greater in the case

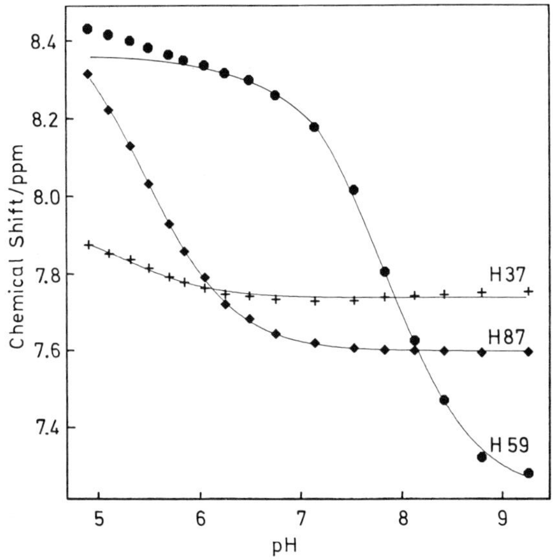

Fig. 11. Proton NMR against pH titration curves for *S. obliquus* plastocyanin PCu(I). Spectra were recorded at 25 °C for solutions in 0.10 M phosfate buffer [49].

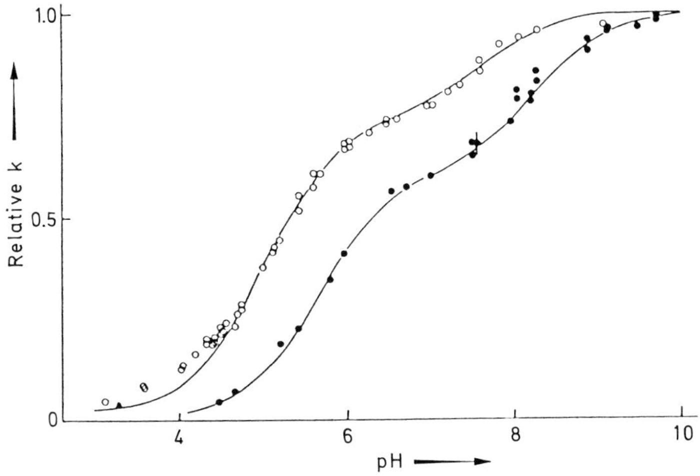

Fig. 12. The variation of second-order rate constants (relative scale) with pH for the $[Fe(CN)_6]^{3-}$ (o) and $[Co(phen)_3]^{3+}$ (●) oxidations of *S. obliquus* PCu(I) at 25 °C, I = 0.10 M(NaCl) [49]

of $[Co(phen)_3]^{3+}$, consistent with an additional contribution from local electrostatic effects.

8.1 Ru-Modification and Intramolecular Electron Transfer

The complex $[Ru(NH_3)_5H_2O]^{2+}$ has proved to be a successful reagent for the modification of metalloproteins because of its affinity to accessible surface histidine residues. The $Ru(NH_3)_5$ moiety has, for example, been attached to specific histidines on ribonuclease A [125], lysozyme [126], horse cytochrome c [127], *Pseudomonas aeruginosa* azurin [128], *Pseudomonas stutzeri* cytochrome c_{551} [129], and *Chromatium vinosum* high-potential Fe/S protein [130]. Not all free histidines are reactive, thus His81 in *Azotobacter vinelandii* cytochrome c_{551}, and His35 (unlike His83) in *Pseudomonas aeruginosa* azurin, are not readily modified. In the case of sperm whale myoglobin, four singly modified derivatives have been prepared by attaching the Ru label to different histidines on the same molecules [131]. In the first two examples [125, 126] the Ru attachment is more a spectroscopic aid. In the others the aim has been to explore intramolecular electron transfer Ru(II) $\rightarrow$ Fe(III), Ru(II)) $\rightarrow$ Cu(II) or Ru(II) $\rightarrow$ Fe_4S_4, as appropriate. Rate constants obtained are for electron transfer over a fixed distance which can be ascertained from X-ray crystallography, see for example, Table 10. The term fixed distance is used guardedly because the act of attaching $Ru(NH_3)_5$ may lead to some change in structure. There is, however, no evidence to suggest that such changes are other than minor, and crystallographic dimensions are believed to be relevant to the distance of electron transfer.

Rate constants for azurin $Ru(NH_3)_5$-modified at His83 have been determined by two methods in which the fully oxidized Ru(III)Cu(II) protein is reduced to Ru(II)Cu(II) by (a) flash photolysis using $^*[Ru(bipy)_3]^{2+}$ as reductant [128], and (b) pulse radiolysis with CO_2^- as reductant [50]. Intramolecular rate constants of $1.9 \pm 1.4\ s^{-1}$ (22 °C) and $2.5 \pm 0.8\ s^{-1}$ (four runs at 17 °C), respectively are in reasonably satisfactory agreement. Both studies were in 0.10 M phosphate, but solutions for pulse radiolysis also have in addition 0.10 M sodium formate present. Ionic strengths were therefore 0.22 and 0.32 M (the value of 0.31 M in Refs. [50] and [129] is closer to and should be adjusted to the latter value; in Ref. [130] an ionic strength of 0.32 M applies to the CO_2^- and O_2 studies, and 0.22 M to the e_{aq}^- study). Although an intramolecular rate constant would not be expected to be affected by ionic strength, reduction potentials do

Table 10. A comparison of rate constants for intramolecular electron transfer in $Ru(NH_3)_5$-modified electron transport metalloproteins, modifications at surface histidine present in native proteins, pH 7. Values of ΔE^0 by determined measurements on modified protein except as indicated

Protein	d/Å	ΔE^0/mV	k/s^{-1}	Ref.
Plastocyanin (*S.o.*)	10–12	276	$<0.26^a$	[50]
Plastocyanin (*A.v.*)	11–9	230^b	$<0.08^a$	[50]
Azurin (*P.a.*)	11.8	240	1.9^c, 2.4^a	[128, 50]
Cytochrome c (horse)	11.8	130^e	30^d, 53^e	[127]
Cytochrome C_{551} (*P.s.*)	7.9	146	13^b	[129]
Hipip (*C.v.*)	7.9	260	18	[130]

[a] 20 °C, 0.10 M phosphate buffer, 0.10 M formate, I = 0.32 M;
[b] From values for native PCu(II)/(I) (*A.v.*) (339 mV), and $[Ru(NH_3)_5\ (imid)]^{3+/2+}$ (109 mV);
[c] 25 °C, 0.10 M phosphate buffer, I = 0.22 M;
[d] 23 °C, 0.10 M HEPES buffer, I = 0.10 M;
[e] 25 °C, 0.10 M phosphate, 0.10 M formate, I = 0.32 M.
$\Delta H\ddagger = 3.5$ kcal mol^{-1}; $\Delta S\ddagger = -39$ cal k^{-1} mol^{-1}

seem to be dependent on this parameter, a point which is returned to below. Similarly, His33 of horse cytochrome c has been Ru-modified, and intramolecular electron transfer studied using the same two techniques to give intramolecular rate constants of 30 and 53 s^{-1} respectively [127].

There are no uncoordinated histidine residues in higher plant plastocyanins. However, some algal plastocyanins, including *S. obliquus* and *A. variabilis* have a histidine at position 59, which is of particular interest since this lies within the remote acidic patch region [132]. Differences between the two are that whereas *S. obliquus* PCu(I) is acidic (charge −9), *A. variabilis* is basic (charge 1+). Also *S. obliquus* has deletions at 57 and 58 with a consequent tightening of the peptide chain in this locality, Fig. 5. Both these plastocyanins were therefore Ru-modified.

Characterization of the modified plastocyanins was by Inductively Coupled Plasma Emission Spectroscopy to analyze for Ru and Cu (1:1 ratio), and by ^{1}H NMR spectroscopy. In the ^{1}H NMR characterization the $C_2\underline{H}$ resonance of His59 at 8.2 ppm is seen to be lost due to paramagnetic line broadening effect of the attached Ru(III), Fig. 13 [50]. In a further test it is known that the His59's of both native plastocyanins react with diethyl pyrocarbonate (DEPC) to give an *N*-ethoxyhistidine derivative, (12), which absorbs strongly at 238 nm (ε 2750 M^{-1} cm^{-1}), Fig. 14 [133].

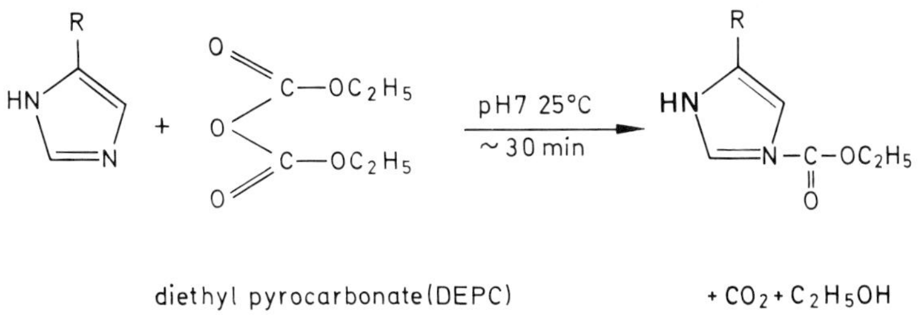

(12)

The Ru-modified proteins give no evidence for DEPC modification, consistent with attachment of Ru at His59. Furthermore, a shoulder is observed in the Ru(III)Cu(II) spectrum at 300 nm (ε 2000 M^{-1} cm^{-1}), which is in agreement with the absorbance for $[Ru(NH_3)_5His]^{3+}$ at 303 nm, (ε 2100 M^{-1} cm^{-1}) [134], and $[Ru(NH_3)_5(Imidazole)]^{2+}$ at 280 nm, (ε 2580 M^{-1} cm^{-1}) 1 [135].

The yield of modified plastocyanin is 30% in the case of *A. variabilis* and 42% for *S. obliquus*. A longer time is required for the modification of *A. variabilis* (4 h) as compared to *S. obliquus* (20 min), consistent with the differences in protein charge. A second 1:1 Ru:Cu modified product is obtained from *A. variabilis* (11%), which is seen as band 2 in the CM52 cation-exchange chromatography [50]. This also gives a negative DEPC test indicating Ru attachment at His59. Characterization is incomplete but a prominent band at 370 nm (ε 2500 M^{-1} cm^{-1}) independent of pH in the range 4.5–9.5 is noted. One possibility is a tetrammine derivative, which completes its coordination by attachment to a second residue on the protein. If alternatively the Ru had

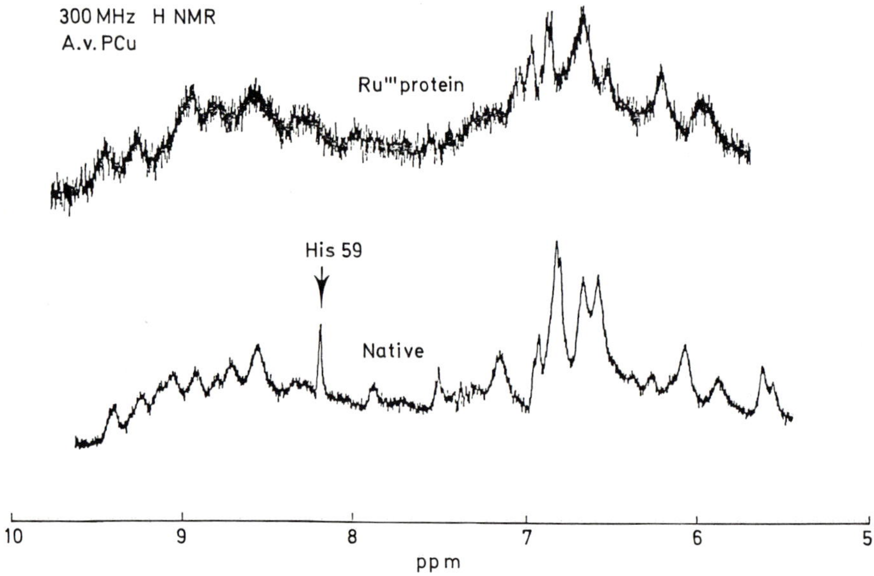

Fig. 13. ^{1}H NMR spectra in the aromatic region for native *A. variabilis* PCu(II) (*lower spectrum*), and the Ru(III)-modified protein [50]

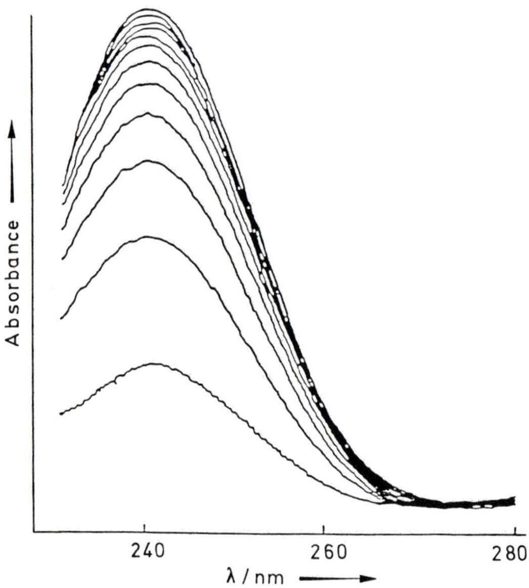

Fig. 14. The increase in absorbance at 238 nm ($\Delta\varepsilon$ 2750 $M^{-1}cm^{-1}$) on DEPC (diethyl pyrocarbonate) modification of *A. variabilis* plastocyanin. Spectra were recorded at 5 min intervals [50]

aquatetrammine coordination it would have a lower E^0, less than that for $[Ru(NH_3)_5(imidazole)]^{3+/2+}$. On present evidence this would seem to be a less likely possibility. Thus E^0 values for derivatives of *cis* and *trans*-$[Ru(NH_3)_4(H_2O)_2]^{3+/2+}$ are in the range 0.12 to -0.10 V [136]. A doubly-modified derivative with 2:1 Ru:Cu is also obtained. The site of attachment of the second Ru has not so far been determined.

Reduction potentials of the *S. obliquus* His59 $Ru(NH_3)_5$-modified protein have been determined by cyclic voltammetry using as electrode the oxidized surface obtained by polishing the edge plane of pyrolytic graphite [137]. The modified protein responds well at the electrode, whereas the native protein requires multi-charged cations, e.g. Mg^{2+} or $[Cr(NH_3)_6]^{3+}$, as mediators to give satisfactory reversibility. Separate reduction potentials at I = 0.10 M(NaCl) for native *S. obliquus* plastocyanin (389 mV) and $[Ru(NH_3)_5$ (imidazole)$]^{3+/2+}$ (109 mV) at 25 °C, become 393 mV and 139 mV respectively (difference 254 mV) for the modified protein. At I = 0.32 M the Ru couple is different (117 mV) indicating an effect of ionic strength. The driving force for the Ru(II)→Cu(II) intramolecular electron transfer is accordingly 254 and 276 mV, respectively at the two ionic strengths. Spectrophotometric titration of PCu(II) with $[Fe(CN)_6]^{4-}$ gives similar results with E^0 for native PCu(II)/(I) of 363 mV and for modified protein 385 mV. No value for the Ru(III)/(II) couple was obtained using this method.

An estimate of the edge to edge S(Cys84) to N^{δ}(His59) distance for intramolecular electron transfer has been obtained for *A. variabilis* plastocyanin, Fig. 15. To do this the Glu59 residue of poplar plastocyanin was replaced by His in the programme FRODO assuming an ideal geometry for the imidazole. The relevant distance is 11.9 Å. Unacceptable contacts with serine 85 (glutamic acid in *A. variabilis*) make a closer approach unlikely. From 2D NMR studies on *S. obliquus* plastocyanin [37] the relevant distance is believed to be less and in the range 10–12 Å due to the two deletions.

On pulse radiolysis using $CO_2^{\cdot -}$ (−2.0 V) to reduce PCu(II)Ru(III) (pH 7, 20 °C) the behavior observed is in both cases very similar [50]. The concentration of $CO_2^{\cdot -}$ was adjusted so that there was <20% reduction of modified protein. As far as can be ascertained, reduction efficiencies in the first stage are about the same, with reaction partitioned between the Cu(II) (72%) and the

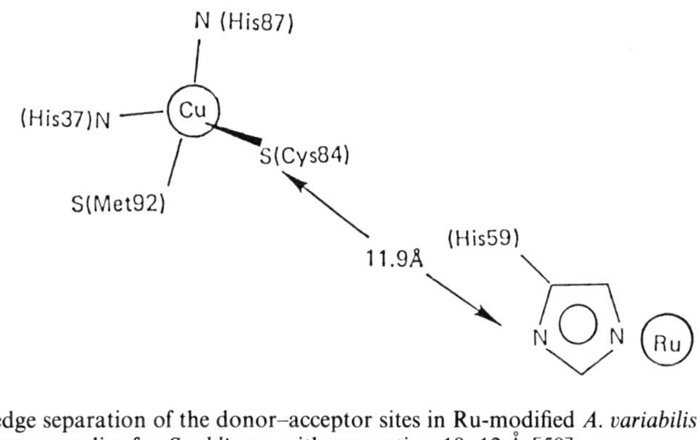

Fig. 15. Showing the edge-edge separation of the donor–acceptor sites in Ru-modified *A. variabilis* plastocyanin. The same diagram applies for *S. obliquus* with separation 10–12 Å [50]

Ru(III) (28%) (13).

$$CO_2^{\cdot -} + PCu^{II}Ru^{III} \longrightarrow \begin{cases} PCu^{I}Ru^{III}\ (72\%) \\ PCu^{II}Ru^{II}\ (28\%) \end{cases} \qquad (13)$$

The combined rate constants obtained by monitoring the Cu(II) decay are $6.7 \times 10^8\ M^{-1}\ s^{-1}$ (*A.v.*) and $7.8 \times 10^8\ M^{-1}\ s^{-1}$ (*S.o.*). The second stage can occur by concurrent intramolecular (k_1) and second-order intermolecular (k_2) steps, (14)–(15),

$$PCu(II)\ Ru(II) \xrightarrow{k_1} PCu(I)\ Ru(III) \qquad (14)$$

$$PCu(II)Ru(II) + PCu(II)Ru(III) \xrightarrow{k_2} PCu(II)Ru(III) + PCu(I)Ru(III) \qquad (15)$$

yielding stable PCu(I)Ru(III). For both plastocyanins, concentration range $(0.6-23) \times 10^{-6}$ M, k_2 is the dominant process $1.2 \times 10^5\ M^{-1}\ s^{-1}$ (*A.v.*) and $3.3 \times 10^5\ M^{-1}\ s^{-1}$ (*S.o.*). Values of k_1 are small $<0.082\ s^{-1}$ (*A.v.*) and $<0.26\ s^{-1}$ (*S.o.*), with zero values not excluded in both studies. Electron transfer across β-sheet from the $Ru(NH_3)_5^{2+}$ attached at His59 through to the Cu does not therefore appear to be a favorable route. Both the distances, 11.9 Å (*A.v.*) and 10–12 Å (*S.o.*), and driving force (276 mV for *S.o.* I = 0.32 M), are within the range or more favorable than for the corresponding reaction of Ru-modified cytochrome c for which intramolecular rate constants of 30 and 53 s^{-1} were obtained.

On the other hand when unattached $[Ru(NH_3)_5(Imidazole)]^{2+}$ (109 mV) is the reductant for *S. obliquus* PCu(II), the rate observed is at the fast limit of the stopped-flow range [50]. With complex and protein at equal concentrations (1.4×10^5 M), and temperature at 8.5 °C, the second-order rate constant is $>4 \times 10^6\ M^{-1}\ s^{-1}$ (pH 7.8, I = 0.10 M NaCl). From studies with $[(NH_3)_5CoNH_2Co(NH_3)_5]^{5+}$ present initial experiments suggested ~50% inhibition (i.e. ~50% of reaction is at the remote site). Such experiments are difficult for a number of reasons and more recently it has not been possible to repeat this result and little or no blocking was observed. It would seem therefore that as in the studies with $*[Cr(phen)_3]^{3+}$ and $*[Ru(bipy)_3]^{2+}$ [103], this very fast reaction may be predominantly at the adjacent site.

8.2 Electron Transfer from the Remote Site

The above experiments suggest the use of a quite specific route or routes for electron transfer to and from plastocyanin. The remote site so far defined is quite large incorporating 42–45 and 59–61 either side of Tyr83. Negative charge at 59–61 appears to assist in increasing association at the remote site, witness the experiments with native *S. obliquus* PCu(I) in which acid dissociation at His59 is seen to give an enhancement in rate with $[Co(phen)_3]^{3+}$ over that observed with

$[Fe(CN)_6]^{3-}$, Fig. 16. However, experiments with the two Ru-modified derivatives clearly indicate that rate constants for electron transfer from His59 are small. Residues 42–45 are more distant from the Cu and are considered less likely lead-in groups. Moreover, attachment of cytochrome c(II) at 42–45 by the carbodiimide method does not lead to a productive electron transfer (see below) [138]. Therefore Tyr83 (Fig. 17) becomes a prime focus as a lead-in group for electron transfer.

The recent X-ray crystal structure of ascorbate oxidase [6] has indicated the relative positions of type 1, 2 and 3 Cu centers. The type 1 center is in a plastocyanin like domain, and is the primary acceptor of electrons from substrate. The shortest pathway for electron transfer from the type 1 to type 3 Cu's is the bifurcated path via Cys508 and either His507 or His509. The two histidines are part of the plastocyanin-like domain, and serve also to coordinate the type 3 Cu's, Fig. 2. The His507 to Cys508 bonding is similar to that of Tyr83

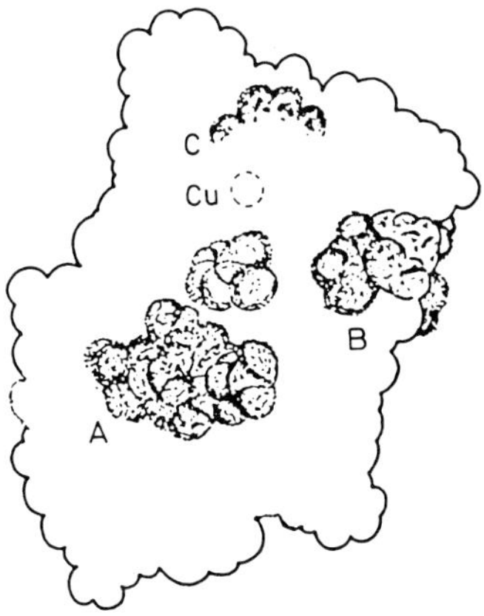

Fig. 16. Model of plastocyanin illustrating the relative positions of the acidic groups 42–45 (*A*) and 59–61 (*B*) either side of Tyr83 (*C*). The position of the buried Cu is also indicated and (*D*) the exposed edge of His87

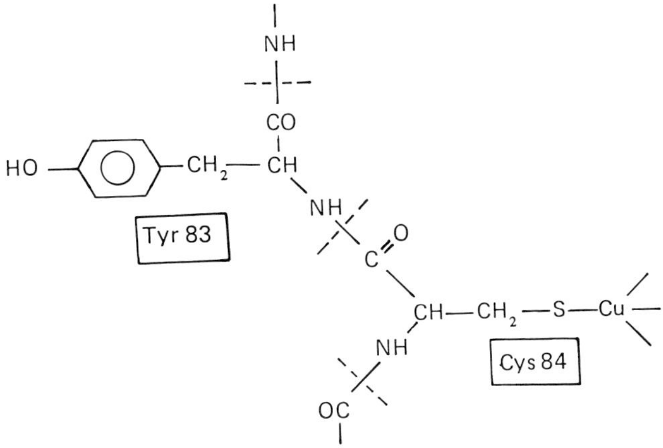

Fig. 17. Showing the coordination of the Cu active site of plastocyanin by Cys84, and the bonding of the latter to Tyr 83 which is a part of the remote site. The aromatic ring of the latter is solvent exposed

to Cys84 in plastocyanin Fig. 15, where the Tyr and His side chains to the aromatic group are of equal length. Precisely how in the two cases electrons are transferred, and indeed how a donor or acceptor reagent interacts with the Tyr83 of plastocyanin has yet to be established. Important features are the aromatic group at one extremity and the cysteinyl S-atom at the other. The carbonyl with its low lying orbitals may be relevant in some way, although the expenditure in energy here has to be balanced against that for a through space electron transfer.

In order to further probe the above route for electron transfer, there is a need to study derivatives modified at Tyr83. One such derivative in which the phenolic group of Tyr83 has been NO_2-modified has so far been prepared [139]. On pulse radiolysis reduction with Cd^+ to give the radical Tyr83$NO_2^{-\cdot}$, fast electron transfer through to the Cu(II) is observed, $k > 10^7\ s^{-1}$ [140].

9 Protein–Protein Reactions

9.1 *Plastocyanin with HIPIP(r)*

Effects of pH and inhibitors have been explored for the reduction of parsley PCu(II) by the high-potential iron-sulfur (HIPIP) protein $Fe_4S_4^{2+}$ (M_r 9500) from *Chromatium vinosum* (reduction potential 350 mV at pH ~7, estimated charge 3−) [97]. In all such cases it is important to have independent evidence (from studies with inorganic complexes), concerning the response of the second protein to changes in, for example pH, which might otherwise give rise to ambiguities in interpretation. Such studies have been carried out. The rate constants obtained for HIPIP(r) reduction of PCu(II) are not significantly affected by variation in pH in the range 5.0 to 8.4 (±7% scatter). There are therefore similarities to the $[Fe(CN)_6]^{4-}$ reduction of PCu(II). Moreover, with increasing amounts of $[Pt(NH_3)_6]^{4+}$ up to 0.36 mM, rate constants increase by a factor of 1.8 at pH 5.8 and 7.5. Again this is similar to the effect of $[Pt(NH_3)_6]^{4+}$ on the $[Fe(CN)_6]^{3-}$ oxidation of PCu(I), which has already been discussed. Therefore, reaction of PCu(II) with negatively charged HIPIP(r) can be assigned to the adjacent His87 site on plastocyanin.

9.2 *Plastocyanin with Cytochrome c(II)*

The kinetics of the reactions of horse cytochrome c(II), M_r 12,400, (charge 8+) reduction potential 260 mV, with parsley and French bean plastocyanins PCu(II) (charges −7 and −8 respectively), have been studied. As in the case of HIPIP, cytochrome c is not a physiologically relevant protein. It is nevertheless important in assessing different approaches prior to investigating the reactions of physiological redox partners. In the case of the reaction of parsley PCu(II) with cytochrome c(II), the rate constant (25 °C) is $1.5 \times 10^6\ M^{-1}\ s^{-1}$ at pH 7.6, I=0.10 M(NaCl) [141]. There is no evidence for saturation kinetics with the concentration of PCu(I) up to 0.15 mM and cytochrome c at $\sim 10^{-6}$ M. The activation parameters are $\Delta H^{\ddagger} = 7.3$ kcal mol^{-1} and $\Delta S^{\ddagger} = -5$, 8 cal K^{-1} mol^{-1}. An upper limit for association prior to electron transfer is estimated at 150 M^{-1}, which suggests that although the charges on the reactants are particularly favorable, association is not observed at concentrations here used. Effects of pH and competitive inhibition have also been studied [97]. On decreasing the pH the percentage decrease in reactivity is more extensive than for the $[Co(phen)_3]^{2+}$, $[Ru(NH_3)_5py]^{2+}$ and $[Co(terpy)_2]^{2+}$ (Fig. 9) [99]. A pK_a of 4.95 is obtained and the inhibition approaches 100% indicating a strong preference of cytochrome c(II) for reaction at the remote site on PCu(II). This is confirmed in the competitive inhibition studies with $[Pt(NH_3)_6]^{4+}$ which indicate at least 70% reaction at this site. The contrast in behavior observed for HIPIP(r) and cytochrome c(II) with PCu(II) is a very good illustration of the different responses which are diagnostic of reaction at the adjacent and remote sites respectively.

For French bean PCu(II) oxidation of cytochrome c(II) the rate constant is 5.1×10^6 M^{-1} s^{-1} and activation parameters $\Delta H^{\ddagger} = 7.6$ kcal mol^{-1}, $\Delta S^{\ddagger} = -2.4$ cal K^{-1} mol^{-1} [115]. Proton NMR line broadening experiments indicate shifts corresponding to the interaction of PCu(II) with cytochrome c(III) [115]. Experiments were at higher protein concentrations than for the kinetic studies, with solutions prepared by adding aliquots of PCu(II) (2–4 mM) plus cytochrome c (0.75–1.0 mM) to a solution of cytochrome c (0.75–1.0 mM) in an NMR tube. From spectra recorded after each addition, the negative patch on PCu(II) and heme edge of cytochrome c were identified as regions of contact. From the first-order rate constant for intramolecular electron transfer and the second-order stopped-flow rate constant, K for the association of cytochrome c(II) and PCu(II) was calculated to be 1.0×10^3 M^{-1} at 25 °C, I = 0.10 M, as compared to the previous value of < 150 M^{-1}. This is a slightly worrying discrepancy which may not be entirely accounted for by the use of different plastocyanins. At lower ionic strengths K is reported to increase to 1.5×10^4 M^{-1}. Preliminary computer graphics docking simulation of the interaction of cytochrome c and plastocyanin predicts a Cu to Fe distance of 18 Å, and a thiolate S(Cys 84) plastocyanin to exposed heme edge separation of ~12 Å. The intramolecular rate constant of 4.8×10^3 s^{-1} is one of the faster rates reported so far for electron transfer over this distance. No evidence was obtained for significant conformational changes from the NMR. Competitive inhibition was observed on addition of Gd^{3+}, which associates at the negative patch on PCu(II) [115].

9.3 Plastocyanin with Cytochrome f

The first cytochrome to be recognised as a component of the photosynthetic electron transport chain was cytochrome f [142]. The properties of cytochrome f have been reviewed [143, 144], and amino-acid sequence information is available for pea, spinach, wheat and tobacco [145]. The axial ligand to the heme-Fe are a histidine and lysine. Based on the 285 amino-acid sequence for pea cytochrome f, determined from the DNA sequence, M_r is 31,000 [146]. A number of cytochrome f's show a tendency to aggregate, a problem related to the presence of a hydrophobic tail consisting of the last 35 of the 285 residues. A protease appears to be retained in the butan-2-one solvent extraction procedure for the isolation of cytochrome f from cabbage leaves which cleaves this tail and results in the isolation of a stable monomeric form [147]. Amino acids 1–250 contain a high proportion of charged residues, consistent with a globular region (Fig. 18), which is pendant within the inner thylakoid where it can react with plastocyanin.

Cytochrome f is one of four proteins present in the so-called b_6f complex. Components are the diheme cytochrome b_6, Rieske's protein, cytochrome f and protein sub-unit IV all of which have been sequenced [145, 148]. The upper section of Fig. 19 shows the individual polypeptides, and the lower half the likely arrangement of the polypeptides in the membrane. The function of plastocyanin, is to shuttle electrons from cytochrome f to P700, which is also membrane-bound. This it presumably does by docking first of all with the exposed heme edge of the globular part of cytochrome f [149]. The reduction potential of cytochrome f is 360 mV, and the isoelectric point close to 5.5 (charlock) [147]. From the sequence the 1–250 protein is estimated to have a small overall negative charge (1 −). The plastocyanin then delivers the electron to P700.

Rate constants for the reaction of cytochrome f with inorganic oxidants are independent of pH when it is between 6.5–8.0 [150]. At pH < 6.5 protonation does have an effect, and with for example $[Co(phen)_3]^{3+}$ there is a decrease and $[Fe(CN)_6]^{3-}$ an increase in rate constants. From the kinetics cytochrome f shows no association with positively charged $[Co(phen)_3]^{3+}$, and no competitive inhibition is observed with $[Pt(NH_3)_6]^{4+}$ or $[(NH_3)_5CoNH_2Co(NH_3)_5]^{5+}$ [150]. It is possible therefore to study the effect of these reagents on the reaction of cytochrome f(II) with PCu(II), which

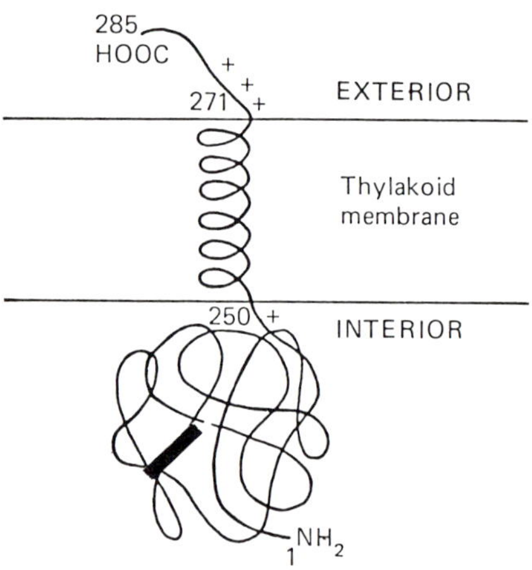

Fig. 18. Schematic representation of the trans-membrane orientation of cytochrome f based on the occurrence of a high proportion of charged residues for 1–250 (represented as a globular region), and hydrophobic residues 250–271 (believed to be helical and trans-membrane) [146]

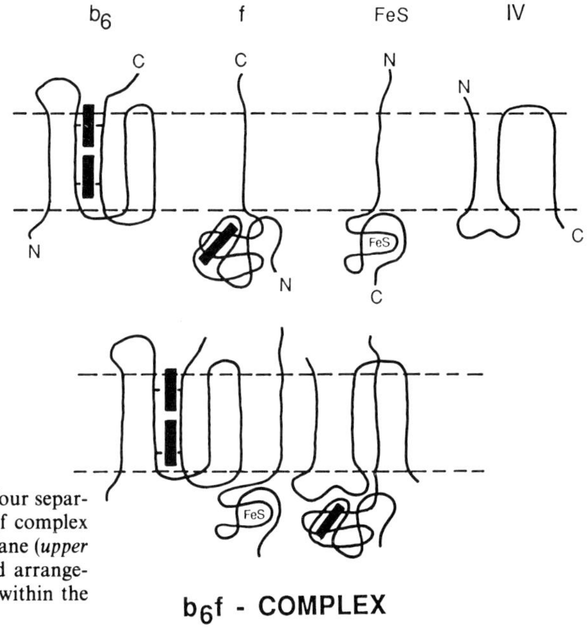

Fig. 19. The orientation of the four separate protein molecules of the b_6f complex relative to the thylakoid membrane (*upper section*), and likely ordering and arrangement of the same polypeptides within the complex [148]

should be diagnostic of the reaction site used on plastocyanin [149]. Competitive inhibition experiments with $[Pt(NH_3)_6]^{4+}$ (at pH 5.8) and $[(NH_3)_5CoNH_2Co(NH_3)_5]^{5+}$ (at pH 7.5) have indicated complete blocking. Also with decreasing pH rate constants for the cytochrome f(II) reduction of PCu(II) decrease substantially. There is a 90% effect (pK_a 5.0) and complete inactivation is not ruled out. Clearly cytochrome f has a high specificity for the plastocyanin remote site. If protonation on cytochrome f had been contributing, then for a negatively charged PCu(II) this would have been expected to give an increase in rate constant with decreasing pH [149, 150]. Studies on the effect of ethylenediamine modification at residues 42–45 of plastocyanin on the reaction with cytochrome f also implicate the remote site [118].

Rate constants for the reaction of cytochrome f with plastocyanin are very fast and at the limit of the stopped-flow range [149]. The reaction can be slowed down by working at a lower temperature (10 °C), and higher ionic strength I = 0.20 M(NaCl). Different sources of the two protein reactants have been explored, with no significant variation in rate constants (see Table V of Ref. [149]). More recently rate constants for the two algal plastocyanins, *S. obliquus*, and *A. variabilis* PCu(II) have been included, Table 11 [124]. That the rate constant for *A. variabilis* is smaller is of interest in view of the differences in charge at the remote site. However, another contributing factor is the smaller driving force, since E^0 is 340 V for this plastocyanin.

For reaction to occur at the remote site of plastocyanin suggests that there is a preponderance of positively charged residues around the exposed heme edge of

Table 11. Rate constants/$M^{-1}s^{-1}$ (10 °C) for the reduction of different plastocyanins, PCu(II), by cabbage (*Brassica oleracea*) cytochrome f(II) at two different pH's, I = 0.20 M(NaCl)[a]

Source	10^{-5}k (pH 5.0)	10^{-5}k (pH 7.5)
Parsley	110	210
S. obliquus	78	
Spinach[b]	115	170
A. variabilis	2.8	4.2
B. komatsuna[c]		450 (20 °C)

[a] Ref. [123];
[b] Christensen HEM, Sykes AG (unpublished work);
[c] *Brassica komatsuna* See Niwa S, Ishakawa H, Nakai S, Takabe T (1980) *J Biochem* 88: 1177

cytochrome f. The redox inactive complex $[Zr(C_2O_4)_4]^{4-}$ inhibits the $[(CN)_5FeCNCo(CN)_5]^{5-}$ oxidation of cytochrome f(II) [150]. At 25 °C, I = 0.10 M (NaCl) K for association of $[Zr(C_2O_4)_4]^{4-}$ with the protein is 530 M^{-1}. This indicates a positively charged locality on cytochrome f. Similarly in the reaction of cytochrome f(II) with plastocyanin PCu(II) at 10 °C, I = 0.10 M (NaCl), $[Zr(C_2O_4)_4]^{4-}$ inhibits the reaction, K ~ 540 M^{-1} [149]. Similar effects have been observed for $[Mo(CN)_8]^{4-}$.

Attention is drawn to the large number (~30) of positive residues in the globular part of cytochrome f, and the occurrence of five pairs of consecutive lysine residues in the sequence. In a recent study on the interaction between the oxidized forms of cytochrome f and plastocyanin (both from spinach), the aim was to identify which regions on cytochrome f are involved in the association process [151]. Solutions of spinach cytochrome f used in these experiments required 0.5% sodium cholate to avoid aggregation. In the first studies fragments of proteoliticly cleaved cytochrome f were allowed to interact with a plastocyanin loaded affinity column. The smallest of the fragments which bound to the column was analyzed, and in this way an ~90 residue (11 kD) fragment containing the *N*-terminal sequence and the heme binding site was identified as showing favorable interaction with the plastocyanin. Secondly, use was made of a modification procedure for the arginines of cytochrome f (there are 10 in all) with the specific reagent hydroxyphenylglyoxal. Plastocyanin has no arginines. The modification procedure was carried out in the presence and absence of plastocyanin. Two peptide fragments containing Arg residues (88 and 154) were found to be modified in the absence of plastocyanin, but not when plastocyanin is present. It is assumed that at the ionic strength used (5 mM) there is significant association between the two reactants (concentrations approaching mM), and that these two residues are protected when cytochrome f is associated with the plastocyanin.

From carbodiimide cross linking experiments it has been suggested that Lys187 of cytochrome f becomes attached to Asp44 of plastocyanin [152].

9.4 *Plastocyanin with P700*

As in the case of Photosystem II the primary process at the Photosystem I reaction center is a light-induced separation of charge. The incidence of light leads to production of an excited state of chlorophyll P700*. An electron is then transferred to an acceptor chlorophyll A_0 to give A_0 and $P700^+$. The A_0 form is a very strong reducing agent (−1.1V) which can bring about reduction of [2Fe–2S] and in turn convert $NADP^+$ to NADPH. At the same time plastocyanin is required to reduce $P700^+$ [153].

Photosystem I is, like b_6f, a transmembrane complex. A procedure for the isolation of Photosystem I 30, the P700-chlorophyll protein complex which has five polypeptides (M_r's of 60–65, 23, 20, 18 and 12 kDa) and 30 chlorophylls per P700 has been described [154]. This is the core complex of Photosystem I depleted of antenna chlorophylls. The preparation of the P700-chlorophyll protein complex containing 180 chlorophylls per P700 (PSI 180) has also been described [155]. The identity of the binding site for $P700^+$ on plastocyanin is of interest but somewhat controversial. A derivative obtained by reduction of PCu(II) with Cr^{2+}, which has a Cr(III) attached at the remote 42–45 site, has been prepared. This is reported to inhibit the reaction with $P700^+$ suggesting that $P700^+$ uses the 42–45 site [156]. Also it is known that addition of low concentrations of $MgCl_2$ increases the effectiveness of PCu(I) as an electron donor to $P700^+$. This could be due to association of Mg^{2+} at the remote site. Alternatively it could be due to the association of Mg^{2+} with the $P700^+$ reactant. At high $MgCl_2$ concentrations the rate is reported to decrease.

As already indicated, plastocyanin can also be modified by reacting remote site carboxylate residues with ethylenediamine [157]. This results in the replacement of a negative carboxylate charge by a positive amine (16).

$$\text{Protein}-CO_2^- + NH_2(CH_2)_2NH_3^+ \rightarrow \text{Protein}-CONH(CH_2)_2NH_3^+ \tag{16}$$

Four products were obtained containing between 2 and 6 moles of amine per plastocyanin. More recently the reaction has been carried out to yield single and double modified products. No effect of modification at 42–45 on the rate of the PCu(I) reduction of $P700^+$ is reported, suggesting that these residues are not involved in the reaction. In a further elaboration the reaction of plastocyanin, cross-linked (by carbodiimide treatment) to cytochrome f at the remote site, with $P700^+$ has been studied and the effects of Mg^{2+} and Na^+ explored [158].

10 Conclusion

Much of this review has been devoted to plastocyanin. Evidence has been presented in support of plastocyanin electron transfer reactivity at adjacent and remote sites. Of the two, reactivity at the remote site is of particular interest in the context of long-distance electron transfer. The physiological reductant cytochrome f appears to react at this site. Evidence for positively charged redox partners (inorganic and protein) using the remote site may be summarised as follows:

(a) Graphs of rate constants with pH (4–8) for the $[Co(phen)_3]^{3+}$ oxidation of PCu(I) are displaced relative to those for negatively charged oxidants (e.g. $[Fe(CN)_6]^{3-}$)
(b) Rate constants for reactions of PCu(II) with positively charged reductants at the remote site decrease with pH.
(c) Saturation kinetics are observed, e.g. with $[Co(phen)_3]^{3+}$, as a result of favorable electrostatic association at the remote site.
(d) Inhibition of cationic redox reagents is observed with redox inactive complexes, e.g. $[Pt(NH_3)_6]^{4+}$ and $[(NH_3)_5CoNH_2Co(NH_3)_5]^{5+}$, due to competitive association at the remote site.
(e) Redox inactive paramagnetic analog complexes such as $[Cr(phen)_3]^{3+}$ give ^1H NMR line broadening of Tyr83 as a result of association at the remote site.

Parsley, spinach, French bean, poplar and *S. obliquus* (but not *A. variabilis*) conform extensively to the above criteria for reaction at the remote site. There is extensive evidence for cytochrome f reacting at the remote site on plastocyanin. The aromatic residue at 83 would seem to be a prime candidate as lead-in group for electron transfer. Desolvation at the surface around 83, and interaction with an aromatic component on the reaction partner, e.g. the porphyrin ring of cytochrome f, may be important. The exact manner of electron transfer has yet to be confirmed. The distance from the aromatic ring of Tyr83 to the Cu for electron transfer is ~ 12 Å.

Last but not least, as this review has attempted to illustrate, there are many other type 1 blue Cu proteins which deserve attention. Azurin has already been fairly extensively studied. As more structural information becomes available, other proteins will no doubt be further investigated, and add to an understanding of this area. The recent ascorbate oxidase structure is extremely important in the further understanding of multi-Cu oxidases.

Acknowledgements The author is indebted to the patience, enthusiasm and efforts of his research group, and to colleagues who have conveyed their own interest and enthusiasm for this area of research. Discussions with Dr G. W. Canters (Leiden) and Dr J. McGinnis (Newcastle) are gratefully acknowledged.

11 References

1. Sykes AG (1985) Chem Soc Rev 14: 283
2. Adman ET (1985) In: Harrison PM (ed) Metalloproteins, Part 1. McMillan, Chapter 1

3. (a) Rydon L (1989) Coord Chem Rev 94: 157 (b) Farver O, Pecht I (1989) Coord Chem Rev 94: 183
4. Seven reviews (1984) In: Lontie R (ed) Copper Proteins and Copper Enzymes, vol 1, CRC, Florida (USA)
5. Messerschmidt A, Rossi A, Ladenstein R, Huber R (1989) J Mol Biol 206: 513
6. Gaykema WPJ, Volbeda A, Hol WGJ (1985) J Mol Biol 187: 255
7. Gray HB (1986) Chem Soc Rev 15: 17
8. Mayo SL, Ellis WR Jr, Crutchley RJ, Gray HB (1986) Science 233: 948
9. Gray HB, Malmstrom BG (1989) Biochemistry 28: 7499
10. Marcus RA, Sutin N (1985) Biochim Biophys Acta 811: 265
11. Peterson-Kennedy SE, McGourty JL, Kalweit JA, Hoffman BM (1986) J Am Chem Soc 108: 1739
12. Miller JR, Calcaterra LT, Closs G (1983) J Am Chem Soc 105: 671
13. McLendon G (1988) Acc Chem Res 21: 160
14. (a) Isied SS, Kuehn C, Worosila G (1984) J Am Chem Soc 106: 1722 (b) Bechtold R, Gardineer MB, Kazmi A, van Hemelryck B, Isied SS (1986) J Phys Chem 90: 3800
15. (a) Deisenhofer J, Epp O, Miki K, Huber R, Michel H (1984) J Mol Biol 180: 385; (1985) Nature (London) 318: 618 (b) Michel H, Epp O, Deisenhofer J (1986) EMBO J 5: 2445
16. Guss JM, Freeman HC (1983) J Mol Biol 169: 521
17. Guss JM, Harrowell PR, Murata M, Norris VA, Freeman HC (1986) J Mol Biol 192: 361
18. e.g. Armstrong WH (1988) In: Que (ed) Metal clusters in proteins, Symposium Series No. 372 Chapter 1
19. Adman ET, Sieker LC, Jensen LH (1973) J Biol Chem 248: 2987
20. Tsukihara T, Fukuyama K, Nakamura M, Katsube Y, Tanaka N, Kakudo M, Wada K, Hase T, Matsubara H (1981) J Biochem 90: 1763
21. (a) Carter CA Jr, Kraut J, Freer ST, Xuong NH, Alden RA, Bartsch RG (1974) J Biol Chem 249: 4212 (b) Carter CW Jr, Kraut J, Freer ST, Alden RA, (1974) J Biol Chem 249: 6339 (c) Carter CW Jr (1977) In: Lovenberg W (ed) Iron-Sulfur proteins, Academic (N.Y.) vol 3 p 158
22. (a) Takano T, Dickerson RE (1981) J Mol Biol 153: 79 (b) Ochi H, Hata Y, Tanaka N, Kakudo M, Sakurai T, Aihara S, Morita Y (1983) J Mol Biol 166: 407
23. Fleischer EB, Jacobs S, Mestichelli L (1968) J Am Chem Soc 90: 2527
24. Wendoloski JJ, Matthews JB, Weber PC, Salemme FR (1987) Science 238: 794
25. Mauk MR, Reid LS, Mauk AG (1982) Biochem 21: 1843
26. McLendon G, Miller JR (1985) J Am Chem Soc 107: 7811
27. Moore GR (personal communication)
28. Yano H, Kamo M, Tsugita A, Aso K, Nozu Y (1989) Protein Seq Data Anal 2: 385. We thank Dr. Y. Nozu for communicating results to us prior to publication
29. Dimitrov MI, Egorov CA, Donchev AA, Atanasov BP (1987) FEBS Lett 226: 17
30. Nielsen PS, Gausing K (1987) FEBS Letter 225: 159 (residues 39–40 in the white campion sequence first reported in Smeekins S, de Groot M, Binsbergen J van, Weisbeek P (1985) Nature 317: 456, should now read Val Val, cited personal communication from Smeekins S)
31. Yoshizaki F, Fukazawa T, Mishina Y, Sugimura Y (1989) J Biochem 106: 282
32. Simpson RJ, Moritz RL, Nice EC, Grego B, Yoshizaki F, Sugimara Y, Freeman HC, Murata M (1986) Eur J Biochem 157: 497
33. Aitken A (1975) Biochem J 149: 675
34. Tan S, Ho K-K (1989) Biochim Biophys Acta 973: 111
35. Ramshaw JAM, Brown RH, Scawen MD, Boulter D (1973) Biochim Biophys Acta 303: 269
36. Jackman MP, Sinclair-Day JD, Sisley MJ, Sykes AG, Denys LA, Wright PE (1987) J Am Chem Soc 109: 6443
37. (a) Moore JM, Case DA, Chazin WJ, Gippert GP, Havel TF, Powls R, Wright PE (1988) Science 240: 314 (b) Chazin WJ, Wright PE (1988) J Mol Biol 202: 623 (c) Moore JM, Chazin WJ, Powls R, Wright PE (1988) Biochem 27: 7806
38. (a) Scott RA, Hahn JE, Doniach S, Freeman HC, Hodgson KO (1982) J Am Chem Soc 104: 5364 (b) Penner-Hahn JE, Murata M, Hodgson KO, Freeman HC (1989) Inorg Chem 28: 1826
39. Segal MG, Sykes AG (1978) J Am Chem Soc 100: 4585
40. Driscoll PC, Hill HAO, Redfield C (1987) Eur J Biochem 170: 279.
41. King GC, Wright PE (1986) Biochem 25: 2364
42. Garrett TPJ, Clingeleffer DJ, Guss JM, Rogers SJ, Freeman HC (1984) J Biol Chem 259: 2822
43. Church WB, Guss JM, Potter JJ, Freeman HC (1986) J Biol Chem 261: 234
44. Collyer CA, Guss JM, Sugimura Y, Yoshizaki F, Freeman HC (1990) J Biol Chem 211: 617.

45. Gewirth AA, Solomon EI (1988) J Am Chem Soc 110: 3811
46. (a) Donovan JW (1979) In: Leach SJ (ed) Physical Principles and techniques in protein chemistry. Academic, N.Y. Part A, p 102 (b) Gross EL, Anderson GP, Ketchner SL, Draheim JE (1985) Biochim Biophys Acta 808: 437
47. (a) Markley JL, Ulrich EL, Berg SP, Krogman DW (1975) Biochemistry 14: 4428 (b) Kojiro CL, Markley JL (1983) FEBS Lett 162: 54
48. Sinclair-Day JD, Sisley MJ, Sykes AG, King GC, Wright PE. J Chem Soc Chem Comm 1985: 505
49. McGinnis J, Sinclair-Day JD, Sykes AG, Powls R, Moore JM, Wright PE (1988) Inorg Chem 27: 2306
50. Jackman MP, McGinnis J, Powls R, Salmon GA, Sykes AG (1988) J Am Chem Soc 110: 5880
51. Armstrong FA, Hill HAO, Oliver BN, Whitford D (1985) J Am Chem Soc 110: 5880
52. (a) Smeekins S, de Groot M, van Binsbergen J, Weisbeck P (1985) Nature (London) 317: 456 (b) Vorst O, Oosterhoff-Teertstra R, Vankan P, Smeekens S, Weisbeck P (1988) Gene 65: 59
53. Adman ET, Stenkamp RE, Sieker LC, Jensen LH (1978) J Mol Biol 123: 35 (b) Adman ET, Jensen LH (1981) Isr J Chem 21: 8
54. Norris GF, Anderson BF, Baker EN (1986) J Am Chem Soc 108: 2784; (1983) J Mol Biol 165: 501
55. Groeneveld CM, Feiters MC, Hasnain SS, van Rijn J, Reedijk J, Canters GW (1986) Biochem. Biophys. Acta 873: 214
56. Pettigrew GW, Leitch FA, Moore GR (1984) Biochim Biophys Acta 764: 339; (see also Ref [89a] and unpublished work)
57. Corin AF, Bersohn R, Cole PE (1983) Biochemistry 22: 2032
58. Karlsson BG, Pascher T, Nordling M, Arvidsson RHA, Lundberg LG (1989) FEBS Lett 246: 211
59. Karlsson BG, Aasa R, Malmstrom BG, Lundberg LG (1989) FEBS Lett 253: 99
60. de Kamp van M, Floris R, Hali FC, Canters GW (1990) J Am Chem Soc 112: 907.
61. Petratos K, Dauter Z, Wilson KS, Lommen A, Van Beeumen J, Canters, GW (1988) J Mol Biol 199: 545
62. Harmel S, Adman E, Walsh KA, Beppu T, Titani K (1986) FEBS Lett 197: 301
63. (a) Murata M, Begg GS, Lambrou F, Leslie B, Simpson RJ, Freeman HC, Morgan JF (1982) Proc Natl Acad Sci (USA) 79: 6434 (b) Guss JM, Merritt EA, Phizackerley RP, Hedman B, Murata M, Hodgson KO, Freeman HC (1988) Science 241: 806
64. Schichi H, Hackett DP (1963) Arch Biochem Biophys 100: 183
65. Morita Y, Wadano A, Ida S (1971) Agric Biol Chem 35: 255
66. Marchesini A, Minelli M, Merkle H, Kroneck PM (1979) Eur J Biochem 101: 77
67 Trost JT, McManus JD, Freeman JC, Ramakrishna BL, Blankenship RE (1988) Biochemistry 27: 7858
68. Zumft W, Gotzmann DJ, Kroneck PMH (1987) Euro J Biochem 168: 301
69. Yamamoto K, Uozumi T, Beepu T (1987) J Bacteriol 169: 5648
70. Lawton SA, Anthony C (1985) Biochem J 228: 719
71. Spira-Solomon DJ, Allendorf MD, Solomon EI (1986) J Am Chem Soc 108: 5318
72. Reinhammer B (1985) Chem Scr 25: 172
73. Calabrese L, Carbonaro M, Musci G (1989) J Biol Chem 264: 6183
74. Adman ET, Turley S, Bramson R, Petratos K, Banner D, Tsernoglou D, Beppu T, Watanabe H (1989) J Biol Chem 264: 87
75. (a) Reinhammar BRM (1970) Biochim Biophys Acta 205: 25; (1972) 275: 245 (b) Bergman C, Gandvik E-K, Nyman PO, Strid L (1977) Biochim Biophys Res Comm 77: 1052
76. (a) Cox JD, Boxer DH (1978) Biochem J 174: 497 (b) Ingledew WJ (1982) Biochim Biophys Acta 683: 89
77. (a) Blake RC, Shute EA (1987) J Biol Chem 262: 14983 (b) Blake RC, White KJ, Shute EA (1988) Biohydrometall Proc Int Symp In: Norris PR, Kelly DP (eds) Sci Technol Lett UK, p 103
78. (a) Blair DF, Campbell GW, Schoonover JR, Chan SI, Gray HB, Malmstrom BG, Pecht I, Swanson BI, Woodruff WH, Cho WK, English AM, Fry HA, Lum V, Norton KA (1985) J Am Chem Soc 107: 5755 (b) Gray HB, Malmstrom BG (1983) Comments Inorganic Chemistry 1983 2: 203
79. Mino Y, Loehr TM, Wada K, Matsubara H, Sanders-Loehr J (1987) Biochemistry 26: 8059
80. Sequence information has been obtained by Ambler RP, Ingledew WJ (unpublished work); (personal communication from Dr. W. J. Ingledew)
81. Paul KG, Stigbrand T (1970) Biochim Biophys Acta 221: 255

82. Bergman C (1980) PhD thesis, Chalmers University of Technology, Goteberg
83. Chapman SK, Orme-Johnson WH, McGinnis J, Sinclair-Day JD, Sykes AG, Ohlsson P-I, Paul K-G. J Chem Soc Dalton Trans 1986: 2063
84. Tobari J, Harada Y (1981) Biochem Biophys Res Comm 101: 502
85. Houwelingen T van, Canters GW, Stobbelaar G, Duine JA, Frank Jzn J, Tsugita A (1985) Eur J Biochem 152: 75
86. Ambler RP, Tobari J (1985) Biochem J 232: 451
87. Lommen A, Canters GW, van Beeymen J (1988) Eur J Biochem 176: 213; and Lommen A, Canters GW (1990) J Biol Chem 265: 2768
88. Lappin AG, Segal MG, Weatherburn DC, Henderson RA, Sykes AG (1979) J Chem Soc 101: 2302
89. Rosen P, Pecht I (1976) Biochemistry 15: 773
90. Sisley MJ, Segal MG, Stanley CS, Adzamli IK, Sykes AG (1983 J Am Chem Soc 105: 225
91. (a) McGinnis J, Ingledew WJ, Sykes AG (1986) Inorg Chem 25: 3730 (b) Lappin AG, Lewis CA, Ingledew WJ (1985) Inorg Chem 24: 1446
92. (a) Groeneveld CM, Canters GW (1985) Eur J Biochem 153: 559 (b) Groeneveld CM, Dahlin S, Reinhammar B, Canters GW (1987) J Am Chem Soc 109: 3247
93. Dahlin S, Reinhammar B, Wilson MT (1984) Biochem J 218: 609
94. Armstrong FA, Driscoll PC, Hill HAO (1985) FEBS Lett 190: 242
95. (a) Chapman SK, Sanemasa I, Watson AD, Sykes AG, J Chem Soc Dalton Trans 1983: 1949 (b) Chapman SK, Samemasa I, Sykes AG, J Chem Soc Dalton Trans 1983: 2549
96. Chapman SK, Watson AD, Sykes AG, J Chem Soc Dalton Trans 1983: 2543
97. Chapman SK, Knox, CV, Sykes, AG, J Chem Soc Dalton Trans 1984: 2275
98. McGinnis J, Sinclair-Day JD, Sykes AG, J Chem Soc Dalton Trans 1986: 2007
99. McGinnis J, Sinclair-Day JD, Sykes AG, J Chem Soc Dalton Trans 1986: 2011
100. Sinclair-Day JD, Sykes AG, J Chem Soc Dalton Trans 1986: 2069
101. Jackman MP, McGinnis J, Sykes AG, Collyer CA, Murata M, Freeman HC, J Chem Soc Dalton Trans 1987: 2573
102. Jackman MP, Sinclair-Day JD, Sisley MJ, Sykes AG, Denys LA, Wright PE (1987) J Am Chem Soc 109: 6443
103. Brunschwig BS, Delaive PJ, English AM, Goldberg M, Gray HB, Mayo SL, Sutin N (1985) Inorg Chem 24: 3743
104. Cookson DJ, Hayes MT, Wright PE (1980) Biochim Biophys Acta 591: 162
105. Armstrong FA, Driscoll PC, Ellul HG, Jackson SE, Lannon AM, J Chem Soc Chem Comm 1988: 234
106. Holwerda RA, Wherland S, Gray HB (1976) Ann Rev Biophys Bioeng 5: 363
107. Bernauer K, Sauvain JJ, J Chem Soc Chem Comm 1988: 353
108. Butler J, Davies DM, Sykes AG (1981) J Inorg Biochem 15: 41
109. King GC (1984) PhD thesis, University of Sydney
110. Holwerda RA, Kuaff DB, Gray HB, Clemmer JD, Crowley R, Smith JM, Mauk AG (1980) J Am Chem Soc 102: 1142
111. Haim A, Sutin N (1976) Inorg Chem 15: 476
112. Farina R, Wilkins RG (1968) Inorg Chem 7: 514
113. Hanford PM, Hill HAO, Lee RW-K, Henderson R'A, Sykes AG (1980) J Inorg Biochem 13: 83
114. Armstrong FA, Driscoll PC, Hill HAO, Redfield C (1986) J Inorg Biochem 28: 171
115. King GC, Binstead RA, Wright PE (1985) Biochim Biophys Acta 806: 262
116. Lappin AG, Segal MG, Weatherburn DC, Sykes AG (1978) J Am Chem Soc 101: 2297
117. Pladziewicz JR, Brenner MS (1987) Inorg Chem 26: 3629
118. Anderson FP, Sanderson DG, Lee CH, Durell S, Anderson LB, Gross EL (1987) Biochim Biophys Acta 894: 386
119. Rush JD, Levine F, Koppenol WH (1988) Biochemistry 27: 5876
120. Sanemasa I, Toda K, Deguchi T, Sykes AG (1985) Anal Chem 57: 2405
121. Christensen HEM, Ulstrup J, Sykes AG, Biochem Biophys Acta (submitted)
122. Rees DC (1980) J Mol Biol 141: 323
123. de Silva DGAH (1987) PhD thesis, University of Newcastle upon Tyne
124. See Fig. 5 in Beoku-Betts D, Sykes AG (1985) Inorg Chem 24: 1142
125. Recchia J, Matthews CR, Rhee M-J, Horrocks WD Jr (1982) Biochim Biophys Acta 702: 105
126. Matthews CR, Erikson PM, Froebe CL (1980) Biochim Biophys Acta 624: 499
127. (a) Yocum KM, Shelton JB, Shelton JR, Schroeder WA, Worosila G, Isied SS, Bordignon E, Gray HB (1982) Proc Natl Acad Sci USA 79: 7052 (b) Nocera DG, Winkler JR, Yocum KM,

Bordignon E, Gray HB (1984) J Am Chem Soc 106: 5145 (c) Isied SS, Kuehn C, Worosila G (1984) J Am Chem Soc 106: 1722 (d) Bechtold R, Gardineer MB, Kazmi A, van Hemelryck B, Isied SS (1986) J Phys Chem 90: 3800
128. (a) Margalit R, Kostic NM, Che C-M, Blair DF, Chiang H-J, Pecht I, Shelton JB, Shelton JR, Schroeder WA, Gray HB (1984) Proc Natl Acad Sc (USA) 81: 6554 (b) Kostic NM, Margalit R, Che C-M, Gray HB (1983) J Am Chem Soc 105: 7765
129. Osvath P, Salmon GA, Sykes AG (1988) J Am Chem Soc 110: 7114
130. (a) Jackman MP, Lim M-C, Salmon GA, Sykes AG, J Chem Soc Chem Comm 1988: 179 (b) Jackman MP, Lim MC, Sykes AG, Salmon GA, J Chem Soc Dalton Trans 1988: 2843
131. (a) Margalit R, Pecht I, Gray HB (1983) J Am Chem Soc 105: 301 (b) Lieber CM, Karas JL, Gray HB (1987) J Am Chem Soc 109: 3778 (c) Crutchley RJ, Ellis WR Jr, Gray HB (1986) In: Xavier AV (ed) Frontiers in Bioinorg Chem, VCH: W Germany 679
132. Jackman MP, Sykes AG, Salmon GA, J Chem Soc Chem Commn 1987: 65
133. Jackman MP, Lim M-C, Osvath P, de Silva DGAH, Sykes AG (1988) Inorg Chim Acta 153: 205
134. Sundberg RJ, Gupta G (1973) Bioinorg Chem 3: 37
135. Sundberg RJ, Bryan RF, Taylor IF, Taube H (1974) J Am Chem Soc 96: 381
136. Che C-M, Margalit R, Chiang H-J, Gray HB (1987) Inorg Chim Acta 135: 33
137. Armstrong FA, Butt JN, Govindaraju K, McGinnis J, Sykes AG Inorg Chem (in press)
138. Peerey LM, Kostic NM (1989) Biochemistry 28: 1861
139. Christensen HEM, Ulstrup J, Sykes AG (1990) Biochim Biophys Acta 1039: 94
140. Govindaraju K, Salmon GA, Sykes AG (1990) J Chem Soc Chem Comm 1003
141. Augustin MA, Chapman SK, Davies DM, Watson AD, Sykes AG (1984) J Inorg Biochem 20: 281
142. Hill R, Scarisbrick R (1951) New Phyt 50: 98
143. Knaff DB (1978) Coord Chem Rev 26: 47
144. Bendall DS (1982) Biochim Biophys Acta 683: 119
145. Hauska G, Nitschke W, Herrmann RG (1988) J Bioinorg Biomembranes 20: 211
146. Willey DL, Auffret AD, Gray JC (1984) Cell 36: 555
147. Gray JC (1978) Eur J Biochem 82: 133
148. Hauska G (1985) Mol Biol Photosynth Apparatus, Cold Spring Harbor Laboratory, USA, p 79
149. Beoku-Betts D, Chapman SK, Knox CV, Sykes AG (1985) Inorg Chem 24: 1677
150. Beoku-Betts D, Sykes AG (1985) Inorg Chem 24: 1142
151. Adam Z, Malkin R (1989) Biochim Biophys Acta 975: 158
152. Monard LZ, Krogmann DW (1988) Abs 14th Ann Midwest Photosynth Conference quoted in Ref [119]
153. See eg. Katoh S (1977) In: Pirson A, Zimmerman MH (eds) Encyclopaedia of plant physiology. vol 5, p 247 Springer-Verlag, Berlin Heidelberg New York
154. Takabe T, Ishikawa H, Niwa S, Itoh S (1983) J Biochem 94: 1901
155. Mullett JE, Burke JJ, Arntzen CJ (1980) Plant Physiol 65: 814
156. Farver O, Pecht I (1981) Proc Natl Acad Sc (USA) 78: 4190
157. Burkey KO, Gross EL (1982) Biochemistry 21: 5886
158. Takabe T, Ishikawa H (1989) J Biochem 105: 98

Author Index Volumes 1–75